Vegetation Dynamics - Ecosystem Management, Conservation, and Protection

Edited by Ana Cristina Gonçalves and Teresa Fidalgo Fonseca

Published in London, United Kingdom

Vegetation Dynamics - Ecosystem Management, Conservation, and Protection
http://dx.doi.org/10.5772/intechopen.1003340
Edited by Ana Cristina Gonçalves and Teresa Fidalgo Fonseca

Contributors
Ana Cristina Gonçalves, Ana Soler, Antía Iglesias, António L. Crespí, Arvid Odland, Damien Hicks, Jaime Sales Luis, José Luís Lousada, João Rocha, Juan Picos, Mallem Hamida, Margarida L.R. Liberato, Maria Emília Silva, Mariana Niculescu, Nuno Silva, Raul Morais, Teresa Fidalgo Fonseca, Teresa Fonseca, Ángeles Cancela

First published in London, United Kingdom, 2025 by IntechOpen
IntechOpen is the global imprint of INTECHOPEN LIMITED, registered in England and Wales, registration number: 11086078, 167-169 Great Portland Street, London, W1W 5PF, United Kingdom

For EU product safety concerns: IN TECH d.o.o., Prolaz Marije Krucifikse Kozulić 3, 51000 Rijeka, Croatia, info@intechopen.com or visit our website at intechopen.com.

British Library Cataloguing-in-Publication Data
A catalogue record for this book is available from the British Library

Vegetation Dynamics - Ecosystem Management, Conservation, and Protection
Edited by Ana Cristina Gonçalves and Teresa Fidalgo Fonseca
p. cm.

This title is part of the Environmental Sciences Book Series, Volume 30
Topic: Ecosystems and Biodiversity
Series Editor: J. Kevin Summers
Topic Editors: Salustiano Mato, Josefina Garrido and Francisco Ramil

Print ISBN 978-0-85014-763-6
Online ISBN 978-0-85014-762-9
eBook (PDF) ISBN 978-0-85014-764-3
ISSN 2754-6713

If disposing of this product, please recycle the paper responsibly.

IntechOpen Book Series
Environmental Sciences
Volume 30

Aims and Scope of the Series

Scientists have long researched to understand the environment and man's place in it. The search for this knowledge grows in importance as rapid increases in population and economic development intensify humans' stresses on ecosystems. Fortunately, rapid increases in multiple scientific areas are advancing our understanding of environmental sciences. Breakthroughs in computing, molecular biology, ecology, and sustainability science are enhancing our ability to utilize environmental sciences to address real-world problems.

The four topics of this book series - Pollution; Environmental Resilience and Management; Ecosystems and Biodiversity; and Water Science - will address important areas of advancement in the environmental sciences. They will represent an excellent initial grouping of published works on these critical topics.

Meet the Series Editor

J. Kevin Summers is a Senior Research Ecologist at the Environmental Protection Agency's (EPA) Gulf Ecosystem Measurement and Modeling Division. He is currently working with colleagues in the Sustainable and Healthy Communities Program to develop an index of community resilience to natural hazards, an index of human well-being that can be linked to changes in the ecosystem, social and economic services, and a community sustainability tool for communities with populations under 40,000. He leads research efforts for indicator and indices development. Dr. Summers is a systems ecologist and began his career at the EPA in 1989 and has worked in various programs and capacities. This includes leading the National Coastal Assessment in collaboration with the Office of Water which culminated in the award-winning National Coastal Condition Report series (four volumes between 2001 and 2012), and which integrates water quality, sediment quality, habitat, and biological data to assess the ecosystem condition of the United States estuaries. He was acting National Program Director for Ecology for the EPA between 2004 and 2006. He has authored approximately 150 peer-reviewed journal articles, book chapters, and reports and has received many awards for technical accomplishments from the EPA and from outside of the agency. Dr. Summers holds a BA in Zoology and Psychology, an MA in Ecology, and Ph.D. in Systems Ecology/Biology.

Meet the Volume Editors

Ana Cristina Gonçalves is an Assistant Professor with Habilitation at the University of Évora, in the Department of Rural Engineering, Portugal and a researcher at the Mediterranean Institute for Agriculture, Environment and Development (MED), Deputy Coordinator of IUFRO subdivision 1. 04. 00 – Agroforestry. She holds a Ph.D. in Forestry, has authored over 100 publications, and has participated in 15 research projects. Her research is focused on silviculture and modelling in pure, mixed, even-aged and uneven-aged stands; biomass; and forest management and planning integrated in a GIS environment.

Teresa Fidalgo Fonseca is an Associate Professor with Habilitation at the University of Trás-os-Montes e Alto Douro (UTAD) in Portugal, an integrated member of the Centre for the Research and Technology of Agro-Environmental and Biological Sciences (CITAB, UTAD) and collaborator at the Centre of Forest Studies (CEF, University of Lisbon). She holds a Ph.D. in Forestry Sciences and conducts research in the scientific domains of her specialization, participating in national and international research projects and authoring scientific and technical publications. Her research focuses on biometrics, forest assessment, silviculture, modelling, and forest management. She is the Co-coordinator of Division 1 (Silviculture) of the International Union of Forest Research Organizations (IUFRO).

Contents

Preface

This book entitled *Vegetation Dynamics – Ecosystem Management, Conservation, and Protection* has eight chapters, four related to vegetation dynamics and four related to monitoring and management.

In the introductory chapter, Ana Cristina Gonçalves and Teresa Fidalgo Fonseca discussed the importance of species traits and sites in the sustainability of the forest systems as well as the importance of monitoring, modelling and management.

Mariana Niculescu discussed the effects of the changes in management practices, invasive species, climate change and mining on meadows and forest systems in Romania. The author pinpointed the effects of unsuitable management on the sustainability of the agroforestry systems. Overall, maintaining agroforestry sustainability, with suitable management practices, would enable regenerating vegetation.

Arvid Odland evaluated the effect of water reduction on forest floodplain wetlands. The species proportion and cover were influenced by its capacity for flowering and fruiting, which were also related to the temperature and water level. This review highlighted the importance of long-term studies as vegetation dynamics evolve slowly over time. Mallem Hamida evaluated the effect of vegetation on soil rehabilitation in pre-desert conditions, particularly in wind erosion mitigation and soil properties improvement in Algeria. Vegetation enhanced dune stabilisation due to its impact on wind erosion, and vegetation improved soil properties. However, variations were found when different species were compared.

António L. Crespí et al. explored the concept of vegetation robustness and proposed a methodological approach to characterise the functional traits of vegetation and their potential correlation with environmental factors. The study indicated that the resilience of these areas was influenced by their resistance and was linked to environmental variability, particularly geomorphological and climatic factors.

Damien Hicks reviewed the influence of sampling vegetation and a set of quantitative tools to analyse data. It included the optimisation protocols to enable the use of data for different analyses, from multivariate analysis to complex statistical analysis of data with network structure, machine learning methods and Bayesian methods.

Fonseca et al. conducted a comprehensive review of the traits, silviculture and management of *Pseudotsuga menziesii* in Portugal. The authors analysed the area potentially suitable for establishing stands of this species in Portugal, as well as its potential growth and wood properties. Iglesias et al. discussed the effects of pure, even-aged plantations of Eucalyptus globulus and invasive species on biodiversity and using an invasive species (*Phytolacca americana*) to produce pulp and paper. The physical, mechanical, and chemical properties of Phytolacca americana confirmed that the species is suitable for producing paper in a pulp blend of the two species.

Ana Cristina Gonçalves
MED – Mediterranean Institute for Agriculture, Environment and Development,
CHANGE – Global Change and Sustainability Institute,
Institute for Research and Advanced Training,
Department of Rural Engineering,
School of Science and Technology,
University of Évora,
Évora, Portugal

Teresa Fidalgo Fonseca
Department of Forest Sciences and Landscape Architecture,
School of Agrarian and Veterinary Sciences,
Centre for the Research and Technology of Agro-Environmental and Biological Sciences,
CITAB, Inov4Agro,
Universidade de Trás-os-Montes e Alto Douro,
Vila Real, Portugal

Section 1

Vegetation Dynamics

Chapter 1

Introductory Chapter: Vegetation Dynamics

Ana Cristina Gonçalves and Teresa Fidalgo Fonseca

1. Introduction

Vegetation covers a large share of the land use in the world and includes forest systems, agroforestry systems, agricultural systems, and other vegetation systems [1, 2]. Vegetation dynamics studies the variability in time and space of the vegetation, whether it is trees, shrubs, or herbaceous plants. It is focused on the number of species, their proportion, and development [3]. Several processes have been studied ranging from physiology to landscape assessments, to enable understanding plants and ecosystem evolution and its impacts on the ecosystems. The high variability of vegetation in time and space brings challenges in the evaluation, monitoring, modeling, and prediction of the evolution of the systems [4, 5], as well as on its management. Sustainable vegetation management is contingent on various factors, primarily the prioritization of key functions such as conservation, protection, and the utilization of goods and services.

Forest ecosystems provide several products (woody and nonwoody) and services (protection, conservation, recreation, and esthetics) [6]. Their contribution to carbon sequestration and storage is due to the trees' large dimensions and longevity [7]. Forest, agroforestry, and agricultural systems are inherently dynamic, exhibiting variability across both space and time. These changes can manifest at different scales, from individual management units to entire landscapes. The dynamics of vegetation can drive the systems to a balanced structure or inversely may jeopardize their resistance and resilience [3, 4]. Maintaining a balanced system of species adapted to local soil and climatic conditions, as well as assuring complementary interactions between different species, is essential to ensure a diversity of products and services [8].

Disturbances, whether natural (e.g., storms, fires, and climate change) or anthropogenic (e.g., derived from management) affect vegetation dynamics [6, 9]. Thus, a comprehensive understanding of ecosystems, their assessment, and monitoring, both at the management unit and landscape level, facilitates the formulation of strategic management alternatives to foster long-term sustainability.

The study of vegetation dynamics is complex and challenging, owing to the high spatial and temporal variability of the systems, as well as to the interactions of the several factors and their relationships [4]. Of importance is the study of the species traits and their suitability to edaphic and climatic factors. This enables to evaluate the relationships between the edaphic and climatic factors and plant and tree development [10–12]. It is also important to select the best suited species to a site [10, 13–16]. The regeneration of the species, particularly, the natural regeneration, is also dependent on the suitability of the species to the site and thus that withstand disturbances and maintain the system sustainability [17, 18].

Climate change can alter the suitability of the species to the sites, resulting in changes in vegetation dynamics, through the death of the plants and preventing a successful regeneration [18–21]. Several authors evaluated the effects of climate change on the potential distribution of forest species [22–24] and on their development [24], highlighting the need to consider species dynamics over the medium and long term.

Management can also change vegetation dynamics, especially in systems with large longevity and slow development of the individuals, such as forest and agroforestry systems, where management can drive systems to a particular stand structure. This is related to the silvicultural system and the production (*sensu lato*, including ecosystem services) goals [15, 25–30]. Ongoing studies compared the existing and new approaches [7, 31]. The main objective is to find or adapt the management of the systems that are flexible and enable to maintain system sustainability, when subject to disturbances [32–35]. While recurrent natural disturbances like storms, fires, and drought, have been studied and management alternatives have been provided [36–39], the intensification and increase of frequency of these disturbances, related to climate change, need further research to enable the development or adaptation of the management tools. For example, in the Mediterranean region prone to droughts and fires, the change in their intensity and frequency demands for the adaptation of the management of the forest stands, to increase their resistance and/or resilience and thus the maintenance of the sustainable systems [17, 31, 40] as well as for the future climate conditions [41].

Vegetation dynamics are influenced by species traits, edaphic-climatic conditions, and topography (Section 2), along with management, making proper monitoring essential (Section 3).

2. Species traits, edaphic-climatic conditions, and topography

There is a thigh relation between species traits and edaphic, climatic, and topographic conditions, regardless of whether they are herbaceous, shrubby or arboreal species. While some species prefer acidic soils (e.g., forest species such as *Quercus suber*, *Pseudotsuga mensiessii*, *Pinus radiata*, and *Pinus pinaster*) others thrive in calcareous soils (e.g., *Quercus faginea*) [42, 43]. Similarly, temperature range for the development and growth of some species is lower than for others. For example, *Pinus pinaster* and *Castanea sativa* prefer cooler sites when compared with *Quercus suber* or *Quercus rotundifolia* [31, 40, 42, 43]. Moreover, some species need higher annual precipitation than others (e.g., *Pinus pinaster* versus *Quercus rotundifolia*) [40, 42]. Altitude and aspect are two other factors that influence species distribution [42, 43].

The variability of traits of the species influences their distribution [44]. Species also vary in their plasticity to adapt to variations in climate. However, under climate change scenarios, some sites will not be suitable for particular species in the future. This will result in the migration for other sites more suitable for their traits and a reduction of the density and/or of their growth when they remain in the site [22, 24, 45]. It is thus of primordial importance to know species traits in detail as this knowledge can facilitate the mitigation of climate change impacts through effective management of vegetation [46]. Such insights have also implications in the conservation and protection of the ecosystems and soils, helping to maintaining their resilience and resistance to both biotic and abiotic disturbances.

Particular care should be taken with invasive or potentially invasive species that might alter composition of the forestry, agroforestry, and agricultural systems [47].

3. Ecosystem monitoring and management

Bearing in mind the species traits and the site conditions, it is possible to maintain and promote ecosystem sustainability by improving their resilience and resistance through proper monitoring and management. Management and monitoring of forest stands are frequently interconnected. Monitoring, through forest inventory with either field plots or remote sensing, enables the characterization of the stand structure [48] and supports modeling [49–51]. In fact, the management prescriptions are usually based on surveys that enable to characterize the stand structure and thus the practices needed to promote or to maintain the sustainability of the system and of their productions [40, 41, 52]. Of particular importance to system perpetuity and sustainability is regeneration [18] and biomass stocks [7].

Models enable to explore different management alternatives that cannot be easily assessed using field plots. However, forest modeling poses challenges due to the variability of tree dimensions. Typically, modeling is done on a per-species basis. Nonetheless, bias may arise due to differences in stand structure, which leads to variations in tree allometry. Thus, several methods and techniques are used to develop growth and production models in order to improve their accuracy and precision and to reduce bias and errors [40, 50, 51]. Accurate models are helpful in exploring management alternatives that are better suited to maintaining sustainability of the ecosystems while preserving their productions.

Management of native and exotic (noninvasive) species can often lead to the development of sustainable systems. For example, in forest stands, thinning can promote growth and may mitigate water stress [53–55]. Mixed stands are described as more resilient and resistant to abiotic and biotic disturbances while having higher diversity [52]. Uneven aged stands enable the sustainability of forest systems due to its regeneration in short time cycles and tend to increase diversity due to the variability of age classes [49, 52].

4. Final remarks

The assessment of species traits is of primordial importance for the sustainability of ecosystems. In addition, short- and long-term monitoring is essential to understand the dynamics of vegetation. This understanding, in turn, contributes to effective ecosystem management, supporting the development of alternative management approaches and thus the resilience and resistance of systems to abiotic and biotic disturbances that are either natural or man-induced.

Conflict of interest

The authors declare no conflict of interest.

Author details

Ana Cristina Gonçalves[1*] and Teresa Fidalgo Fonseca[2,3,4]

1 MED – Mediterranean Institute for Agriculture, Environment and Development and CHANGE – Global Change and Sustainability Institute, Institute of Research and Advanced Education (IIFA), Department of Rural Engineering, University of Évora, Apartado, Évora, Portugal

2 Department of Forest Sciences and Landscape Architecture (CIFAP), University of Trás-os-Montes and Alto Douro, Vila Real, Portugal

3 Centre for the Research and Technology of Agroenvironmental and Biological Sciences, CITAB, Inov4Agro, Universidadede Trás-os-Montes e Alto Douro, UTAD, Quinta de Prados, Vila Real, Portugal

4 Forest Research Centre (CEF), School of Agriculture University of Lisbon, Lisboa, Portugal

*Address all correspondence to: acag@uevora.pt

References

[1] Schulze K, Malek Ž, Verburg PH. Towards better mapping of forest management patterns: A global allocation approach. Forest Ecology and Management. 2019;**432**:776-785

[2] van Asselen S, Verburg PH. A land system representation for global assessments and land-use modeling. Global Change Biology. 2012;**18**:3125-3148

[3] Hufnagel L, Mics F. Introductory chapter: Vegetation dynamics, basic phenomena, and processes. In: Hufnagel L, El-Esawi MA, editors. Environmental Sciences [Internet]. London, United Kingdom: IntechOpen; 2023. Available from: https://www.intechopen.com/chapters/83412 [Accessed: October 8, 2024]

[4] Nakhoul J, Fernandez C, Bousquet-Mélou A, Nemer N, Abboud J, Prévosto B. Vegetation dynamics and regeneration of Pinus pinea forests in Mount Lebanon: Towards the progressive disappearance of pine. Ecological Engineering. 2020;**152**:105866

[5] Pausas JG. Mediterranean vegetation dynamics: Modelling problems and functional types. Plant Ecology. 1999;**140**:27-39

[6] Bolte A, Mansourian S, Madsen P, Derkyi M, Kleine M, Stanturf J. Forest adaptation and restoration under global change. Annals of Forest Science. 2023;**80**(1):7. DOI: 10.1186/s13595-022-01172-6

[7] Gonçalves AC. Stand structure and biomass. In: Gonçalves AC, Malico I, editors. Forest Bioenergy [Internet], Green Energy and Technology. Available from: https://link.springer.com/10.1007/978-3-031-48224-3_3. Cham: Springer International Publishing; 2024. pp. 53-90 [Accessed: January 5, 2024]

[8] Zheng L, Barry KE, Guerrero-Ramírez NR, Craven D, Reich PB, Verheyen K, et al. Effects of plant diversity on productivity strengthen over time due to trait-dependent shifts in species overyielding. Nature Communications. 2024;**15**(1):2078

[9] Jandl R, Spathelf P, Bolte A, Prescott CE. Forest adaptation to climate change—Is non-management an option? Annals of Forest Science. 2019;**76**(2):48

[10] Bentrup G, Dosskey MG. Tree advisor: A novel woody plant selection tool to support multifunctional objectives. Land. 2022;**11**(3):397

[11] Grossman JJ, Vanhellemont M, Barsoum N, Bauhus J, Bruelheide H, Castagneyrol B, et al. Synthesis and future research directions linking tree diversity to growth, survival, and damage in a global network of tree diversity experiments. Environmental and Experimental Botany. 2018;**152**:68-89

[12] Kunstler G, Falster D, Coomes DA, Hui F, Kooyman RM, Laughlin DC, et al. Plant functional traits have globally consistent effects on competition. Nature. 2016;**529**(7585):204-207

[13] Diaci J. Nature-Based Forestry in Central Europe Alternatives to Industrial Forestry and Strict Preservation. Ljubljana: Univ. Ljubljane, Biotech. Faculty, Dept. of Forestry and Renewable Forest Resources; 2006

[14] Mason WL, Diaci J, Carvalho J, Valkonen S. Continuous cover forestry

in Europe: Usage and the knowledge gaps and challenges to wider adoption. Forestry: An International Journal of Forest Research. 2022;**95**(1):1-12

[15] Pommerening A, Murphy ST. A review of the history, definitions and methods of continuous cover forestry with special attention to afforestation and restocking. Forestry. 2004;**77**(1):27-44

[16] Schütz JP. Historical emergence and current application of CCF. In: Pukkala T, Gadow KV, editors. Continuous Cover Forestry. Dordrecht: Springer Science+Business Media B.V; 2012. pp. 1-28

[17] Gonçalves AC. Impact of continuous cover forestry on Forest systems. In: Raj A, Jhariya MK, Banerjee A, Nema S, Bargali K, editors. Land and Environmental Management through Forestry [Internet]. 1st ed. Hoboken, USA: Wiley; 2023. pp. 133-159. DOI: 10.1002/9781119910527.ch6 [Accessed: November 27, 2023]

[18] Gonçalves AC, Fonseca TF. Influence management and disturbances on the regeneration of forest stands. Frontiers in Forests and Global Change. 2023;**6**:1123215

[19] Kremer K, Promis Á, Bauhus J. Natural advance regeneration of native tree species in *Pinus radiata* plantations of South-Central Chile suggests potential for a passive restoration approach. Ecosystems. 2022;**25**(5):1096-1116

[20] Löf M, Castro J, Engman M, Leverkus AB, Madsen P, Reque JA, et al. Tamm review: Direct seeding to restore oak (*Quercus* spp.) forests and woodlands. Forest Ecology and Management. 2019;**448**:474-489

[21] Malcolm DC, Mason WL, Clarke GC. The transformation of conifer forests in Britain Đ regeneration, gap size and silvicultural systems. Forest Ecology and Management. 2001;**151**:7-23

[22] Akyol A, Örücü ÖK, Arslan ES. Habitat suitability mapping of stone pine (*Pinus pinea* L.) under the effects of climate change. Biologia. 2020;**75**(12):2175-2187

[23] López-Tirado J, Hidalgo PJ. Predicting suitability of forest dynamics to future climatic conditions: The likely dominance of holm oak [*Quercus ilex* subsp. Ballota (Desf.) Samp.] and *Aleppo pine* (*Pinus halepensis* mill.). Annals of Forest Science. 2018;**75**(1):19

[24] Paulo JA, Palma JHN, Gomes AA, Faias SP, Tomé J, Tomé M. Predicting site index from climate and soil variables for cork oak (*Quercus suber* L.) stands in Portugal. New Forests. 2015;**46**(2):293-307

[25] Biber P, Borges J, Moshammer R, Barreiro S, Botequim B, Brodrechtová Y, et al. How sensitive are ecosystem services in European Forest Landscapes to silvicultural treatment? Forests. 2015;**6**(12):1666-1695

[26] Bugalho MN, Dias FS, Briñas B, Cerdeira JO. Using the high conservation value forest concept and Pareto optimization to identify areas maximizing biodiversity and ecosystem services in cork oak landscapes. Agroforestry Systems. 2016;**90**(1):35-44

[27] O'Hara KL. Multiaged Silviculture Managing for Complex Forest Stand Structures. Oxford: Oxford University Press; 2014. p. 213

[28] Plieninger T, Hartel T, Martín-López B, Beaufoy G, Bergmeier E, Kirby K, et al. Wood-pastures of Europe: Geographic coverage, social–ecological values, conservation management,

and policy implications. Biological Conservation. 2015;**190**:70-79

[29] Puettmann KJ, Coates KD, Messier CC. A Critique of Silviculture: Managing for Complexity. Washington, DC: Island Press; 2009. 189 p

[30] Schütz JF. Opportunities and strategies of biorationalisation of forest tending within nature-based management. In: Diaci J, editor. Nature-Based Forestry in Central Europe Alternatives to Industrial Forestry and Strict Preservation. Ljubljana: Univ. Ljubljane, Biotech. Faculty, Dept. of Forestry and Renewable Forest Resources; 2006. p. 37

[31] Gonçalves AC. Challenges to management of evergreen oak forest systems in the Mediterranean basin. In: Jhariya MK, Meena RS, Banerjee A, Kumar S, Raj A, editors. Agroforestry for Carbon and Ecosystem Management [Internet]. London, United Kingdom: Academic Press, Elsevier; 2023. pp. 295-310. Available from: https://shop.elsevier.com/books/agroforestry-for-carbon-and-ecosystem-management/jhariya/978-0-323-95393-1

[32] Bončina A. Conceptual approaches to integrate nature conservation into forest management: A central European perspective. The International Forestry Review [online]. 2011;**13**(1):13-22. DOI: 10.1505/ifor.13.1.13 [Accessed: March 27, 2025]

[33] O'Hara KL. What is close-to-nature silviculture in a changing world? Forestry. 2016;**89**(1):1-6

[34] Puettmann KJ, Messier C. Simple guidelines to prepare forests for global change: The dog and the Frisbee. Northwest Science. 2020;**93**(3-4):209

[35] Spathelf P, Bolte A, van der Maaten ECD. Is close-to-nature Silviculture (CNS) an adequate concept to adapt forests to climate change? Landbauforschung - Applied Agricultural and Forestry Research. 2015;**65**(3-4):161-170

[36] Fahey RT, Alveshere BC, Burton JI, D'Amato AW, Dickinson YL, Keeton WS, et al. Shifting conceptions of complexity in forest management and silviculture. Forest Ecology and Management. 2018;**421**:59-71

[37] Jandl R, Ledermann T, Kindermann G, Freudenschuss A, Gschwantner T, Weiss P. Strategies for climate-smart Forest Management in Austria. Forests. 2018;**9**(10):592

[38] Thom D. Natural disturbances as drivers of tipping points in forest ecosystems under climate change – Implications for adaptive management. Forestry. 2023;**96**(3):305-315

[39] Thom D, Keeton WS. Stand structure drives disparities in carbon storage in northern hardwood-conifer forests. Forest Ecology and Management. 2019;**442**:10-20

[40] Fonseca TF, Gonçalves AC, Lousada J. Maritime pine, its biological and silvicultural traits for the basis of natural resources: An overview. In: Gonçalves AC, Fonseca TF, editors. Conifers - Recent Advances [Internet]. London, United Kingdom: IntechOpen; 2022. pp. 1-35. Available from: https://www.intechopen.com/chapters/80840 [Accessed: October 28, 2024]

[41] Fonseca TF, Costa RNM, Marques CP, Louzada JL, Gonçalves AC. Exploring the biometric traits and potential of Radiata pine (*Pinus radiata* D. Don) as a non-native species for sustainable Forest Systems in Portugal. In: Fonseca TF, Gonçalves AC, editors. Conifers - From Seed to Sustainable

Stands [Internet]. London, UK: IntechOpen; 2024. pp. 51-68. Available from: https://www.intechopen.com/books/1002680

[42] Correia AV, Oliveira AC. Principais espécies florestais com interesse para Portugal: Zonas de influência mediterrânica. Lisboa: Direcção-Geral das Florestas; 1999. 318 p

[43] Correia AV, Oliveira AC. Principais espécies florestais com interesse para Portugal: zonas de influência atlântica. Lisboa: Direcção-Geral das Florestas; 2003. 322 p

[44] Caudullo G, Welk E, San-Miguel-Ayanz J. Chorological maps for the main European woody species. Data in Brief. 2019;**12**:662-666

[45] Alegria C, Roque N, Albuquerque T, Gerassis S, Fernandez P, Ribeiro MM. Species ecological envelopes under climate change scenarios: A case study for the main two wood-production forest species in Portugal. Forests. 2020;**11**(8):880

[46] Espeland EK, Kettenring KM. Strategic plant choices can alleviate climate change impacts: A review. Journal of Environmental Management. 2018;**222**:316-324

[47] Miniat CF, Fraterrigo JM, Brantley ST, Callaham Jr MA, Cordell S, Jeffrey S, et al. Impacts of invasive species on forest and grassland ecosystem processes in the United States. In: Poland TM, Patel-Weynand T, Finch DM, Miniat CF, Hayes DC, Lopez VM, editors. Invasive Species in Forests and Rangelands of the United States: A Comprehensive Science Synthesis for the United States Forest Sector [Internet]. Cham: Springer International Publishing; 2021 [cited 2025 Feb 19]. Available from: https://link.springer.com/10.1007/978-3-030-45367-1

[48] Gonçalves AC. Effects of Forest stand structure in biomass and carbon. In: Shukla G, Chakravarty S, editors. Forest Biomass and Carbon [Internet]. Rijeka, Croatia: IntechOpen; 2018. pp. 1-21. Available from: http://www.intechopen.com/books/forest-biomass-and-carbon/effects-of-forest-stand-structure-in-biomass-and-carbon [Accessed: August 9, 2019]

[49] Sandim A, Emília Silva M, Cristina Gonçalves A, Tomé M, Fonseca TF. Management of *Pinus pinaster* Aiton for wood and resin production: A technical-financial feasibility analysis. In: Gonçalves CA, Fonseca TF, editors. Conifers - Recent Advances [Internet]. London, United Kingdom: IntechOpen; 2022. pp. 165-183. Available from: https://www.intechopen.com/chapters/81743 [Accessed: October 28, 2024]

[50] Gonçalves AC. Stand structure impacts on Forest modelling. Applied Sciences. 2022;**12**(14):6963

[51] Gonçalves AC. Modelling Biomass. In: Gonçalves AC, Malico I, editors. Forest Bioenergy [Internet], Green Energy and Technology. Cham: Springer International Publishing; 2024. pp. 121-146. Available from: https://link.springer.com/10.1007/978-3-031-48224-3_5

[52] Gonçalves AC. Influence of stand structure on forest biomass sustainability. In: Jhariya MK, Meena RS, Banerjee A, Meena SN, editors. Natural Resources Conservation and Advances for Sustainability [Internet]. Cambridge, United States: Elsevier; 2022. pp. 327-352. DOI: 10.1016/B978-0-12-822976-7.00007-7

[53] Sohn JA, Saha S, Bauhus J. Potential of forest thinning to mitigate drought

stress: A meta-analysis. Forest Ecology and Management. 2016;**380**:261-273

[54] Sohn JA, Gebhardt T, Ammer C, Bauhus J, Häberle KH, Matyssek R, et al. Mitigation of drought by thinning: Short-term and long-term effects on growth and physiological performance of Norway spruce (*Picea abies*). Forest Ecology and Management. 2013;**308**:188-197

[55] Gonçalves AC. Thinning: An overview. In: Gonçalves AC, editor. Silviculture [Internet]. London: IntechOpen; 2021. pp. 41-58. Available from: https://www.intechopen.com/books/silviculture/thinning-an-overview

Chapter 2

Aspects Regarding Vegetation Dynamics in Forests and Meadows in Romania

Mariana Niculescu

Abstract

Along with climate changes, the expansion of urbanization, the reduction of pastoral practices at the rural level, and the expansion of invasive species, the vegetation, are facing very great threats and pressures. They affect the physiognomy, the floristic structure of the phytocoenoses within the plant communities of forests and meadows with repercussions on habitats and ecosystems. Vegetation dynamics and maintaining the state of conservation of ecosystems are in close correlation with the natural landscape. The complex mosaic landscape characteristic of ancestral rural landscapes ensures, in addition, a sustainable exploitation of natural resources and exceptional biodiversity. In this chapter, we have tried to include important aspects regarding the syndamics of meadows and forest vegetation in several areas of conservation interest in Romania, an integral part of the Balkan region. These data on the evolution of the vegetation in these areas are the result of the studies we have carried out over several years, at the level of different types of ecosystems, habitats, subject to different ecological, climatic, and anthropic factors. These research studies had well-defined objectives in establishing the type, the degree of syndynamics of the vegetation, and the importance of its evolution, in close correlation with the conservative management measures of the entire natural capital.

Keywords: vegetation dynamics, natural heritage, ecosystem, habitat, phytocoenosis, climate change, anthropogenic factors, conservation, protection

1. Introduction

The vegetal carpet is an essential component of the natural capital, the volume of vital resources, which has a special role in the existence and development of human society. The concept of vegetation includes important indicators regarding the importance and conservation of biological diversity in close correlation with the current eco-pedo-climatic changes, the production of phytomass, the source of oxygen necessary for life, the protection and conservation of water and soil. Resources, conservation of natural landscapes in the fight against aggression, and anthropogenic pressure resulting from human activities, agriculture, industry, tourism, and urban development. Vegetation is a key element that plays a role in mitigating global climate change, particularly by sequestering carbon [1].

Vegetation is a basic component of the landscape of a region, a complex structured, dynamic component, conditioned ecologically, anthropically, and historically, performing important eco-protective, productive, and socio-economic functions [2, 3].

Vegetation, in all its complexity, has the role of regulating the energetic, climatic and hydrological regime of the earth's surface. Vegetation is also a good ecological and geographical indicator because the plant species from which it is formed show special adaptations to climate and soil, and the spread of a certain type of vegetation delimits territories in which a special climate is manifested; there is a special covering of soils and a special complex of biocenoses and natural ecosystems is formed [4].

The vegetal carpet, as an integral part of the living world and implicitly of natural landscapes, has undergone important transformations over time as a result of the interaction with ecological, climatic, and anthropic factors, all of which affect the inheritance of the entire natural capital from which future generations must also benefit. The vegetal carpet is an inexhaustible source of natural wealth, which began to be exploited since the time of the most primitive human communities. However, the rational exploration of the potential resources of the vegetation cover is not possible without a deep knowledge of the elements that make it up, with the reflection of all the mutual links and their own legitimacy [5].

In recent decades and especially in the last decade, together with global climate changes, manifested both by the increase in temperatures at the planetary level, the appearance of phenomena characterized by strong fluctuations in temperature and precipitation, the appearance of very strong winds and tornadoes in regions where did not exist in the past, they caused major changes in the physiognomy and floristic structure of the entire plant cover. All these climatic disturbances have determined the occurrence of extreme weather phenomena, such as floods or the outbreak of strong fires, which have a particularly negative impact on the entire biodiversity. The reality of the last decades and the last years, in particular, has alarmingly proven that humanity does not have infinite natural resources to cover all its needs.

The climate changes and the action of anthropic pressures (overexploitation of biological resources, agriculture and forestry, pollution, industrialization and urbanization, residential and commercial development, human intrusions and disturbances on biodiversity, invasive non-native species, geological events, natural disasters) are in close correlation with the expansion, contraction, degradation, and fragmentation of the distribution area of plant communities, influencing in one way or another the structure and dynamics of the vegetation [6].

Biodiversity around the world is under threat, and the entire vegetation cover is at risk, agricultural and forestry productivity is in decline due to increased pollution, continued soil degradation, and loss of fertility [6, 7].

Forest vegetation is often devastated by the increasingly polluted atmosphere, acid rain, natural catastrophes such as wind blows, heavy floods or fires, and most of the time by its poor management.

The vegetation of the meadows underwent profound changes from a phytocenotic and syndamic point of view, the state of conservation of plant communities being influenced by several factors, their action influencing the evolution of meadow ecosystems.

Studies on the current state of the grassy or woody plant carpet, from a structural and productive point of view, have a decisive role in the conservation of the diversity of plants, animals, and biocoenoses/ecosystems and are of particular importance in establishing the level of its degradation and the possibilities of the application of ecological reconstruction and conservation measures. From here derives the need to know and analyze the vegetation from multiple aspects, as well as to decipher and understand its structural and functional mechanisms and its dynamics in close connection with the optimal actions and methods for its management. Vegetation monitoring, both qualitatively and quantitatively, has the role of providing important data regarding the quality of life and its sustainability.

In order to better manage the vegetal carpet in all its complexity, we have carried out phytocoenotic, syndamic studies in close correlation with protection and conservative management studies in several areas of Romania, an integral part of the Balkan peninsula. The objectives that formed the basis of the studies included in this chapter sought to provide a more faithful image of the dynamics of vegetation in certain areas of Romania, in the context of the action of various disturbing factors and the current eco-pedo-climatic conditions.

2. Materials and methods

2.1 Study area

The paper presents the results of the investigations carried out over several years in several areas of Romania, an integral part of the Balkan Peninsula, from the plain area to the mountain floor, at the level of both forest and meadow vegetation. The studied area included important areas from the so-called "Sahara of Oltenia" but also many areas from the Danube basin and the Romanian Carpathians.

2.2 Methods

Vegetation dynamics research consisted of the analysis of the changes in the physiognomy and floristic composition of the species with an impact on the deep transformations regarding the vegetation cover. In order to provide the most accurate image of the process of vegetation dynamics, in South-Eastern Europe, especially in Romania, we have carried out studies on the progressive and regressive development processes of plant communities under the action of several eco-pedo-climatic factors and anthropic. For the study of the vegetal carpet, we have used methods of phyto-sociologic research characteristic of the Central European phyto-sociologic School, which was based on the principles and methods elaborated by Braun-Blanquet [8] and adapted by Borza [9] to the particularities of our country's vegetation. The plant communities were identified and distinguished according to their characteristics, edifying, dominant, and differential species. For the identification of the plant communities and the coenotaxonomic classification, synthesis works of Mucina et al. [10] and Rodwell [11] were used.

Regarding the studies carried out on vegetation dynamics, were taken into account according to the methodology described in the works developed by Cristea et al. [2, 3].

Following the research on the vegetal carpet, we will take into account its analysis and characterization from a structural, physiognomic, and syndamic point of view in close correlation with the profound eco-pedo-climatic changes of recent years.

Direct observations were made of the processes of phytocoenogenesis manifestations – the formation of new plant communities and the loss of old plant communities. The regression or degeneration of plant communities, as well as the structural transformation of phytocoenoses, the decrease in the number of individuals within a phytocoenosis, as well as the modification of the cenotic nucleus of a plant community are important including for the assessment of the conservation status of biodiversity and natural landscapes. Following the studies carried out in different areas, the direct observation of vegetation dynamics was based on periodic inventories in permanent sample areas to establish the structure of phytocoenoses in a plant community or the extent of changes in the floristic composition during these observations. Thus, in this work, we sought to present concrete models of vegetation dynamics under the action of various anthropogenic factors and especially the action of global climate changes in recent decades. Regarding the dynamics of meadow vegetation in the context of the abandonment of agro-pastoral practices, we took into account the complexity of the vegetal carpet, its functions and characteristics, the evaluation of the dynamics and the pastoral value based on the analysis of the physiognomy and floristic composition, the degree of coverage (%), of the annual evolution of the phytocoenotic structure, as well as the direction of evolution over time of the vegetal carpet, all of which are based on the principles of the management of the pastoral fund.

3. Results and discussions regarding vegetation dynamics

The changes occurring in the structure of the vegetation under the action of different disturbing factors result in certain morpho-functional dynamics of it.

Vegetation dynamics can be expressed in two broad categories: the internal dynamics of vegetation and its succession. The processes of the internal dynamics of the vegetation are manifested at the phytoceonosis level and involve quantitative and reversible changes. The internal dynamics of the vegetation are characterized by a frequency of occurrence, as well as a variable duration of manifestation of these processes in close connection with the variation of abiotic factors, but also by the biology of the component populations [2, 3].

The internal dynamics can be a dynamic of the effective populations of short duration, which appear as a result of seasonal, periodic, regular or irregular and unpredictable fluctuations with direct action, especially on the physiognomy of the phytocenoses and, to a certain extent, the specific wealth. Likewise, the internal dynamics of the vegetation can also be expressed through degeneration-regeneration processes, the value of which depends on the intensity of the factors that act in a shorter or longer time interval on the phytocenoses. The greater the amplitude of changes, as a result of which phytocoenesis is able to regenerate, the higher its elasticity or resilience [5].

The phenomenon of degeneration affects the floristic structure as well as the stability of the phytocenoses within a plant community subject to anthropogenic (defective agro-pastoral practices, incorrect management of forest habitats) or natural (floods, wind blows, explosion of non-living species, and excessive species attack) disturbing factors of phytosanitary quarantine).

Regeneration is the reverse phenomenon of degeneration and occurs when the disturbing factors listed above significantly reduce their action until complete stagnation, so that the phytocenoses restore their structure and floristic composition.

Vegetation succession is a dynamic, directional, irreversible process that is characterized by essential transformations in the composition and physiognomy of a phytocenosis that lead to its replacement with another type of phytocenosis. The succession that has as its starting point a bare land, without vegetation, is a primary succession, and the one that starts from a pre-existing plant grouping is a secondary succession [12].

Such a primary succession can occur, for example, in arid conditions, when the soil goes through deep transformations, becoming sandy and favoring the establishment of a specific, psammophilous vegetation. The secondary succession is much faster and unfolds in a shorter time under climatic and geomorphological conditions comparable to those characteristic of the primary succession [2, 3].

Vegetation dynamics through succession are influenced by biotic and abiotic environmental factors, but especially by certain characteristics of the competing species such as vitality, temperament, growth rate, longevity, and regeneration capacity [13].

The complex process of vegetation dynamics must be approached both in the context of the structural and functional changes that occur both at the level of grassy and woody vegetation.

Forests constitute one of our main natural resources and are of particular importance from an economic, ecological, and esthetic point of view, representing, at the same time, an invaluable genetic fund for flora and fauna and a real source of health. Forests contribute to the preservation of landforms and the environment, having an extremely important role in the phenomena of soil erosion and landslides. In the last decades, forests all over the globe have undergone profound changes from a structural and functional point of view, on the one hand, due to their faulty management or lack of sustainable management. Correct management of forests and stopping the process of degradation and degeneration of the forest fund in all its complexity is of particular importance for the protection of the environment, and the preservation and conservation of the entire natural heritage.

Since ancient times, natural meadows have represented an extremely important element from a botanical point of view, but also from a socio-economic point of view, having an essential role in ensuring the current well-being of mankind. From an esthetic point of view, pastoral landscapes delight the eye. In recent years, there has been a sustained struggle at the European level for the application of pastoral management measures that take into account that the meadows are very well maintained and exploited, so that their conservation status remains within favorable limits in close correlation with the capacity maximum grazing and mowing.

Both the forest vegetation and the grassy vegetation have undergone profound changes from a structural, functional, conservative and productive point of view due to the action of numerous disturbing factors.

The data presented below refer to concrete situations of vegetation dynamics, encountered in Romania in recent years.

3.1 Vegetation dynamics in the context of global climate change

In recent years, major climate changes have produced major disruptions that have affected all types of ecosystems, and all biodiversity.

The excessive temperatures and the prolonged drought of recent years, to which added the irrational deforestation of the *Robinia pseudoaccacia* plantations, have accentuated the phenomenon of desertification in the southwestern area of Romania, in the Oltenia region. The phenomenon of desertification in this area of Romania has been reported since the 1960s by Alexandru Buia [14]. This area was named "Sahara Oltenia" (**Figure 1**) and is individualized both by the presence of sandy soils but also by the very high temperatures and the low amount of precipitation recorded, its surface increasing more and more every year, reaching up to currently over 80,000 hectares, growing by about 1000 hectares every year.

Figure 1.
Aspect with the sandy soil of "Sahara Oltenia," Romania (photo: M. Niculescu).

Due to the current type of vegetation, the wind that blows from the west during summer and autumn produces a strong eolian deflation of the sands, forming strips of sand arranged parallel to the direction of the wind. Thus, the increase of sandy surfaces from year to year has a very large impact on the vegetation, amplifying its dynamics. The influence of all these factors produced major changes in the spontaneous vegetation of this region. These changes aim both at expanding and narrowing the distribution area of certain plant species, as well as the disappearance or appearance of certain plant species in this area.

Thus, the herbaceous plant communities specific to this area are: *Molluginetum cervianae* Borza 1963 [15, 16, 17], *Plantaginetum arenariae* Buia et al. 1960 [15, 16, 18], *Brometum tectori* (Kern. 1863) Soó (1925) 1939 [15, 16, 19], *Vicio-Polygonetum arenarii* Timár (1953) 1957 [15, 16, 20, 21], *Tribulo-Tragetum* Soó et Timár 1954 in Timar 1954 [15, 16, 22], *Artemisio austriacae – Poëtum bulbosae* I. Pop 1970 [15, 16, 22], *Festucetum vaginatae* (Rapaics 1923) Soó 1929 [15, 16, 23]. In this area, large areas of the *Robinia pseudoaccacia* plant community were deforested 30–40 years ago, establishing these types of herbaceous plant communities, which previously occupied much smaller areas and which could not stabilize the sandy soil.

As the desertified surface and the area of these plant communities increases, the physiognomy and floristic composition changes in a very short time, and at the same time, the abundance-dominance of psammophilous species within the phytocoenoses increases to the disadvantage of other species.

Regarding the forest vegetation, the forest plant communities in the Danube basin and its tributaries in the "Sahara Oltenia" and reduced the area, some of them undergoing profound changes in terms of floristic composition, including in terms of building species, for example, in the last five decades, most of the time and on large surfaces the edifying species of *Populus alba* has been replaced by *Populus x canadensis* (Canadian poplar), a lover of higher temperatures, and a much more productive species, affecting the entire biodiversity.

The forests of *Quercus robur* have also suffered, considering the type of soil on which this species has optimal development, the sandy soil being unfavorable to the entire forest plant community that it builds up and which goes through a dynamic process of vegetation succession. In this way, the plant community underwent changes, both from the point of view of the surface, it shrinking considerably, and of the physiognomy and floristic composition, some species disappearing completely or meeting less and less, on small surfaces and affecting the cenotic nucleus of the vegetational community *Convalallario-Quercetum roboris* Soó (1934 n.n) 1975 [15, 24] specific to the area in southern Romania, and implicitly of the forest habitat – 91I0* *Euro-Siberian steppe woods with Quercus spp.* [25, 26], if we refer from a conservative point of view.

The edifying species in the grassy layer *Convallaria majalis*, which gives the vegetal community a characteristic note and has considerably reduced its abundance-dominance and, at the same time, its area, due to ecological conditions with profound changes determined by climate changes.

Due to prolonged drought and high temperatures from the period 2020–2024, the aquatic and swamp vegetation has suffered in the Danube riverside region, in the southwest of Romania. Thus, on Copănița Island on the banks of the Danube, which is an integral part of a protected area, some plant communities have regressed to the point of extinction, bringing with them the entire biodiversity.

The plant communities *Lemno-Salvinietum natantis* Miyawaki et J. Tx. 1960 (**Figure 2**) [16, 27], *Spirodelo-Salvinietum natantis* Slavnic 1965 [16, 28], *Trapo-Nymphoidetum* Oberd. 1957 [16, 29], *Trapetum natantis* (V. Karpati 1963) Th. Muller et Gors 1960 [16, 30], *Lemno-Azolletum carolinianae* Nedelcu 1967 (**Figure 3**) [16, 31], *Myriphyllo verticillati-Nupharetum luteae* Koch 1926 [16, 32] almost disappeared or were replaced by some synanthropic plant communities such as *Xanthio strumarii – Bidentetum tripartitae* Timar 1947 (**Figure 4**) [16, 33]. At the same time, in the south-west of Romania and beyond, the plant community that harbors the species *Marsilea quadrifolia* has disappeared in proportion to 90%, this rare, protected species being almost non-existent due to several factors including climate changes and water pollution, with industrial residues, agricultural and domestic. Following the studies we

Figure 2.
Lemno-Salvinietum natantis *plant community deeply affected by drought, Copănița island, Romania, July 2022 (photo: M. Niculescu).*

Figure 3.
Azola filiculoides *in drought conditions (photo: M. Niculescu).*

Figure 4.
Installation of synanthropic vegetation, Copănița island (photo: M. Niculescu).

have carried out in the last 5 years, this species was no longer found in the locations where it had been cited years ago [34].

Due to the prolonged drought in the Danube Valley, the beach expanded and new islands were formed on which pioneer vegetation built up by poplar took root (**Figure 5**). In the willow plant communities of the Danube Valley, due to the prolonged drought, the grassy layer has regressed to the point of despair (**Figure 6**).

3.2 Vegetation dynamics under the influence of natural disasters

Natural catastrophes, such as fires, floods, and windstorms, which have registered a fairly high frequency in the last decade, deeply influence the state of vegetation conservation. In recent years, although Romania has a temperate continental climate, which would not predispose to wildfires, in the southwest of this country, in the Cernei, Almajuli, Mehedinți-Domogled Mountains, on the calcareous rocks, the degree of dryness of the vegetation and the prolonged drought, caused by extremely high temperatures, with significant deviation from normal, the lack of precipitation in the hot and dry season, to which is often added the negligence of people, caused

Figure 5.
Poplar pioneer vegetation, Danube Valley (photo: M. Niculescu).

Figure 6.
Plant community white willow, Danube Valley (photo: M. Niculescu).

the appearance of such phenomena, which had a special impact particularly serious on forest vegetation. For example, in 2022, over 160,000 hectares were affected by wildfires in Romania.

In the Almaj Mountains, in the Portile de Fier Natural Park in recent years, including this year 2024, in these hot days of July, a large fire broke out that affected a large area of *Pinus nigra* ssp. *banatica* forest (**Figures 7** and **8**). This species is found spontaneously only in the southwestern part of Romania, more precisely in the southern area of Banat and western Oltenia (Cernei, Almăjului, Mehedinți-Domogled, Danube-Tricule Mountains), on calcareous cliffs, between 150 m (Tricule) and 1500 m altitude (Mehedinți Mountains) being an endemism of the Southern Carpathians. In the southwest of Romania, the Banat black pine, together with the rare herbaceous species, *Genista radiata*, builds the habitat 9530* *Sub-Mediterranean pine forests with endemic black pines* [25, 26], Natura 2000 habitat, with a special structure.

Following the fires that took place in recent years, not only the layer of trees suffered, but also the grassy carpet in the forest was scorched; *Genista radiata*, a rare species, was no longer found following the studies we carried out in the last years, the entire floristic structure of the plant community being disturbed, the dynamics of the vegetation being inevitable.

Figure 7.
Forest habitat with Pinus nigra *ssp.* banatica *"Iron Gates" Natural Park (photo: M. Niculescu).*

Figure 8.
Fire started in the forest habitat of Pinus nigra *ssp.* banatica*, "Portile de Fier" Natural Park, July 2024 (photo: M. Niculescu).*

These fires also affected the habitats of protected meadows in the area, and the hot temperature, the wind, and the rough terrain, difficult to access, rocky, and with steep slopes, favored the spread of the fires and made it very difficult to extinguish them.

In this way, the entire biodiversity is affected in this area, which also has direct implications for the physiognomy and structure of natural landscapes.

Strong winds, to which is added the exposure and the degree of inclination of the slopes in the mountainous region, can cause trees to fall over large areas. In many areas affected by massive falls, due to the difficulty of access, fallen trees cannot be completely removed, using anti-ecological methods to collect them, considering this as a consequence of anthropogenic impacts (**Figure 9**).

Countless times, in the Romanian Carpathians, such a windthrow phenomenon has had a direct influence on the dynamics of the vegetation; in some hard-to-reach places, the forest vegetation being replaced by shrub and underbush vegetation, such as that built by *Rubus iadeus, Sambucus racemosa or Vaccinium myrtillus.*

The overflow of the main rivers and their tributaries caused by the sudden melting of the snow and the heavy rains of the last decades have produced floods, which have repeatedly had a devastating impact both on the entire biodiversity but also from a socio-economic point of view.

Most of the floods produce ecological effects, such as the destruction of vegetation and ecosystems. Heavy flooding can result in the formation of silt-covered areas or the formation of swamps.

Following strong floods, there can be changes in the structure of the soil due to excess moisture, the appearance or disappearance of islands, the extension of concave

Figure 9.
Spruce forests affected by wind in the Apuseni Mountains, Romania (photo: M. Niculescu).

banks through erosion and the formation of steep banks, the washing of nutrients from the soil, the physical influence of water on vegetation, all of these intervening in the vegetation dynamics process.

Some plant communities of tall grasses, included in the habitat 6430 *Hydrophilous tall-herb fringe communities of plain and of the montane to alpine levels* [25, 26], habitat Natuara 2000, which meet on the banks of the waters following strong floods undergo changes in physiognomy and the floristic composition, the habitat of some species being destroyed. For example, the plant community *Telekio-Petasitetum hybridi* (Morariu 1967) Resmeriţă et Raţiu 1974 [16, 35], which is an integral part of the habitat of the Natura 2000 species *Tozzia carpathica* Woll. (species present in Poland, Romania, Slovakia and Bulgaria) has suffered many times, the distribution area of this species being more and more restricted, and the conservation status being unfavorable.

3.3 Mining quarries in close correlation with vegetation dynamics

Surface mining exerts a smaller or larger anthropogenic impact on the vegetation and implicitly on the entire biodiversity, depending on the capacity and management of the natural resources, altering the natural landscape in the respective area [36].

The forested lands that are on the border of mining quarries in Romania, as well as in the entire Balkan region and beyond, are subject to high atmospheric pollution, showing important signs of degradation. Since 2012, we have carried out such research studies on the forest vegetation in the area of several mining quarries that exploit limestone in the Carpathian Mountains: Bistrita – Vâlcea, Lespezi, Băiţa – Crăciuneşti, Bicaz Chei, Hoghiz, Malnaş and Mateiaş. The vegetation that is on the border or even inside a mine pit is primarily affected by the dense dust from the mine pits, which prevents the normal development of physiological processes. The phytocoenotic structure of a plant community undergoes changes, the vegetation being in a dynamic stage of degeneration (**Figure 10**). Landslides affect the entire vegetation that is in the immediate vicinity of the exploited area. Following the mining operations, the succession process of the vegetation is present in these areas, and primary succession can be found on the waste heaps but also in the steps of the quarry where the mining operation was carried out (**Figure 11**).

Figure 10.
Forest vegetation in the context of limestone mining (Mining quarries Bistrita – Vâlcea and Lespezi, Romania) (photo: M. Niculescu).

Figure 11.
Installation of synanthropic vegetation (primary succession) – after mining, Băița-Crăciunesti Mine, Romania (photo: M. Niculescu).

3.4 Overexploitation of the forest vegetation

The current needs of our society regarding wood materials are far beyond the natural potential of the existing forest stock.

The irrational deforestation that took place recently worldwide had dire consequences, especially in terms of the climate, leading to the appearance of negative phenomena of aridization and desertification. Also, the negative anthropogenic impact exerted on the forest vegetation in our country led and lead to the emergence of erosion phenomena and landslides. In this sense, it is necessary to impose measures to protect, conserve, and improve the forest vegetation and implicitly, the soils, an essential condition for the sustainable development of forestry in our country.

The industrial overexploiting of the wood has a negative impact, leading to the degradation of the forests and the forest soil, allowing some erosion processes to

Figure 12.
Anthropogenic impact on forest vegetation (photo: M. Niculescu).

start. Usually, forest exploitation tends to exhaust, to fragment and to homogenize the forests. In order to resist the impact of such a challenge, it is necessary to increase not only the productivity of the already existent forests but also the reforesting of the areas where a lot of trees were cut down. On the other hand, we must avoid artificial reforesting, under the form of monocultures, for this thing could lead to continuous diminishing of the flora and fauna diversity [6]. The exploitation of the forests led to the appearance of a dense network of forest roads, which itself supposes important deforesting. These ways of access, in the regions with pronounced slopes, strongly favor soil erosion and sediment formation in the river basins. The heavy machines, used for log transportation, also have a negative impact because of the soil strong settling (**Figure 12**) [6].

3.5 The evolution of vegetation in the context of migration and invasion of adventitious species and undesirable species

Another very important aspect that intervenes in the process of vegetation dynamics is also represented by the appearance of some invasive plant species, adapted to the current eco-pedo-climatic conditions. Adventitious invasive species exert a particular anthropogenic impact on the natural vegetation, being present both in the forest floor and meadows. Most of these species are introduced directly, intentionally, for ornamental purposes or through agricultural or forestry practices by not complying with phytosanitary quarantine rules. It should be noted that every spring, some areas on the banks of the Danube are affected by the increase in the levels and flows of this river, especially the riparian areas located in the low, undammed areas, prone to flooding in the event of the Danube overflowing, being flooded pastures and hayfields natural areas, arable land, fields, forest floor, a series of invasive species being brought in this way.

Of the total species that have a high degree of invasiveness and aggressiveness and influence the dynamics of plant communities, we can mention *Ailanthus alissima* (Miller) Swingle, *Sicyos angulatus* L., *Asclepias syriaca* L., *Amorpha fruticosa* L., *Ambrosia artemisiifolia* L., *Erigeron canadensis* L., *Echinocystis lobata* (Michx) Torr. et A. Gray, *Symphyotrichum lanceolatum* (Willd.) G.L. Nesom., *Phytolacca americana* L., *Acer negundo* L., *Oenothera biennis* L., *Reynoutria × bohemica* Chrtek & Chrtková, *Reynoutria japonica* Houtt., *Solidago canadensis* L., *Rudbeckia laciniata* L. [37].

In the last decades, *Albizia julibrissin* Durazz., which represents a new threat from an invasive point of view, was introduced in Romania for ornamental purposes.

Following the observations made in the last years, this species has an extremely fast growth rate, low water requirements and the ability to withstand hot temperatures and atmospheric pollution.

In Romania, *Ailanthus alissima* (**Figure 13**) is an invasive phytosanitary quarantine species that was initially introduced for ornamental purposes, and then it was used for the reconstruction of habitats and degraded, eroded lands. This species has an extremely high capacity to reproduce vegetatively, through budding, but also generatively, producing an extremely large number of fruits and seeds annually. Following the studies, we found that this species also has a very high sprouting capacity after the stems were cut at a distance of 20–30 cm above the ground and injected with certain types of herbicides. From a morphological point of view, fruits and seeds show adaptations that greatly favor dissemination and germination. In recent years, this species has invaded the forest, practical and ruderal habitats, intervening in the structure and floristic composition of the phytocenoses, greatly influencing the dynamics of the vegetation. In 2011 in Romania, Sîrbu and Oprea [38] described a new plant community – *Balloto nigrae-Ailanthetum altissimae* Sîrbu and Oprea [38] This plant community built up by *Ailanthus altissima* and *Ballota nigra* was described from sunny lands, on roadsides, railway embankments, railway stations, steep slopes, ravines, abandoned fields, abandoned orchards or vineyards on terraces, ruderal lands, sandy lands, degraded pastures, or on the edge of *Robinia pseudacacia* plantations, in disturbed habitats, originally occupied by different groups of ruderal weeds (*Stellaritea mediae* and *Artemisietea vulgaris*), or by some degraded pastures (*Festuco-Brometea*), in various successional stages [39].

Sicyos angulatus (**Figure 14**) had a very large explosion in Romania, in the Danube basin, after the natural floods of 2005, having a very large anthropogenic impact, especially on the willow and poplar plant communities [40]. Following the studies carried out on the degree of invasion and the current eco-pedo-climatic conditions, we found that there is a direct relationship between the density of individuals of *S. angulatus* and the structure and dynamics of plant communities.

Asclepias syriaca (**Figure 15**) is frequently found in Romania in open habitats, most often in meadows, on uncultivated agricultural land, at the edge of the forest, and on fertile, sandy soils. Following the completed studies, we encountered abundant populations with over 650 individuals/100 m^2, in the south and southwest of Romania, in the Danube Valley, and on Old Moldova Island. Following the studies carried out in

Figure 13.
Ailanthus altissima *Moldova Veche Island (photo: M. Niculescu).*

Figure 14.
Sicyos angulatus, *Copănița Island (photo: M. Niculescu).*

Figure 15.
Asclepias syriaca, *Danube Valley (photo: M. Niculescu).*

the Danube Valley, it was found that the majority of individuals in a population can produce 5–6 capsules, each with a very large number of viable seeds, which are spread very easily due to their specific morphological characters and which have a capacity of extremely high germination. All this results in the clearing of the invaded surfaces and the damage to the native vegetation and its strong syndynamics [37]. These populations invade the following types of meadows: *Festuco vaginatae-Corynephoretum* Soó 1935 [16, 41], *Bassio laniflorae-Brometum tectorum* (Soó 1938) Borhidi 1996 [16, 42], *Agrostio-Festucetum pratensis* Soó 1949 [16, 43], *Potentillo arenariae-Festucetum pseudovinae* Soó (1939) 1950 [16, 44].

Amorpha fruticosa (**Figure 16**) is an adventive species, frequently spread on floodplains in river valleys and on sandy, oligotrophic soils. In Romania, in the valleys of the main rivers and especially in the Danube Valley, they form dense populations, in the form of strips of varying widths, sometimes quite large and very dense, so that they are difficult to cross, having an extremely negative impact on the structure of the vegetation and the microclimate in the invaded area [37]. The meadow forest

Figure 16.
Amorpha fruticosa *in forest and meadow habitats (photo: M. Niculescu).*

plant communities of the Danube Valley and the main tributaries subject to major disturbances from a physiognomic, structural and implicit syndamic point of view are *Salici-Populetum* Meijer-Drees 1936 [16, 45], *Stellario nemori-Alnetum glutinosae* (Kastner 1938) Lohmeyer 1957 [16, 46], *Carici remotae-Fraxinetum* Koch ex Faber 1936 [16, 47], *Fraxino danubialis-Ulmetum* Soó 1936 corr. 1963 [16, 48], *Salicetum fragilis* Passarge 1957 [16, 49], *Salicetum albae* Issler 1924 [16, 50]. *Amorpha fruticosa* has a massive development destroying the herbaceous and shrub layer within these plant communities. It also negatively influences the natural regeneration of the edifying, characteristic, and dominant species, suffocating including the saplings. Many times *Amorpha fruticosa* also appears in the habitats of mesophilic meadows, intervening in their floristic structure. In order to restore the forest habitats in the Danube Valley, we carried out experiments for the mechanical and chemical control of this species [37]. The best yields were given by the application of some combined mechanical and chemical methods (**Figure 17**).

Phytolacca americana (**Figure 18**) grows a lot in forest habitats, sometimes together with other invasive adventive species *Erigeron canadensis, Echinocystis lobata* intervening in the modification of the floristic structure of the grassy and subshrub layer due to the abundance of high dominance [37].

Reynoutria japonica produces major disturbances, especially in riparian or synanthropic vegetation on roadsides, on sandy, alluvial, marshy soils, and in ruderalized areas [37, 39]. The massive development of this species produces profound changes in the floristic structure of phytocoenoses, reducing the development and regeneration

Figure 17.
Mechanical control measures of the Amorpha fruticosa *species (Moldova Veche Island) (photo: M. Niculescu).*

Figure 18.
Phytolacca americana.

of native species and, at the same time, their diversity, implicitly intervening in the dynamics of plant communities built by *Salix alba, S. fragilis, S. purpurea, Populus alba*, or communities plants of tall grasses on the banks of the waters. This species is considered one of the most dangerous invasive plants, acting negatively on the entire biodiversity, habitats, and populations of fauna species. It can form the plant community *Reynoutrietum japonicae* Hilbig 1995 [16, 51] (Syn.: *Polygonetum cuspidati* (Moor 1958) Oberd. et al. 1967) [16, 52].

Acer negundo was intentionally introduced by foresters, being planted as an auxiliary species in forest habitats, especially in poplar, black alder, and willow groves, developing massively and having the capacity to substitute native vegetation due to its great power of adaptation, spread, and resistance to less favorable environmental conditions, leading to the reduction of local biodiversity and changes in the structure and functionality of the invaded ecosystems.

Following the geobotanical studies carried out in Romania, we found that some species are undesirable species and affect the floristic composition and favorable state of conservation of both the forest and meadow vegetation communities.

Some species such as *Rubus hirtus, R. canescens, Salvia glutinosa, Pteridium aquilinum, Glechoma hederacea, G. hirsuta,* and *Alliaria petiolata* have a massive development in some phytocoenoses exceeding 20% coverage, suffocating the other species in the grassy and subshrub layer.

Rubus hirtus (**Figure 19**) grows in forest habitats, being often found in brambles, carpino-brambles, and gorunets. It has a very negative impact on forest regeneration, suffocating seedlings and saplings of edifying tree species. In the southwest of Romania, this species has been identified in several types of forest, of which it exerts a very large impact both on the grass layer and on the saplings within a rare forest plant community: *Castaneo-Quercetum* Horvat 1938 [53]. This plant community edified by *Castanea sativa* edifies the forest habitat 9260 *Castanea sativa* woods [25, 26], Natura 2000 habitat. Extremely rare in Romania, from a chronological point of view, meeting only on the hilly floor in the north, northwest and southwest of the country. The purest and most valuable chestnut forests of this type are found in southwest Romania: Pocruia Hill, Tismana Valley, Cracul Gorgului, Dl. Pădurea, Poiana Măgurei, Dealul Cioclovina, Valea Rachițeaua, Valea Bistricioara, Runcu [54].

Figure 19.
Rubus hirtus *in the forest in the* Alnus glutinosa *forest with* Fagus sylvatica *(photo: M. Niculescu) (photo: M. Niculescu).*

Rubus canescens is an undesirable species with a massive development in the meadows included in the following plant communities: *Medicagini minimae-Festucetum valesiacae* Wagner 1941 [16, 55, 56]; *Botriochloëtum ischaemi* Pop 1977 [16, 55, 57]; *Festucetum valesiaco-rupicolae* Csüros et Kovacs 1962 [16, 55, 58]; *Agrostio-Festucetum valesiacae* Borisavljevic et al. 1955 [16, 55, 59]. The high abundance-dominance of this species leads to the change of the floristic composition of these plant communities until their complete replacement. This process of degradation and at the same time of dynamics of meadow vegetation under the action of the species *Rubus canescens* (**Figure 20**) was found in some hilly areas in the north of Oltenia, Romania.

3.6 Poor forest management and use

Intensive grazing also has a negative impact on the forest phytocoenoses because of the decrease in the vegetal biomass and in the number of species with a fodder value. The forests near the sheepfolds, which actually are cow sheepfolds, are characterized by a very poorly degraded grassy layer because of the intensive grazing in these areas and the large number of animals in these sheepfolds, and also due to the too many sheepfolds.

Figure 20.
Rubus canescens *invading the meadow habitat (photo: M. Niculescu).*

So said, within what limits can grazing be admitted to the forest fund? Does it have an important role in forest ecosystems or not? Perhaps if it is completely prohibited, it may lead to an excessive development of the shrubby and subshrub layer of *Crataegus monogyna, Rosa canina, Prunus spinosa, Ligustrum vulgare, Euonymus sp., Rubus caesius, R. fruticosus,* and *R. hirtus*. This excessive development affects the state of conservation of the forest vegetation and implicitly of the entire ecosystem and the natural landscape. Animals can consume the saplings of these species, if grazing is controlled and carried out within certain limits, so that the shrub layer does not develop excessively so that a deciduous forest is not impenetrable. Such situations were found especially in *Quercus cerris* and *Q. frainetto* forests in southern Romania, where the shrub layer developed excessively.

In the last decade, on the desertified lands found in the "Oltenia Sahara," an extensive process of ecological forest reconstruction and restoration of vegetae communities began by restoring the forest curtains of *Robinia pseudoaccacia* [15]. The restoration of forest plant communities is based on the phenomenon of regeneration. The best way to regenerate the forest vegetation is the natural regeneration from the seed, which is established naturally after cutting. If this is not possible and the restoration of forest plant communities is done through artificial regeneration, it is very important to use native forest saplings, produced in nurseries, to restore small areas resulting from definitive cuttings, razed or in those that are not have regenerated naturally through the application of intensive, rational and multifunctional management methods of the degraded forest stock.

From a sequential point of view, due to the current eco-pedo-climatic conditions, it was found that there is a general tendency of altitudinal descent of the area of the main forest species. As regards the dynamics of the succession after the exploitation by clear cutting or the deforested lands following strong floods or strong wind blows, we can first speak of a pine vegetation edified by birch, aspen, red elder, etc. Within natural forests, in the wake of forestry operations, successions can be produced by replacing the native edifying species that can be replaced by fast-growing species for example after the exploitation of *Quercus robur*, it can be replaced by *Tilia cordata*, it is a fast-growing species [60]. In the current stage of forest evolution, it is appreciated that there is a general tendency for altitudinal descent of the area of the main forest species, a descent that also imprints the current character of the succession of forest vegetation. In order to ensure the conservation of the forest fund in close connection with the vegetation succession processes, certain silvicultural norms must be respected: for example, the creation of a regeneration composition with native species corresponding to the specific ecological conditions, the natural regeneration to be realized by using the seeds of high-value species.

3.7 The dynamics of meadow vegetation in the context of the abandonment of agro-pastoral practices

The abandonment of agro-pastoral practices in the rural area had a negative impact on the vegetation of meadows and also on the sindynamic and the conservation of the traditional archaic rural landscape. Vegetation dynamics and maintaining the state of conservation of ecosystems are in close correlation with the natural landscape. The complex mosaic landscape characteristic of ancestral rural landscapes ensures, in addition to a sustainable exploitation of natural resources and an exceptional biodiversity. Rural revitalization in accordance with the traditional practices of revitalizing meadows and maintaining natural landscapes.

With the abandonment of traditional agro-pastoral practices in the last decades, not only has the meadow vegetation undergone profound transformations to the point of degradation, but also the natural landscape has also changes irreversibly. The thickets of *Crataegus monogyna, Rosa canina, R. gallica, Prunus spinosa, P. insititia, P. mahaleb, Rubus caesius, R. fruticosus, R. hirtus, Cotinus coggyria,* and *Berberis vulgaris* (**Figure 21**) have started to invade pastures and hay fields as a result of delaying their correct management.

Sometimes, the meadows in the hill or mountain region are found on the border of the forest or inside them, so that some forest species can migrate to these meadows. In order to avoid the transformation, afforestation of meadows, or harm to the plant communities of meadows, their abandonment and the application of their cleaning works must be avoided. Woody vegetation quickly settles in meadows abandoned or incorrectly managed by the owners. Agricultural practices must be revitalized, their regular mowing and the cleaning and removal of woody species must be carried out according to the norms imposed by the pastoral arrangements, so that the meadows are permanently in a favorable state of conservation. In the area with low financial potential, it is necessary to find possibilities for applying some agro-pedo-ameliorative measures or land improvements (drainage, drying, irrigation, and fertilization) that will lead to the revitalization of plant communities of shepherd and avoid their syndinamic in terms of anthropic view.

Intensive grazing and the excessive stationing of animals lead to the appearance of some species that lead to the deterioration and dynamics of vegetation, especially around sheepfold. Thus, for example, there can be an explosion of species such as *Veratrum album* (**Figure 22**), *Rumex obtusifolim, Urtica dioica*, and *Pteridium aquilium* (**Figure 23**) [60]. Thus, the floristic structure of these meadow plant communities

Figure 21.
Invasive shrub species in meadows (photo: M. Niculescu).

Figure 22.
Veratrum album *invading the meadow habitat, Căpățânii Mountains (photo: M. Niculescu).*

Figure 23.
Pteridium aquilium *invading the meadow habitat, Cernei Mountains (photo: M. Niculescu).*

was modified led to their evolution into plant communities such as *Rumicetum alpini* Szaf., Kulcz. and Pawl. 1923, 1927 [16, 56], *Rumici obtusifolii-Urticetum dioicae* Kornaš 1968 [16, 56, 61], *Clinopodio vulgaris-Pteridietum aquilini*i Dihoru 1975 [16, 56, 62], *Veratretum albi* Puşc. et al. 1956 [16, 56, 63].

Following the phytocoenotic research and the assessment of the pastoral value in some areas of the Romanian Carpathians, it was found that the meadows overgrazed by *Festuca rubra, F. nigrescens, F. airoides, Nardus stricta,* and *Poa media* go through a significant syndamic process. The phytocoenoses of these plant communities can return to *Festuca nigrescens* or *Nardus stricta* meadows, if intensive grazing and long-term stocking of animals are eliminated in those areas [60]. The phytocoenoses of these plant communities can return to *Festuca nigrescens* or *Nardus stricta* meadows, if intensive grazing and long-term livestock stationing are eliminated on the respective surfaces.

In order to maintain the favorable state of conservation of meadow vegetation, grazing with a mixture of animal species (cows, sheep, horses, and goats) on the same pasture area should be avoided. Grassland plant communities are subject to changes in

Figure 24.
Strips of Pinus mugo *in the Căpătânii Mountains (photo: M. Niculescu).*

phytocenotic and physiognomic composition and due to loss of surfaces by changing the category of use and/or the destination of the land, as well as by the expansion of grassland surfaces over land with other uses.

In Romania, in the last decades, in order to increase the grazing areas, important areas of *Pinus mugo* were deforested and sometimes burned in the form of strips, so that the plant community built by this species in the Romanian Carpathians has reduced the area, and the areas left after deforestation or the arsonists suffered from a syndamic point of view, in some places these evolved into plant communities of *Nardus stricta* and in others *Bruckenthalia spiculifolia*, *Vaccinium mirtyllus* or *V. gaultherioides* were established. In the Căpăţânii Mountains, such actions were encountered, the entire biodiversity and, implicitly the natural landscapes being subjected to deep transformations (**Figure 24**) [60].

4. Conclusions

Confirmation of the expansion of the degree of general biological degradation also comes from the decrease in the diversity of life: in the Balkan area but also in the whole of Europe, a large number of species of flora and fauna are increasingly rare, vulnerable or threatened, this state of fact being caused by the onslaught of industrial development, continuous and chaotic urbanization, as well as by profound climate changes that affect the entire natural heritage.

The revitalization of the pastoral fund by applying management measures regarding changes in the structure of meadows through the expansion of invasive species, and shrub species, but also by limiting anthropogenic activities with a major impact on practical habitats such as mining, continuous urbanization, tourist activities, and global climate changes.

The conservation of the agro-pastoral landscape and biodiversity must be in close accordance with the degree of expansion, the dynamics of the floristic composition and the dynamics of the distribution area.

The biodiversity of these landscapes represents the natural heritage of a region, and conservation has an essential role in European agriculture. Conservation of the agro-pastoral landscape and biodiversity is a current problem that requires the implementation of appropriate management measures. Conservation preserves the regional and global ecological balance and ensures the regeneration of biological resources and the maintenance of a quality environment necessary for society, at present and for future generations in the context of sustainable development.

Acknowledgements

This paper contains many of the results obtained in the following research that I have carried out on many works, several years for the realization of the PhD thesis "Flora and vegetation in the upper basin of the Luncavăţ River" ("Babeş-Bolyai" University, Faculty of Biology and Geology, Department of Taxonomy and Ecology, Cluj-Napoca, 2006), as well as numerous studies from the research projects in which I have worked: "Study of monitoring for the migration and invasion species: *Ambrosia artemiisifolia* and *Phytolacca americana* in the forest habitats from the lower basin of Jiu river. Identification of the active measures to combat" (Project no. 10C/05.10.2018, financially support by University of Craiova and Terra Silva Company, Romania), "The scientific study assessment of Natura 2000 habitats and species to substantiation ecological reconstruction in the Lespezi Quarry "(project won the competition-Quarry Life Award launched by HeidelbergCement Group, 2012), "Services for the Application of active conservation measures to maintain the distribution area of priority habitats and improve their conservation status" (Project POIM, 2020-2022, financially support by Integra Trading, Bucharest, EPMC Consulting, Cluj-Napoca) and "Revision of the Management Plan of the Natural Protected Areas ROSCI0045 Jiului Corridor, ROSPA0010 Bistreţ and the Drănic Fossil Site-2391 and Zăval Forest Nature Reserves- IV.33" (Project POIM 150549 coordinated by Dolj County Council and University of Craiova).

Author details

Mariana Niculescu
Faculty of Agronomy, Department of Botany, University of Craiova, Craiova, Romania

*Address all correspondence to: mniculescum@yahoo.com

References

[1] Le Quéré C, Andrew RM, Friedlingstein P, Sitch S, Pongratz J, Manning AC, et al. Global carbon budget. Earth System Science Data. 2017;**10**:405-448. DOI: 10.5194/essd-10-405-2018

[2] Cristea V. Fitocenologie Şi Vegetaţia României. Cluj-Napoca: Univ. "Babeş-Bolyai"; 1993. 180 p

[3] Cristea V, Gafta D, Pedrotti F. Phytosociology. Cluj-Napoca: Editura Presa Universitară Clujeană; 2004. 164 p

[4] Donita N. Etape în cartografia pădurilor şi a învelişului vegetal din România. Bucovina Forestieră. 2014;**14**(1):117-120

[5] Borza A, Boşcaiu N. Introducere în studiul covorului vegetal. Bucureşti: Ed. Acad; 1965. 338 p

[6] Niculescu M, Duță A. The anthropic impact of the wooden vegetation in the Căpățânii Mountains (Romanian Carpathians). In: Proceedings of the XXXIV ESNA Annual Meeting, Novi Sad, Serbia and Muntenegru. Ed. Mondograf; 2004. 415-418 p

[7] Tivy J. Biogeography: A Study of Plantsin the Ecosphere. Essex, Routledge: Longman Scientific & Technical, Taylor & Francis Group; 1993. 440 p

[8] Braun-Blanquet J. Pflanzensoziologie. Springer, Berlin: Grundzüge der Vegetationskunde; 1928

[9] Borza A. Studii fitosociologice în Munții Retezatului, Bul. Grăd. In: Boot. și al Muz. Bot. Vol. 14. Cluj-Napoca: De la Univ. Cluj; 1934. 1-2, 1-84 p

[10] Mucina L, Bültmann H, Dierßen K, Theurillat J-P, Raus T, Čarni A, et al. Vegetation of Europe: Hierarchical floristic classification system of vascular plant, bryophyte, lichen, and algal communities. Applied Vegetation Science. 2016;**19**(Suppl. 1):3-264. DOI: 10.1111/avsc.12257

[11] Rodwell JS, Schaminée JHJ, Mucina L, Pignatti S, Dring Moss JD. The Diversity of European Vegetation - an Overview of Phyto-Sociological Alliances and their Relationships to EUNIS Habitats, Report No. EC-LNV 2002 (054). Wageningen, The Netherlands: National Reference Centre for Agriculture, Nature and Fisheries; 2002

[12] Ivan D. Fitocenologie şi vegetaţia Republicii Socialiste România. Bucureşti: Ed. Didactică şi pedagogică; 1979. 331 p

[13] Talmaci I., Miron A., Managementul durabil al pădurilor și pajiștilor deținute de autoritățile publice locale, Chișinău [Internet] 2016. 46 p. Available from: https://faolex.fao.org/docs/pdf/mol224797.pdf [Accessed: August 12, 2024]

[14] Buia A. Există stepă naturală în Oltenia? Vol. 1. Bucharest, Romania: Com. de botanică, Soc. de Şt. Nat. şi Geografie din R.P.R; 1957;**1**:1-10

[15] Niculescu M, Grecu F, Popescu V. The corology, ecology and phytosociology of the sands plant communities and natural habitats in the sandy area from Dabuleni, Dolj County – Romania. In: Proceedings of the 14th SGEM GeoConference on Ecology, Economics, Education and Legislation, Iulie 2014; Albena, Bulgaria. Vol. 2. 2014. pp. 697-702

[16] Sanda V, Popescu A, Barabaş N. Cenotaxonomia şi caracterizarea grupărilor vegetale din România. Is

Stiinte si Comunicari Journal (St. Com.). 1997;**14**:5-366

[17] Borza A. Pflanzengesellschaften der Rumänischen Karpathen, Biologia, Casopis Slovenschej Académie Vied. Bratislava. 1963;**18**(11):856-964

[18] Buia A, Păun M, Maloş C. Pajiştile naturale din regiunea Craiova şi îmbunătăţirea lor (I). Probleme Agricole. Craiova, Romania; 1960;**12**(12):31-40

[19] Soó R. Homokpustai és sziki növényszövetkezetek a Nyírségen. Dand- und Alkali-steppenassoziationen des Nyírség [Sandy- and alkali-steppe associations of Nyírség]. Botanikai Közlemények. 1939;**36**:90-108

[20] Timár L. Zonációtanulmányok szikes vizek partján, Bot. Közlem. 1957;**47**:157-163

[21] Timár L. A Tiszazug növényföldrajza. (Pflanzengeographie des Tiszazug.) Földrajzi Értesítő. 1954;**2**(3):554-567

[22] Pop I. Adnotaţiuni la studiul comparativaupra vegetaţiei litoralului Mării Negre din împrejurimilelocalităţii Vama Veche (Dobrogea). Cluj: Contrib. bot; 1970. pp. 99-110

[23] Soó R. Revision der Orchideae-Ophrydineae von Ostasien und dem Himalaya. Annales Musei Historico-Naturalis Hungarici. 1929;**26**:339-384

[24] Soó R. Synopsis Systematica-Geobotanica. In: Florae Vegetationsque Hungariae. Budapest: I-IV, Académiai Kiadó; 1975. pp. 1964-1980

[25] Gafta D, Mountford O. Romanian Manual for Interpretation of Eu Habitats. Cluj-Napoca: Risoprint; 2008. 101 p

[26] Mountford O, Gafta D, Anastasiu P, Barbos M, Nicolin A, Niculescu M, et al. Natura 2000 in Romania. In: Habitat Fact Sheets. Bucharest, Romania: Romanian Ministry of Environment and Sustainable Development; 2008. 252 p

[27] Miyawaki A, Tüxen J. Über Lemnetea-Gesellschaften in Europe und Japan. Vol. 8. Stolzenau Weser, Germany: Mitteilungen der floristisch-Soziologischen Arbeitsgemeinschaft. N.F. Heft Stolzenau/Weser, Germany; 1960. pp. 127-135

[28] Slavnić Ž. O intraspecijskim oblicima vrste Euphorbia carniolica Jacq. u zapadnoj Jugoslaviji. Glasnik Zemaljskog muzeja u Sarajevu – Prirodne nauke. 1965;**III/IV**:193-196

[29] Oberdorfer E. Süddeutsche Pflanzengesellschaften. Vol. 10. Ed. Fischer. XXXVIII + 564 pp., Jena, 1957

[30] Muller T, Görs S. Pflanzengesellschaften stehender Gewässer in Baden Württemberg. Beitr. Z. Naturk. Forschung I Südwestdeutschland. 1960;**19**(1):60-100

[31] Nedelcu AG. Flora acvatică şi palustră a câtorva locuri din Câmpia Română, cu unele consideraţii morfologice [Phd thesis]; Bucureşti. 1967

[32] Koch W. Die Vegetationseinheiten der Linthebene unter Berücksichtigung der Verhältnisse in der Nordostschweiz. Jahrbuch der St. Gallischen Naturwissenschaftlichen Gesellschaft. 1926;**61**(2):1-146

[33] Timár L. Szeged és környéke vetési gyomvegetációja. (Kézirat.) (Ünkrautvegetation der Saaten von Szeged und ihrer Umgebung.) Handschrift. 1947

[34] Niculescu M, Florea E. Monografia ariilor naturale protejate ROSAC(ROSCI)0045 Coridorul Jiului,

ROSPA0023 Confluența Jiu-Dunăre, ROSPA0010 Bistreț și Rezervațiile Naturale - Locul fosilifer Drănic și Pădurea Zăval. Craiova: Sitech; 2023. 215 p

[35] Resmeriță I, Rațiu O. Vegetația higro-și hidrofilă din Maramureș. Cluj Napoca: Contrib. Bot; 1974. pp. 115-119

[36] Niculescu M, Nuță IS. Cariera minieră Bicaz-Chei – biodiversitate, reabilitare ecologică. Craiova: Sitech; 2019. 201 p

[37] Niculescu M. The alluvial forest vegetation disturbed by the invasive alien plants, in the Danube valley, between Cetate and Calafat. Scientific Papers, Series A, Agronomy. 2023;**LXVI**(1):787-792

[38] Sîrbu C, Oprea A. Contribution to the study of plant communities dominated by Ailanthus altissima (Mill.) Swingle, in the Eastern Romania (Moldavia). Cercetări Agronomice în Moldova. 2011;**XLIV**(3(147)):51-74. Available from: https://repository.iuls.ro/xmlui/handle/20.500.12811/2431

[39] Sîrbu C, Oprea A. Plante adventive in flora României. Iasi: Ion Ionescu de la Brad; 2011. 733 p

[40] Niculescu M, Iancu P, Păniță OF. Invasion of Sicyos angulatus in riparian habitats in the Jiu and Danube area (Romania). Diversity. 2024;**16**(7):400. DOI: 10.3390/d16070400

[41] Soó R. A pusztuló bátorliget. Természettudományi Közlöny. 1935;**67**:14-21

[42] Borhidi A. Critical Revision of the Hungarian Plant Communities. Pécs: Janus Pannonius University; 1996. 95-138 p

[43] Soó R. Revue systématique des associations végétales des environs de Kolozsvár (respectivement de la Mezöség et de la région de la Szamos, en Transylvanie). Acta Geobotanica Hungarica. 1949;**6**(1):1-50

[44] Soó R. A korszerű növényföldrajz kialakulása és mai helyzete Magyarországon (Development and recent state of modern phytogeography in Hungary). Ann. Biol. Univ. Debrecen. 1950;**1**:4-26

[45] Meijer-Drees E. De bosvegetatie van de Achterhoek en Enkele aangrenzende Gebieden. Wageningen, NL: Veenman & Zonen; 1936

[46] Lohmeyer W. Der Hainmieren-Schwarzerlenwald. Mitteilungen der Floristisch-Soziologischen Arbeitsgemeinschaft 1957;**6**:247-257

[47] Koch W. Die Vegetationseinheiten der Linthebene. In: Jahrb. St. Gall. Naturwiss. Ges. 61, II Tl. St. Gallen. 1926

[48] Soó R. Systematische Übersicht der pannonischen Pflanzengesellschaften VI. Die Gebirgswälder II. Acta Bot. Acad. Sci. Hung. 1963;**9**:123-150

[49] Passarge H. Waldgesellschaften des nördlichen Havellandes. Wiss. Abh. Deutsch. Akad. Landwirtschaftwiss. 1957;**26**:1-139

[50] Issler E. Les associations végétales des Vosges méridionales et de la plaine rhénane avoisinante. Première partie: Les forêts (fin). Bulletin de la Société d'Histoire Naturelle de Colmar. N. S. 1924;**19**:1-109

[51] Hilbig W. The Vegetation of Mongolia. Amsterdam, NL: SPB; 1995

[52] Oberdorfer E, Görs S, Korneck D, Lohmeyer W, Müller T, Philippi G,

et al. Systematische Übersicht der westdeutschen Phanerogamen- und Gefäßkryptogamen-Gesellschaften. Schriftenreihe Vegetationsk. (Bad Godesberg). 1967;**2**:7-62

[53] Horvat I. Biljnosociološka istraživanja šuma u Hrvatskoj [Phytosociological studies of forests in Croatia]. Glasnik Za Šumske Pokuse. 1938;**6**:1-279

[54] Niculescu M. Natura 2000 in Romania: Habitat Fact-Sheets, 9260 Castanea sativa Woods Habitat Fact Shett. Bucharest, Romania: Romanian Ministry of Environment and Sustainable Development; 2008

[55] Niculescu M, Iancu P, Pănițặ OF, Făgăraș MM. Evaluation of the Spiranthes spiralis population's phenotypic density in the plant community of Medicagini minimae–Festucetum valesiacae meadows in the Natura 2000 Jiului corridor site, Romania (ROSCI0045). Diversity. 2024;**16**(1):1-13. DOI: 10.3390/d16010001

[56] Wagner H. Die Troekenrasen gesellschaften an Alpenostrand. Vegetation. 1941;**104**:81

[57] Pop I. Studiu comparative asupra pajiștilor de Botriochloa ischaemum din România. Cluj-Napoca, Romania: Contrib. Bot; 1977. pp. 111-120

[58] Csüros Ș, Kovacs A. Cercetări fitocenologice în raioanele Sighișoara și Agnita. Cluj: Contrib. Bot; 1962. pp. 167-188

[59] Borisavljević L, Dunjić-Jovanović R, Mišić V. Vegetacija Avale (Vegetation auf der Avala). Zbornik Radova- Institut za Ekologiju i Biogeografiju, Knjiga. 1955;**6**:1-43

[60] Niculescu M. Flora and Vegetation in the Upper Basin of the Luncavat River. Craiova: Sitech; 2009. 357 p

[61] Kornas J. A tentative list of recently introduced synanthropic plants (kenophytes) established in Poland. Mat. Zakl. Fitosocjol. Stos. UW. 1968;**25**:43-53

[62] Dihoru G. Învelişul vegetal de pe muntele Siriu. Bucharest, Romania: Ed Acad. R.S.R; 1975

[63] Puşcaru D, Puşcaru-Soroceanu E, Păucă A, Șerbănescu I, Beldie A, Ștefureac T, et al. Pășunile alpine din Munții Bucegi. Editura Academiei Republicii Socialiste Romania. 1956;**9**(4):365-369

Chapter 3

Long-Term Floodplain Vegetation Dynamics after a Permanent Water Level Regulation

Arvid Odland

Abstract

This chapter is a review of studies on vegetation dynamics on a freshwater delta ecosystem after a permanent water level drawdown. Floodplain wetlands are globally threatened and have long been recognized as sensitive to changes in the hydrological regime. The original floodplain vegetation types were left "hanging" over the new water level, and secondary successions were immediately initiated both in the original vegetation and on the newly exposed sediments. Permanent transects were established running from the original vegetation toward the new water's edge. Repeated analyses were conducted to document plant dynamics over a period of 33 years. The focus has been to analyze the dynamics of the decomposition of the original vegetation, the development of new vegetation on exposed sediments, and the varying fertility of plants. The plants' responses are discussed in relation to effects of important environmental factors and why the original vegetation has not been reestablished after the regulation.

Keywords: littoral vegetation, water level drawdown, secondary succession, coverage dynamics, vegetation degradation, flower production, rate of vegetation change

1. Introduction

A permanent regulation of a lake water level will have irreversible effects on the nearby floodplains. I have investigated vegetation dynamics after the regulation for 33 years, and the results are reviewed and discussed in the present article. Data collection was carried out in different parts of a freshwater delta in the Vestland County, SW Norway. The studies have focused on several aspects of vegetation changes both within the original vegetation and on the exposed sediments where new plants immigrated. Dynamic long-term trends have been studied with a focus on both individual species and on the development of secondary successions initiated after the regulation.

The near-shore areas of lakes and rivers are biologically important interfaces between terrestrial and aquatic ecosystems. The floristic structure of the floodplains is primarily regulated by the hydrological regime. Numerous studies have identified flood events as an important factor influencing plant distribution along elevation gradients on floodplains. Floods processes shape and maintain the spatial and temporal heterogeneity of floodplain areas and their plant communities [1]. Floodplain wetlands have long been recognized as sensitive to disturbances such as land reclamation and river regulation, which are globally threatened ecosystems. For example, 80–90% of the floodplains in Europe have been lost to agriculture and urban development [2].

Areas influenced by variable degrees of flooding have frequently been separated into three main zones: sublittoral, eulittoral, and epilittoral (supralittoral) [3, 4]. The littoral zones of undisturbed lakes may form narrow or broad fringing wetlands, with helophytes sorted by their tolerance to variations in the water level.

Two different succession processes were initiated after the regulation: a retrogressive (degradation of the original vegetation) and a progressive one (development of new vegetation on exposed sediments). In all areas, the plant populations will become dynamic and initiate long-time directional changes [5].

Littoral vegetation types usually include populations of many different species, which are dynamic because the individual plants have different physiologies, life cycles, and reproduction strategies. The vegetation in any area is naturally dynamic, with annual (seasonal) variations in growth, reproduction, and abundance. In areas influenced by environmental changes, long-term directional trends in the species composition will also be initiated. Dynamics of the behavior of plants in a community or study plot can include changes in species size, number of taxa (biodiversity), abundance (cover percentage), distribution, and fertility over time [6, 7]. To my knowledge, studies of plant dynamics over a long-term are few, with the exception for studies on tree species. The dynamics of herbs and graminoids are highly different because all aboveground biomass decay during the autumn and new shoots must be developed during spring and summer. Therefore, shoot biomass (cover) can be highly variable both from year to year and on the long term. My previous studies have quantified changes in the vegetation with focus on the following topics:

- The dynamics of species during the degradation of the original vegetation types
- The dynamics of the establishment of species on newly exposed sediments
- The dynamics of flower establishment during the secondary succession
- The dynamics of flower coverage in the degraded original vegetation
- The dynamics of the development of new vegetation types on exposed sediments
- The dynamics of changes in total species coverage and diversity during the development of new vegetation
- The dynamics of the development of new vegetation in areas where the original vegetation had been degraded
- The dynamics in species turnover and rate of floristic changes during the secondary successions

The results are discussed in relation to essential environmental variables and previous studies.

2. Study area

The investigations have been conducted on the Myrkdal delta (60° 40′N, 6° 28′E) located at the northern end of lake Myrkdalsvatn (229 masl.) in SW Norway. The delta extends approx. 700 m inward from the north end of the lake. The part of the delta closest to the lake is approx. 600 m wide and is divided into sections by several rivers and canals (**Figure 1**). The inner parts of the delta had for several 100 years been used for agricultural purposes. Most of these areas were more than 1 m above the original summer water level. The lower parts of the delta were originally dominated by *Equisetum fluviatile* and *Carex vesicaria* growing both above and below the water level.

More than 90% of the delta sediments consisted of particles smaller than 125 µm. The pH was mostly around 5.0, the loss on ignition was around 3%, and the base saturation was mostly between 14 and 25% [8].

2.1 The water level regulation

The summer lake water level was permanently lowered by approx. 1.4 m in August 1987. In addition to the lowering, the outlet was extended to reduce water level fluctuations. The main reason for the regulation was to improve and expand the areas used for grass production. Frequently the water level increased to more than 3 m above the normal water level before the regulation. After the regulation, the area of cultural fields increased, and forests gradually developed dense stands close to the new lake shoreline (**Figure 2**).

Figure 1.
Photo (taken in 1978) of the lower parts of the Myrkdal delta before the regulation.

Figure 2.
Recent map of the lower parts of the Myrkdal delta after the regulation. Yellow color indicates cultural fields and light green color indicates forested areas (From Norgeskart.no, the National Map Agency).

2.2 Hydrology and climate after the regulation

Hydrological data were made available from NVE (Norwegian Water Resources and Energy Directorate). A limnigraph had recorded water levels continuously in the lake. Based on daily measurements, mean June water level, mean summer water level (defined as mean level during May, June, and July), and mean annual water levels were calculated (**Figure 3**). The durations of the flood periods in percentage of the year were estimated at different elevations both before and after the regulation. The elevation of all study plots was related to the elevation scale on the limnigraph using a leveling instrument.

Water level data and the vegetation conditions before the regulation have shown that populations dominated by *E. fluviatile* grew in areas where the flood duration lasted more than approx. 26% of the year. *C. vesicaria* grew where the flood duration was originally between approx. 10 and 26% of the year. Shrubs, trees, and grass species dominated where the flood duration was less than approx. 10%. After the regulation, the flood duration at mean summer water level was approx. 25%, and at the mean June water level, the flood duration was approx. 12%.

Mean annual rainfall in the area was approx. 1300 mm. Mean annual temperature was 4.4°C, mean July temperature was 13.5°C, and mean December temperature was −5.0°C.

Temperature data during the study period were interpolated from nearby meteorological stations recorded at Norwegian Meteorological Institute from 1987 to 2020. Summer temperatures (mean of June and July) varied between years from 12.4 to 17.6°C, with a mean of 14.5 ± 1.3°C. Highest values were in 2018 and lowest in 1993, 1996, and 2015. There was a significant linear increase of 2.1°C ($r = 0.8$, $p < 0.01$) in the summer temperature during the study period.

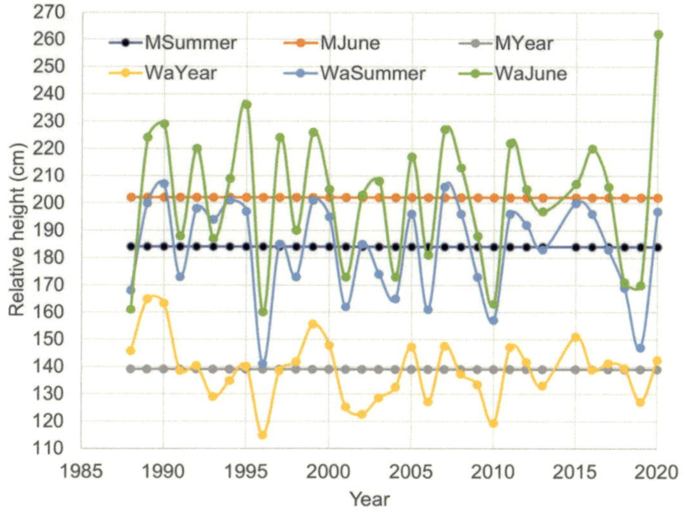

Figure 3.
Variation in the lake water level during the study period shown in relation to the new mean summer water level (184 cm). Mean water levels for the whole period was denoted M. The variation in mean water levels were estimated each year and denoted Wa.

3. Review of the sampling methods

This article summarizes the results of investigations conducted between 1987 and 2020, where the sampling methods have been described. Several distinct aspects of vegetation ecology in different parts of the delta have been studied. The delta was first visited in August 1987, 1 month after the regulation. Three permanent transects were established for long-time monitoring in different parts of the delta. The transects covered different elevations above the new summer water level, and the survey periods could vary because different study topics were initiated and ended at different times. The study plots (1 m long and 0.5 m broad) were analyzed in late July or early August. In some analyses, data from four nearby plots were combined into ten 4 x 0.5 m large plots where the mean species coverage and the mean elevation were calculated. The species' coverage was visually estimated as a percentage. The number of species recorded in the study plots (species richness) was counted, and the total coverage of all species in each of the plots was summed.

The main transect (T1) covered an area from the original *E. fluviatile* vegetation toward the new water level [8, 9]. T1 included 40 plots situated at elevations from 7 to 97 cm above the new mean summer water level (**Table 1**). In time, plants were only established above the new summer water level. T1 therefore does not include the lower parts of a potential full littoral zonation. T1 also included eight plots in the lower part of the original *E. fluviatile* vegetation, which was situated 30–75 cm below the original summer water level. The transect was surveyed a total of 19 times between 1987 and 2020, most often at one-year intervals. Variations in the number of flowers produced by *C. vesicaria* and *Carea rostrata* were counted from 1993 until 2020.

Transect T2 was established in an original *C. vesicaria*-dominated vegetation zone [10]. T2 included 15 plots situated from 24 cm above to 29 cm below the original summer water level. Variation in species coverages was investigated from 1988 until 2002,

Limn	Z1	H1	H2	F% 1	F% 2
394		62			
	Epilittoral				
		32		<10%	
364		24	174	10%	
	Eulittoral				
332	WL 1	0	147		
		−29	121	26%	
304		−30	120	>26%	
	Hydrolittoral				
		−57	97		< 1%
		−75	75		
255					<10%
	Exposed				
			15		
					12%
			7		13%
184	WL 2		0		25%
			−5		>25%
	Aquatic				
124			−60		

WL1 indicates the original summer water level and WL2 indicates the new summer water level. Elevation (H) and duration of flooding (F) of the zones were estimated both in relation to the situation before (H1) and after the regulation (H2). Duration of flooding (in percentage of the year) associated with the elevational distribution of the littoral zones before (F% 1) and after (F% 2) the regulation.

Table 1.
Schematic overview of the elevation (cm) of the littoral zones and mean summer water level on the delta before (Z1) the regulation.

and the variation in the number of flowers produced by *C. vesicaria* was counted from 1993 until 2002.

Transect T3 was established in the lower part of the exposed delta sediments. T3 included 14 plots situated from 15 cm above to 5 cm below the new summer water level. Variation in species coverages was investigated from 1988 until 2002, and the variation in the number of flowers produced by *C. vesicaria* and *C. rostrata* was counted from 1993 until 2001 [11].

The vertical distribution of the studied vegetation zones was estimated in relation to the scale of the lake limnigraph. Elevations of the transect plots were given in relation to both the old and the new mean summer water level (**Table 1**).

Data collected in the various investigations were analyzed using statistical methods. Long-term trends in species cover and flower production changes were analyzed by regression analyses [10]. Trends in the development of new vegetation types were explored by ordination/gradient analyses. Ordination techniques are used in ecology to describe long-term relationships between gradients in species composition patterns and underlying environmental factors. Variation in species study plot composition during the study periods will be related to time (number of years) or elevation. DCA (Detrended Correspondence Analysis) is an unconstrained ordination method where environmental factors were included as passive variables. DCA can be used to find trends in data with large standard deviation [9]. The rate of change (often denoted ROC) or species compositional floristic turnover is a measure of the speed at which a variable changes over a specific period.

Long-term trends in floristic change in data with relatively little standard deviation were explored by PCA (Principal Component Analysis). DCA and PCA are multivariate statistical techniques widely used by ecologists to find long-term trends or gradients in ecological studies. The various methods that were used are described in the results section.

The degree of plant fertility is generally assumed to be more sensitive to the environmental conditions than their vegetative structures. To elucidate this topic, the number of *C. vesicaria* and *C. rostrata* shoots producing flowers was counted in study plots from 1993 onward. The plants produce new shoots from underground rhizomes each year, and a varying proportion of these may produce flowers. The number of shoots with flowers was counted in the 0.5 m^2 plots at each survey.

4. Results and comments

During the study period, the frequency and coverage of all species were highly dynamic. The number of occurrences of the most common species recorded in T1 during the study period is shown in **Table 2**. All species can be regarded as common in most freshwater wetlands.

4.1 Decomposition dynamics of the original vegetation types

As a result of the regulation, the original vegetation zones were left "hanging" above the new water level. It had to be expected that these zones in time would disappear on site. Before the regulation, the eulittoral zone was dominated by *C. vesicaria* (T2), and the hydrolittoral zone was dominated by *E. fluviatile* (T1). The zones had wide distributions in the lower parts of the delta. Both species are clonal, and the populations were monocultures. Interesting questions would then be how long the species would survive in their original positions, and how the dynamics of the plants' decomposition developed. T1 and T2 were monitored for several years [10–12].

The *C. vesicaria* zone was situated between a zone dominated by *Phalaris arundinacea* on the upper part and *E. fluviatile* on the lower part. T2 consisted of 15 plots situated between 24 cm above and 29 cm below the original summer water level and 174–121 cm above the new summer water level (**Table 1**). The plots were surveyed from 1988 until 2002.

The coverage reduction of *C. vesicaria* started from the edges (upper and lower part) of the original zone [10]. Mean plot coverage decreased linearly from 94.4 to 1.3% (R^2 = 0.96) for 15 years. The coverage reduction was, however, variable between the study plots. In the middle part, *C. vesicaria* maintained a coverage of 100% until 8 years after the regulation. Fourteen years after the drawdown, the coverage was 40% in the middle part of the zone, while it had disappeared or had a coverage lower than 20% at the edges. After 15 years, only a few scattered shoots were recorded [10, 11].

Species	Abb.	No
Agrostis stolonifera	*A.sto*	117
Alopecurus geniculatus	*A.gen*	13
Calamagrostis phragmitoides	*C.phr*	316
Callitriche palustris	*C.pal*	22
Calttha palustris	*C.pal*	52
Carea rostrata	*C.ros*	474
Carex canescens	*C.can*	95
Carex echinata	*C.ech*	12
Carex leporina	*C.lep*	79
Carex nigricans	*C.nig*	35
Carex vesicaria	*C.ves*	548
Deschampsia cespitosa	*D.ces*	336
Eleocharis acicularis	*E.aci*	7
Filipendula ulmaria	*F.ulm*	41
Galeopsis bifida	*G.bif*	8
Galium palustris	*G.pal*	339
Juncus filiformis	*J.fil*	263
Phalaris arudinacea	*P.aru*	592
Ranunculus repens	*R.rep*	477
Ranunculus reptans	*R.rept*	35
Salix nigricans	*S.nig*	21
Sparganium angustifolium	*S.ang*	23
Subularia aquatica	*S.aqu*	26

No = number of occurrences in the 760 plots during the study period.

Table 2.
The most common species recorded in transect 1, with full names and abbreviations.

The coverage of *E. fluviatile* was studied in eight permanent plots situated at the lower part of the original *E. fluviatile* zone [11]. The plots were situated between 30 and 75 cm below the original summer water level and 120–75 cm above the new summer water level. The coverage reduction was variable from year to year, but the mean coverage decreased in general linearly from 39.8% in 1988 to 3.3% after 14 years (R^2 = 0.81) when *E. fluviatile* had disappeared almost completely. However, the occurrence of scattered shoots showed that the rhizomes were still alive 33 years after regulation [13].

The results show that the coverage reduction of the two species was dynamic, but both had disappeared almost completely 15 years after the regulation. *C. vesicaria* retained a 100% coverage for 8 years after drawdown in the center of the population. The mean coverage of *E. fluviatile* was higher than expected in 1995 (year 8), probably because of a high water level that year (**Figure 3**).

This may indicate that *E. fluviatile* was more dependent on periods with high water level than *C. vesicaria* and therefore more sensitive to the drawdown. This is surprising since the regulation represented a greater water level change on *C. vesicaria* than on *E. fluviatile* since the former was originally situated at a higher elevation.

4.2 Dynamics of species immigration on the exposed sediments

The immigration of plants on the exposed sediments was highly dynamic both in relation to time and space (elevation above the water level). Already 1 month after the regulation, several plants were established on the exposed sediments. During the first few years, the sediments were dominated by bryophytes [8, 14]. The bryophyte richness was high, especially with many acrocarpous species. The coverage of the dominant bryophytes was highly dynamic during the first years. *Blasia pusilla* dominated the first year, frequently with a plot coverage of 70%. The next years, *Polytricum commune* increased gradually until it reached a mean coverage higher than 40% after year 4. Also, the *Drepanocladus* species increased gradually and reached a peak during year 5. Later, the bryophyte coverage decreased strongly, and most disappeared as the coverage of vascular plants increased.

Several vascular plants were also quickly established on the sediments, but their coverage was low the first month after the regulation in 1987 [8, 9, 14]. Dominants were *Alopecurus geniculatus, Subularia aquatica*, *Ranunculus reptans*, *Callitriche palustris*, *Eleocharis acicularis,* and *Sparganium angustifolium*. These species were mainly found in the six first years after the regulation. *Agrostis stolonifera* was also common until year 8, and *Juncus filiformis* was common until year 10.

The coverage of the tall plants gradually increased, and most bryophytes and the small vascular plants disappeared after year 5. Frequency of vascular plants in the 40 plots in T1 during the study period is shown in **Table 2**, and their coverages have previously been described [9].

Mean coverage of the most common species established along T1 was highly dynamic during the study period (**Figure 4**). The most rapid invader was *Deschampsia cespitosa*, which had a plot coverage between 0 and 80% the first year after the regulation. *C. vesicaria* and *C. rostrata* needed 4–5 years to reach high coverages. *C. vesicaria* had high coverage during two periods: between years 3–10 and years 29–33. *P. arundinacea* and *C. phragmitoides* reached 100% cover after approx. 10 years. The coverage of *P. arundinacea* increased over the whole period and had a mean coverage of approx. 50% from year 13 until year 33. *P. arundinacea* and *C. vesicaria* dominated during the last study period, but they were mainly separated to elevations above or below elevations of 50 cm, respectively (**Figure 5**). The mean coverage of *C. phragmitoides, D. cespitosa,* and *Ranunculus repens* decreased during the last study period. *F. ulmaria* was sporadically present and reached a coverage of 40–70% between year 20 and 26.

The lowest part of the original zone dominated by *E. fluviatile* was situated at a 31 m distance from the lake summer water level. After the regulation, the rhizomes began to migrate downward toward the new water level. After 13 years, it had moved 22 m, which gave a mean migration of approx. 1.7 m/year. *E. fluviatile* had almost completely disappeared in T1 after 18 years [13].

The 40 plots in T1 were divided into 10 elevation zones above the mean summer water level. Mean coverage of the species was estimated in each of these zones. The species' coverages were dynamic also in relation to elevation, indicating that the species responded differently to the distance above the mean summer water level during the study period (**Figure 5**). The coverage of *C. vesicaria* increased toward low elevations, and it became more stable at elevations lower than approx. 40 cm. The coverage of *P. arundinacea* was the highest between 50 and 70 cm, while the coverage of *C.*

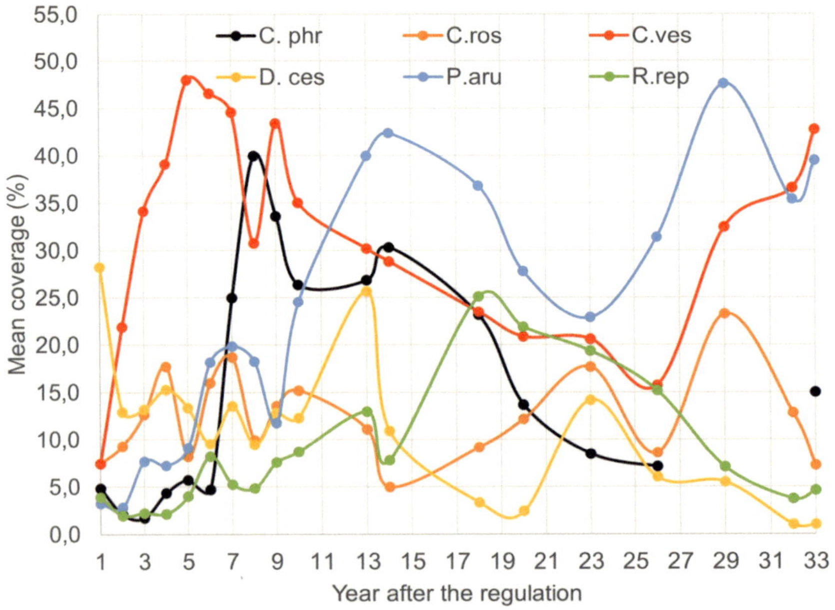

Figure 4.
Mean coverage of the most common species in all the 40 study plots estimated each study year. Species abbreviations are explained in **Table 2**.

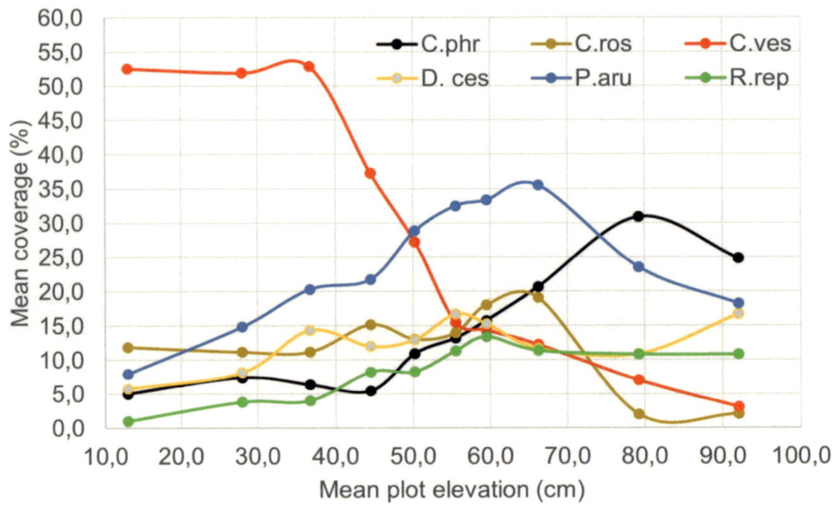

Figure 5.
Mean coverage of the most common species in the 10 study plots with different elevation above the mean summer water level estimated during the whole study period. Species abbreviations are explained in **Table 2**.

phragmitoides was the highest around 80 cm. *A. geniculatus, J. filiformis, R. reptans, E. acicularis*, and *S. aquatica* were all most common below 40 cm.

The study shows that the invasion of species on the exposed sediments was rapid. A prerequisite for this was that the sediments consisted of fine-grained material and had good moisture conditions and that large seed banks were normally present in floodplain sediments [15]. The pioneer plants were mostly a mixture of annuals, aquatic plants, and helophytes. These were over the first years distributed over most of all transect plots. Later, clonal helophytes and grasses became dominants. *D. cespitosa, C. vesicaria*, and *C. rostrata* were the first dominants followed by *C. phragmitodes* and *P. arundinacea*. Over most of the study period, only *C. vesicaria* and *P. arundinacea* retained a high coverage. But the coverage of all species was highly dynamic. Most species responded in time differently to the distance above the mean summer water level, becoming separated to specific elevations. After 2001 (year 14), the species distribution had become stable in T1 with the following weighted elevation optima: *C. vesicaria* at 31 cm, *C. rostrata* at 32 cm, *P. arundinacea* at 60 cm, *C. phragmitoides* at 76 cm, and *D. cespitosa* at 87 cm [11]. Before the regulation, *C. vesicaria* populations were found from 24 cm above to 29 cm below the summer water level (**Table 1**). This shows that *C. vesicaria* had not developed populations in relation to the water level as before the regulation.

4.3 Dynamics in flower production

The number of *C. vesicaria* and *C. rostrata* shoots producing flowers was counted in 26 plots in T1 from 1993 to 2020. During this period, both water levels and air temperatures varied strongly [12]. The study plots were situated between 7 and 59 cm above the mean summer water level. In plots situated higher than 60 cm, almost no shoots produced flowers.

In most plots, *C. vesicaria* produced more flowers than *C. rostrata*. The largest mean number of shoots with flowers in the plots was 43 for *C. vesicaria* and 30 for *C. rostrata*. The largest numbers found in all 26 plots throughout the study period plots were 344 (2019) and 212 (1994). The variation in the mean number of flowering shoots was highly dynamic during the study period (**Figure 6**). Except for the first 4 years (1993–1996), *C. vesicaria* produced more flowering shoots than *C. rostrata*. The mean annual number of flowering shoots during the study period was 2.1 for *C.*

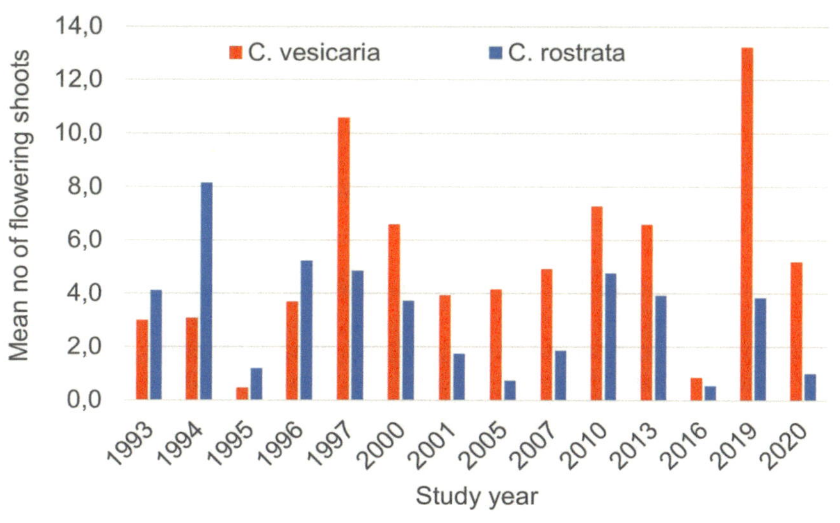

Figure 6.
Dynamics in the mean number of C. vesicaria *and* C. rostrata *shoots producing flowers in the eight plots between 1993 and 2020.*

rostrata and 3.5 for *C. vesicaria*. There was no significant trend in flower production during the study period. Both species produced almost no flowers in 1995 and 2016, which were years of high water levels (**Figure 3**). The species showed almost identical relative annual variations in flower production, with a significant correlation ($r = 0.62$) [12]. This indicates that both species were affected by the same environmental factors although the degree of flowering was different.

The significance of water level and air temperature for the variation in flower production was investigated statistically. Different values for annual water levels and temperatures were tested. The flowering of *C. vesicaria* was significantly related to the mean summer water level (average for June and July) and the summer temperature (average for June and July), both estimated on data from the current year and the previous year. The flowering of *C. rostrata* showed no significant relationship to these factors. The production of *C. vesicaria* flowers was not significantly related to its coverage degree [12].

The number of flowering shoots of *C. vesicaria* and *C. rostrata* was also counted at low elevations (T3) from 1993 until 2001 [11]. T3 included 10 plots situated from 15 cm above to 5 cm below the mean summer water level. Also, in these plots, the production of flowering shoots was highly dynamic, but the results were different from the data in T1. In T3, *C. rostrata* produced more flowers than *C. vesicaria* most years. The mean number of flowering *C. vesicaria* shoots was 10.1, and it was 14.5 for *C. rostrata*. During the study period, the number of *C. rostrata* shoots producing flowers increased significantly, while *C. vesicaria* showed no significant trend.

The number of flowering *C. vesicaria* shoots was also counted during the degradation period in T2 from 1993 until 2002 [11]. The average coverage of *C. vesicaria* decreased with time, but the number of flowering shoots was dynamic. Relatively few flowering shoots were recorded in 1995, which was a year with high water level (**Figure 3**). The number of flowering shoots was highest in 1996 (the highest number in a plot was 35). Ten flowering shoots were found in the middle of the transect, but almost none in either the upper or the lower plots in 1996. After 2002, no flowering shoots were found. This shows that shoots with flowers were recorded up to 15 years after the regulation.

Comparisons between the number of flowering shoots in the three investigated transects were different with major differences from year to year. The mean numbers of flowering *C. vesicaria* and *C. rostrata* shoots between 1993 and 2001 were 4.5 and 4.1 in T2 and 10.1 and 14.5 in T3, respectively. Obviously, the number of flowering shoots of both species varied with distance to the summer water level. The results indicate that drawing conclusions from a few years of research and from only one area can give contradictory results.

4.4 Long-time population dynamics of *C. vesicaria* and *P. arundinacea* coverage

Population dynamics is a field within ecology where changes in species occurrence (quantity or number) and distribution patterns are investigated over time. Whether there are significant long-term trends can be investigated using statistical methods.

C. vesicaria and *P. arundinacea* dominated the transect study plots over most of the study period, but their coverage varied both with elevation and time after the regulation. Their long-term coverage dynamics was therefore interesting to study by statistical methods [16].

Variations in mean coverage of *C. vesicaria* and *P. arundinacea* in the 40 study plots over 33 years are shown in **Figure 7**. The boxplots show that the coverage of

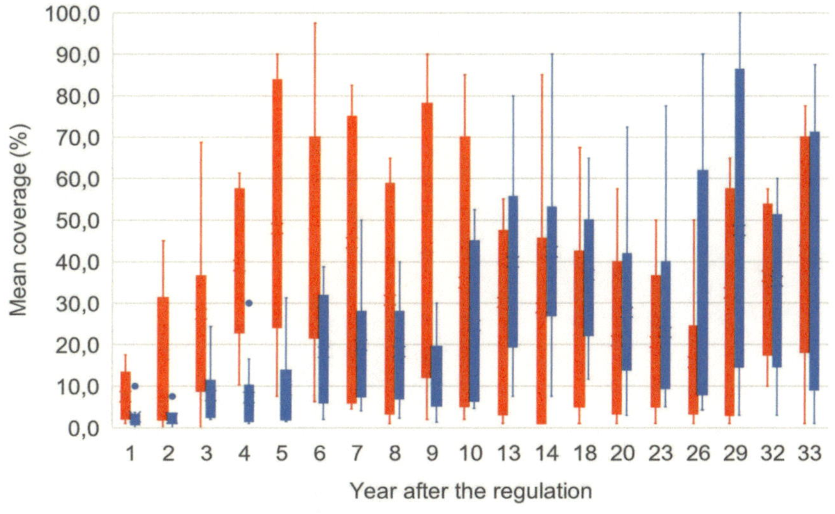

Figure 7.
Boxplots showing the variation in the mean coverage of C. vesicaria *(red columns) and* P. arundinacea *(blue columns) in the forty transect plots during the study period.*

both species was highly dynamic. The increase of *C. vesicaria* coverage started earlier than for *P. arundinacea*. During the last 4 years, both species had high coverage but at different elevations (**Figure 5**).

Over time, the species' coverage was gradually separated to specific elevations [16]. The 40 plots in T1 were divided into 10 consecutive elevation zones. The study plots were in this case 4 x 0.5 m large (2 m^2), with mean elevations from 13 to 93 cm above the summer water level. The species' mean coverages were estimated in each of these zones. During the study period, the coverage was highly dynamic as shown in **Figure 8**. *P. arundinacea* coverage was the highest at a mean elevation of 66.3 cm, while *C. vesicaria* had the highest coverage at low elevations. Over the two last years,

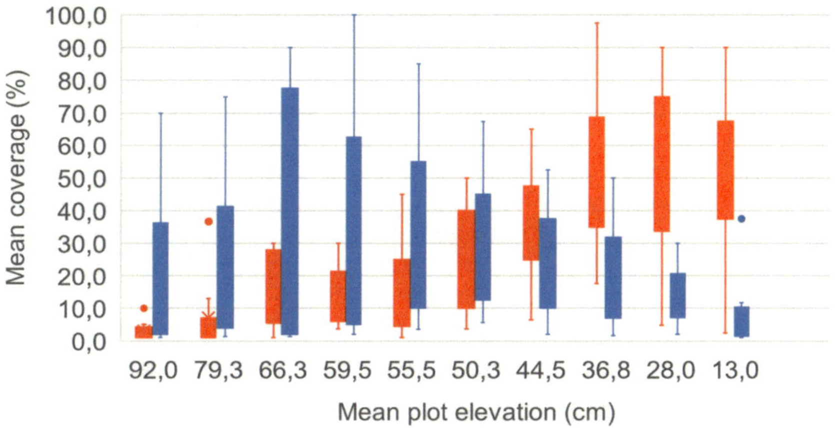

Figure 8.
Boxplots showing the variation in mean coverage of C. vesicaria *(red columns) and* P. arundinacea *(blue columns) in ten transect plots with mean elevations between 92 and 13 cm above the mean summer water level during the study period.*

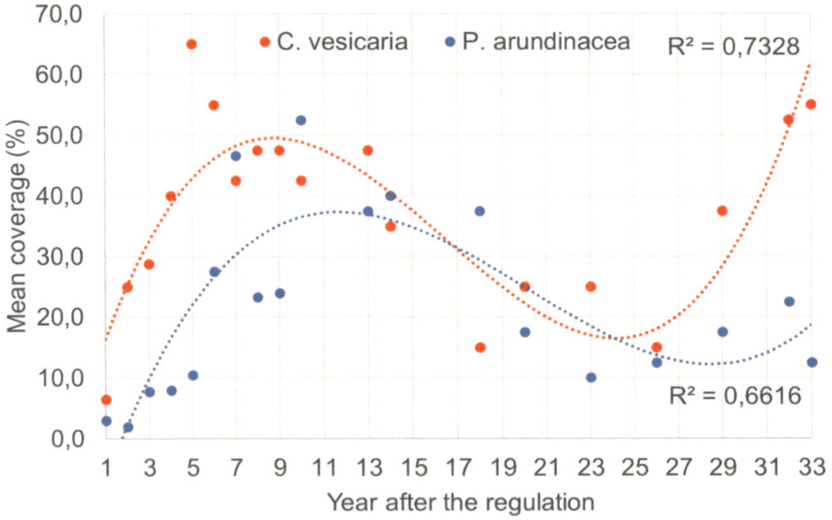

Figure 9.
Mean coverage of C. vesicaria *and* P. arundinacea *in plots situated at a mean elevation of 44.5 ± 1.1 cm from 1988 to 2020. The data show significant third-degree (cubic) distributions.*

their coverages were more equal, but they were mainly separated to below or over elevations of approx. 50 cm.

Coverage variations from 1988 to 2020 at a mean elevation of 44.5 cm are shown in **Figure 9**. Trends in their dynamic mean coverages were statistically tested by different polynomial regression models. The long-term trends were significantly explained by third-degree functions [16]. At elevations between 37 and 56 cm, both species significantly followed third-degree (cubic) functions. In most of the other elevation levels, the trends were also significantly explained by third-degree functions, but the significance levels were often lower. The long-term dynamics mostly explained by third-degree functions are difficult to explain in terms of plant biology and ecology (see chapter 5.1). The exposed sediments situated less than 70 cm above the mean summer water level were dominated by *P. arundinacea* and *C. vesicaria* as shown in **Figure 10**.

4.5 Development of new vegetation on the exposed sediments

The exposed sediments were rapidly invaded by species, as shown in **Figure 4**. It could be expected that both the floristic vegetation composition, the total coverage, and the species richness in the plots should be dynamic but become more stabilized during the last periods of the study period. The long-term trends during the secondary succession have previously been described by Refs. [8, 9].

Twenty-six species were recorded in the transect plots during the study period (**Table 1**), but many of them were sparse and disappeared after short periods. The mean coverage of species established in all study plots was estimated during the study period, and the coverage sum of the species is shown in **Figure 11**. The sum of the mean coverage of all species increased gradually during the first 13 years. The highest coverages were found during year 7–14, and thereafter, it decreased. This shows that

Figure 10.
Vegetation development on the exposed sediments 33 years after drawdown. Mainly C. vesicaria *and* P. arundinacea *dominated areas less than 20–30 m from the summer water level.* S. nigricans *developed most areas situated more than 20–30 cm above the water level. The photo was taken in 2020, but later the changes were small. Vegetation development out into the water had been limited below the mean summer water level.*

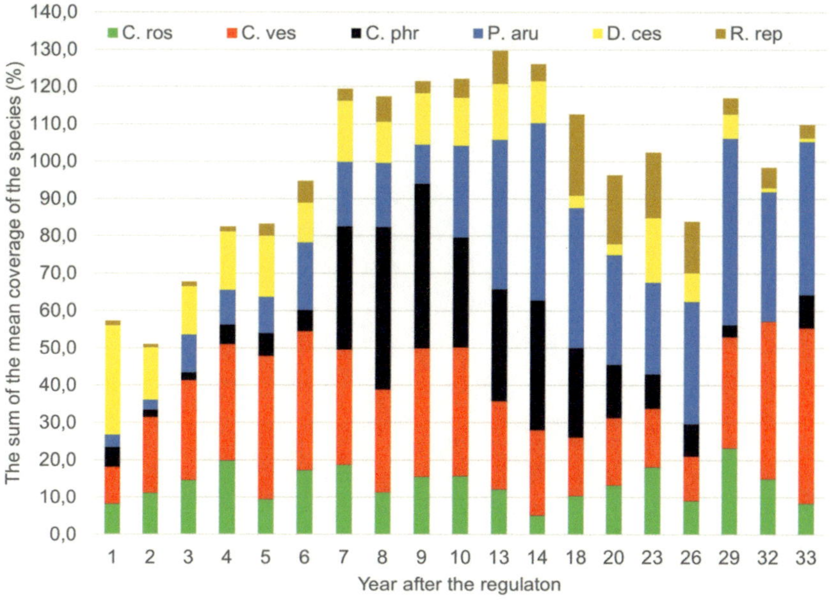

Figure 11.
Variation in mean coverage of the six most common species in transect T1 from 1988 until 2020. The sum of the mean species coverage in the forty plots has been estimated each study year. Species abbreviations are explained in **Table 2**.

the total plot coverages during the succession were highly dynamic. The total coverage was higher during the first year after the regulation than during year 2. This was mainly a result of the high coverage of the small pioneer plants along T1. During the study period, species richness was the highest during year 2 with 4–13 species in all T1 plots. After 13 years, the richness varied mostly between 1 and 6, with a mean of 4.2 ± 1.4. A quadratic model significantly described the long-term trend [9].

Previous studies have shown that the variation in number of species can be highly variable during secondary successions [8]. Biodiversity may increase, decrease, or be the highest at intermediate succession stages, but there are few available long-term studies on the variation in species richness during a secondary wetland succession.

4.6 Rate of change during the secondary succession on exposed sediments

Changes in the floristic composition of the T1 plots were analyzed by DCA [8, 9]. The units are expressed in standard deviations (SD) of species turnover rate along floristic gradients [17].

The variation in floristic composition of the study plots along T1 each year and throughout the entire study period is shown in **Figure 12**. The vertical axis (DCA axis 1) shows the main difference between the plot floristic composition along T1 each study year. The compositional floristic turnover in relation to time showed a significant non-linear trend. During the first 15–16 years after the regulation, there was a strong change in the species composition. Over the last years, however, the mean DCA axis 1 plot scores showed relatively small floristic changes.

Previous DCA analyses have shown that the time (number of years after regulation) and the elevation above the summer water level explained equal parts of the succession development [8, 9, 18]. DCA axis 2 showed a significant linear compositional floristic turnover along the elevation gradient.

The variation in the species composition in relation to the water level appears to be constrained. After year 7, the floristic composition in all study plots situated in the lower part of T1 had almost the same DCA axis 1 plot scores (**Figure 12**).

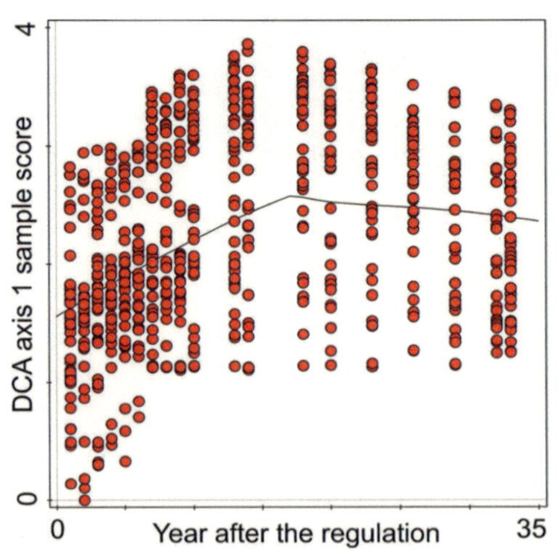

Figure 12.
Result of the DCA analysis. Variation in DCA axis 1 sample scores plotted against the study years after the regulation. Result of the lowess smoothing is shown.

The variation in the floristic composition along DCA axis 1 showed a significant polynomial change in the floristic composition in relation to time [8, 11]. After 15–16 years, the mean DCA axis 1 plot scores showed almost no long-term trend.

4.7 The dynamics of plant immigration in the original vegetation zones

It had to be expected that new vegetation types would gradually be developed as the original vegetation was degraded. Interesting questions would be to investigate the dynamics of the invading species and how long it would take before new stable vegetation types were established.

Species invasion in the 15 study plots originally dominated by *C. vesicaria* is shown in **Figure 13**. *C. vesicaria* retained a high coverage for several years, and the invasive species had a low coverage. After 15 years, *C. vesicaria* rarely occurred. *P. arundinacea* was established with a coverage of 30% after year 3. After 14 years, its coverage was higher than 70%. *C. phragmitoides* had a 60% coverage in year 9, but thereafter, it decreased. *E. fluviatile* was sporadically recorded with low coverage from year 2 to year 14. After year 12, *P. arundinacea* was dominant, and other species had low coverage [10].

Variation in coverage of species immigrating in the original *E. fluviatile* zone is shown in **Figure 14**. *D. cespitosa* was established the year after the regulation, and it was recorded until year 29. *C. phragmitoides* was also quickly established, and after year 7, it became dominant for a period. After year 29, it was not recorded, and later *P. arundinacea* took over as the dominant species. *C. vesicaria* was recorded with low coverage from year 8 but with somewhat higher coverage in years 32 and 33.

The development of new vegetation types in the original littoral zones was highly dynamic. The species richness was higher in the original *E. fluviaile* zone, but after 29 years, *P. arundinacea* became dominant in both zones.

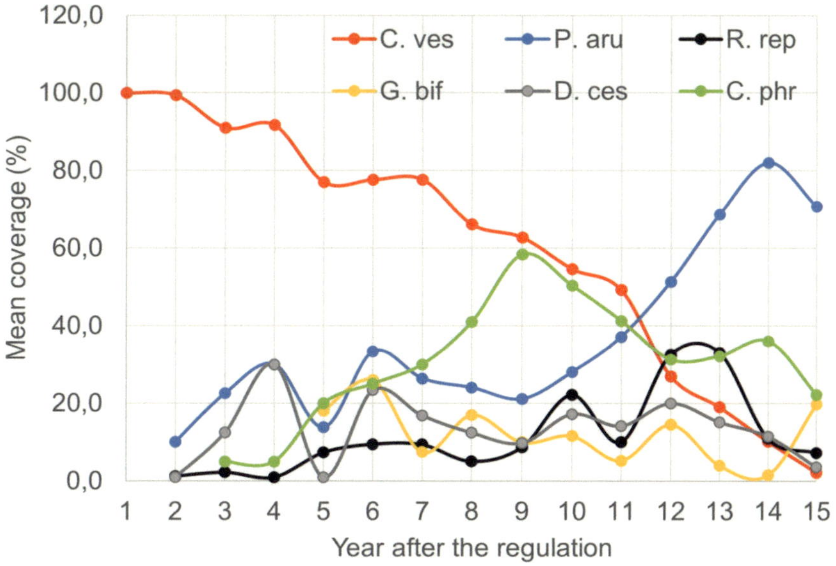

Figure 13.
Dynamics of the plant invasion in the original C. vesicaria *zone. Species abbreviations are explained in* **Table 2**.

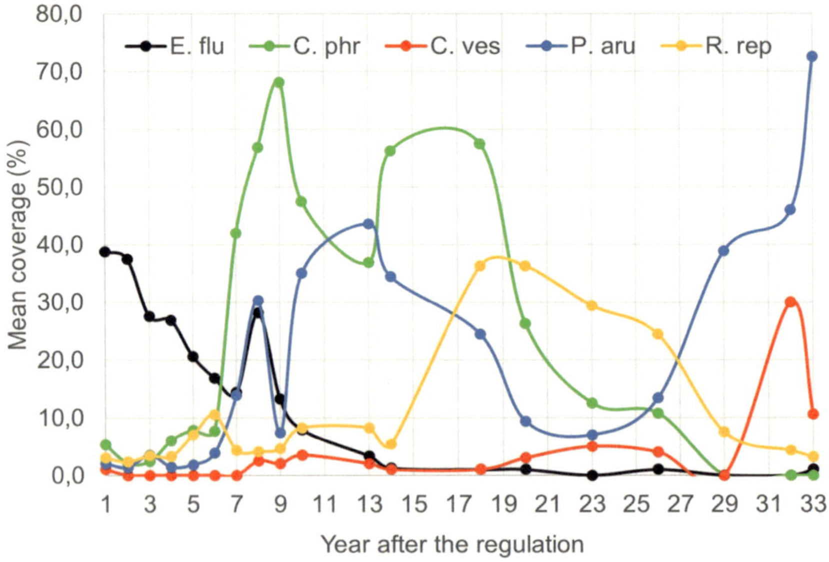

Figure 14.
Dynamics of the plant invasion in the original E. fluviatile *zone. Species abbreviations are explained in* ***Table 2****.*

4.8 Rate of vegetation change after the decomposition of the original vegetation types

Invasion of new species in the original vegetation zones was highly dynamic (**Figures 13** and **14**). Variations in the mean cover of all species along the transects in relation to time were analyzed by PCA (Principal Component Analysis) [17] with time as passive variables. The analyses describe the development during the secondary successions in the original vegetation types.

The rate of change of the mean coverage of species in all the transect plots during the development of new vegetation in the *C. vesicaria* zone is shown in **Figure 15**. The analysis shows a gradual change in the floristic composition during 15 years after the regulation. The plots were not studied after 2002 (year 15), but *P. arundinacea* had by then become the dominant species.

The rate of floristic change in all the transect plots during the development of new vegetation in the original *E. fluviatile* zone is shown in **Figure 16**. The analysis showed a gradual change in the floristic composition during the first 10–15 years after the regulation. Thereafter, the changes decreased, but there were still some variations.

The original coverage of *C. vesicaria* was 100%, while in the plots dominated by *E. fluviatile*, the coverage was only 40%. The invasion of other species in the *E. fluviatile* plots was therefore faster and more species-rich than in the *C. vesicaria* plots. In the *C. vesicaria* vegetation, the coverage sums of all plants increased to 200% in year 6 but thereafter decreased to 130% in year 15. The species richness varied mostly between 6 and 8 during the study period. In the *E. fluviatile* zone, the coverage sums of all plants varied between 70 and 130%, and the species richness varied between 19 and 7. In both populations, *P. arundinacea*, *C. phragmitoides*, and *R. repens* became common and often dominant. These trends appeared to be the same over most of the lower parts of the delta after the regulation. Although the vegetation and ecological conditions in the two types were different initially, both showed the same long-term trends after regulation, and new stable conditions were reached after approx. 15 years.

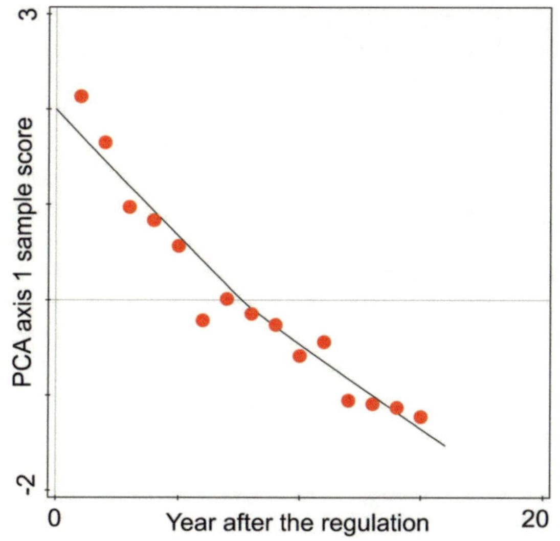

Figure 15.
Dynamics of vegetation development in the original C. vesicaria *zone. The PCA analysis shows major species change in the fifteen study plots during the first 15 years after the regulation. The analysis was based on the mean coverage of all species in the 15 study plots from 1988 until 20. Later, the floristic changes were small. Result of the lowess smoothing is shown.*

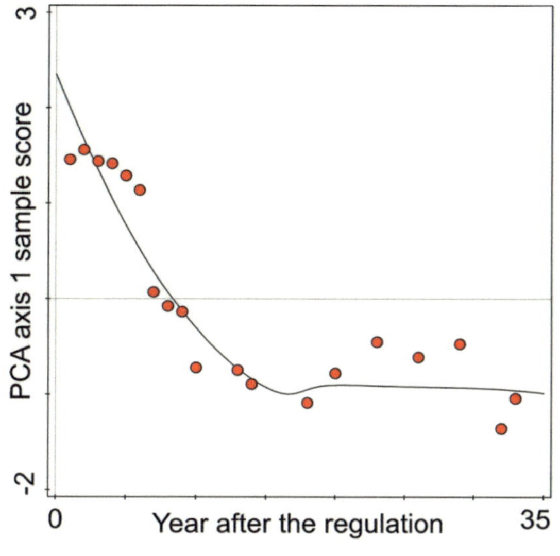

Figure 16.
Dynamics of vegetation development in the eight plots in the original E. fluviatile *zone. The analysis was based on the mean coverage of all species in the eight study plots from 1988 until 2020. After 10–15 years, the floristic changes were relatively small. Result of the lowess smoothing is shown.*

5. General discussions

Effects of seasonal water level drawdowns frequently employed for hydropower and flood control have widely been studied. Long-term vegetational effects of permanent lake water level drawdown have, however, hardly been studied previously.

The dynamic vegetation changes described in these investigations are effects of interactions between hydrological and plant physiological factors. It is easy to describe changes in plant distribution after a regulation, but to explain the results in terms of plant biology is difficult. The underlying mechanisms governing large-scale species turnover are ambiguous and poorly studied [19].

The changes in species' coverage and distribution have also resulted in changes during the secondary vegetation successions. Patterns of vegetation succession after a drawdown depend on factors such as sediment characteristics, topography, short- and longtime changes in the water level fluctuation, and effects of wind and wave erosion [20]. These factors may all influence the biological and physical structures of the littoral zones. The primary factor that distinguishes wetlands from terrestrial landforms is that the wetland vegetation is dominated by helophytes adapted to the unique anoxic hydric soils. The plants have structures both above and below the soil surface, and the biomass is dynamic in both places.

5.1 Growth factors associated with plant coverage dynamics

Most plant species of wetland habitats expand through clonal growth, often forming dense, monospecific stands. Underground plant rhizomes continuously produce new shoots and root systems. Clonal propagation is the main mechanism of macrophyte reproduction and population growth, while sexual reproduction is often a subordinate strategy [21]. Neither water levels nor temperatures could explain the coverage variation in the study plots [16]. It is therefore likely that the plants' growth responses control the cyclical variation in coverage.

The capacity of clonal plants to exchange resources and growth substances between interconnected shoots has long been held responsible for the ability of clonal plants to prevent shoot overproduction and for the general lack of density-dependent mortality and self-thinning [18, 22]. The long-term variation in plant coverage is therefore probably mainly a result of rhizome growth dynamics, and not of aboveground self-thinning [23].

Previous studies have shown that *P. arundinacea* produces dense crowns and prominent networks of vigorous underground rhizomes, allowing for aggressive vegetative spread on disturbed and moist sites [24]. Tillering young clones could cover an area of one square meter consisting of 100 tillers by the end of the first growing season. During the fourth year of its life, the partial clump did not form aboveground shoots and died out during the fifth year.

Aerial growth dynamics of *P. arundinacea* and *C. vesicaria* have been studied for 2 years [25]. Major differences were found, probably because of differences in underground clonal growth. The lifespan of the dominant species may explain their long-term coverage dynamics, but there are few studies on the life history of plants. The average lifespan of *C. rostrata* shoots varied from 1.5 to 6 years, with an annual shoot survival rate of 60% [26–28]. There are few available studies on *C. vesicaria*, but it can have an intensive long-term tillering [29]. Consequently, it is not immediately easy to explain why the coverage dynamics of *P. arundinacea* and *C. vesicaria* were significantly described by cubic functions.

5.2 Effects of the water level amplitude

Water level amplitudes and the duration of flooding periods have large effects on the ecological processes and patterns of littoral vegetation zonation [30, 31]. Flooding duration, which is a function of the water level amplitude, can be an important stress factor for plants, but this proposition has never been tested [32].

The number of days with flood impact decreases with increasing elevation above the normal water level [33]. Based on this, elevation can be used as a proxy for flood frequency. Water level fluctuation can be estimated in number of days or quantified on different time scales: daily, seasonal, during months, and during the study periods. The mean annual flooding duration can be estimated for the mean June water level (1 month) to be 8.2%, and the mean summer water level (3 months) to be 24.7%.

Harmful effects of reduced water level amplitudes have previously been demonstrated [34]. The flooding tolerance of angiosperms species determines their distribution along flooding gradients in river forelands [32, 35, 36]. The low diffusion rate of gases (and more particularly carbon dioxide and oxygen) leads to problems of carbon and oxygen supply [37, 38]. A major problem for many species will therefore be to survive long periods of flooding.

These studies indicate that short flooding periods are not critical; it is the long-term mean water levels that are most important. Experiments have shown that species inhabiting river foreland survived both 3 and 6 weeks of flooding and were classified as tolerant to flooding duration [37, 39]. The production of flowering shoots was significantly reduced during years with a high water level.

C. vesicaria, *D. cespitosa*, and *P. arundinacea* growing on a floodplain have their optima at different elevations relative to the water level [40]. *C. vesicaria* was found to be confined to the lowermost parts that were flooded for more than 40 days during the study period.

P. arundinacea has been found to be sensitive to summer flooding, and after 3 weeks, rhizome growth was reduced [41]. A 50-day flood period (ca. 14% of the year) adversely affected its biomass production [42, 43]. It was restricted to elevations less than 1 m above normal high water level and only 11.2 days of flooding during the growing season [44].

Results of the present investigation are consistent with previous results, which have shown that flooding determines the distribution of most floodplain plants. The prevalence of *P. arundinacea* and *C. vesicaria*, which were dominants, was found above or below approx. 50 cm above the summer water level, where the flood duration was around 10%. As indicated in **Table 1**, the transition between the epilittoral and the eulittoral zones had a flooding duration around 10%, and the transition between the eulittoral and the hydrolittoral zones had a flooding duration around 26%. The studies show that the elevational distribution of the littoral zones is associated by the average long-term water levels. The mean flooding duration had negligible effect on the species' coverage, but it can have a decisive effect on the development of flowering shoots.

5.3 Effects of wind and wave erosion

Transect T1 was initially supposed to describe the vegetation development in the new sublittoral zone, but below the mean summer water level, only scattered shoots were periodically established. Previous studies have also shown that wind and wave actions can restrict the distribution of the littoral vegetation zones by regulating plant development on the within-lake shorelines [45–48].

The reduced water level amplitude increased the erosion at the mean summer water level. This is probably because the water level fluctuations have become much smaller and more stabilized at the mean summer water level. Before the regulation, there were well-developed populations dominated by *E. fluviatile* down to approx. 60 cm below the summer water level (**Table 1**). Studies have shown that the new sublittoral vegetation was limited to small areas protected from erosion after the regulation. In delta areas that were protected from erosion, vegetation dominated by *C. vesicaria* was found below the mean summer water level, and *E. fluviatile* developed populations down to a water depth of approx. 1 m [11, 49].

5.4 The value of long-term studies

This review demonstrates the importance of studies in permanent plots over several years to document the dynamics of both individual plant coverage and the overall floristic composition of the study plots. Previous investigations have also emphasized needs for surveys of permanent plots. Any study of vegetation dynamics at the species level is impossible without permanent plots [34, 50, 51]. Long-term time series are necessary to better understand population dynamics because brief time series may lead to wrong conclusions, given large natural year-to-year variability [52].

Several ecologists have called for more studies on plant population dynamics because they are at the core of the ecological sciences but poorly represented in the current literature [53]. Explanation of why and how plant dynamics change in space and time is important from a fundamental scientific point of view and has become a cornerstone of modern ecological research and conservation biology [54, 55].

An essential question has been raised as to how long of a time series is necessary because brief time series may lead to wrong conclusions given large natural year-to-year variability. It has been shown that to identify long-term changes in abundance, on average 15.91 years of continuous monitoring were often required, but there was a wide distribution of estimated minimum times [52]. It was also shown that 72% of time series required at least 10 years of continuous monitoring to achieve a high level of statistical power.

Most of the plants examined in the delta studies showed major annual variations in their biomass (coverage). An essential question would then be how many study years would be necessary to describe the different long-term trends. The several topics investigated in these studies required different lengths of time to fully describe the long-term dynamics.

A period of 33 years appeared to be necessary to show that the long-term coverage dynamics of *C. vesicaria* and *P. arundinacea* significantly followed third-order functions.

The development of secondary successions should, in theory, over time result in stable floristic compositions. In the present studies, the species compositions in the study plots were stabilized after approx. 15 years. The degradation of the original vegetation types also required a survey period of 15–18 years.

The number of *C. vesicaria* and *C. rostrata* shoots producing flowers showed no significant trend during the whole survey period. Both species responded, however, with decreased flower production during years with high water levels. *C. vesicaria* responded also with significantly increased flower production during years with high summer temperatures. This shows that a study period of 27 years was necessary to establish significant data on factors of importance for *C. vesicaria* flower production.

5.5 Have new littoral zones been developed in accordance with the new hydrological regime?

It would be interesting to investigate if the secondary succession was stabilized during the study period and if a new vegetation zonation was recreated in accordance with the original conditions after the regulation. These questions are particularly important where rare or vulnerable habitat types are affected.

The regulation has changed the distribution of several vegetation types. Usually, during a secondary succession, the vegetation will gradually stabilize [56–59]. The DCA analyses of the vegetation in T1 have shown that after approximately 15 years, no long-term trends in the distribution of the species were recorded. Another fact indicating that the ecological conditions were stabilized was that biodiversity no longer showed a long-term trend. The species richness in T1 was stabilized after approx. 15 years [8, 9].

But the littoral zonation has not been recreated as it was before the regulation. Wind and wave erosion over large areas have prevented eulittoral and sublittoral vegetation types to develop and reduced the extent of the vegetation dominated by *E. fluviatile*. Before regulation, it could dominate areas more than 100 m from the water edge, while at present, only scattered stands occur in protected areas. In undisturbed lakes, *E. fluviatile* may extend to sites more than 100 cm below the summer water level, while *C. rostrata* and *C. vesicaria* are often found 60 cm below the mean summer water level [60, 61]. The distribution of *C. vesicaria* was also reduced, but stands have developed in areas situated around the mean summer water level.

The stabilized water level fluctuations have decreased the areas affected by frequent flooding, which has benefitted the distribution of *S. nigricans*. Most tree species do not tolerate long periods of flooding. *S. nigricans* has largely expanded its range and now covers large parts of the lower part of the delta where forest did not exist before the regulation (**Figures 1** and **2**). Large parts of the delta situated 20–30 cm above the summer water level and approx. 30 m inland from the water edge are at present dominated by *S. nigricans* forests [11]. The forest field layer is mainly dominated by *C. phragmitoides*, *P. arundinacea*, and *D. cespitosa*, which do not tolerate long flooding periods (**Figure 10**).

Dense stands dominated by *E. fluviatile* were gradually developed in areas protected from wave erosion 10 years after the regulation. They grew from the mean summer water level down to a depth of more than 60 cm [11].

6. Conclusions

The investigations have shown major dynamics of both individual plant species coverage and the floristic composition of the new vegetation types after the regulation. Due to favorable substrates, the changes have proceeded very quickly. The distribution of both species and vegetation types was permanently changed compared to the state before the regulation. The littoral zones have shifted downward toward the new summer water level. The distribution of sublittoral and eulittoral vegetation has been reduced, and forests dominated by *S. nigricans* have become more widespread.

All the investigated species showed major coverage dynamics both annually and across long terms. The plants had periods of high coverage interspersed with periods of low coverage. But the trends showed little agreement between the species, and the lengths of periods with high and low cover were mostly variable. More specific results of the investigations are:

- The original delta vegetation dominated by *E. fluviatile* and *C. vesicaria* was gradually degraded and disappeared almost completely after approx. 15 years.
- After the regulation, *E. fluviatile* began to migrate downward toward the new water level from its original lowermost site.
- The number of *C. vesicaria* shoots producing flowers was reduced over time in the original vegetation. After 15 years, no flowering shoots were recorded.
- Sediments exposed after the regulation were completely dominated by vascular plants within 3 years. Study plot total coverage became higher than 100% after 7 years.
- Over time, most plants achieved their greatest coverage at specific levels above the summer water level.
- Most species, except *C. vesicaria* and *C. rostrata*, had their highest coverage 50 cm above the summer water level.
- At the end of the study period, *C. vesicaria* and *P. arundinacea* became dominants but at different elevations above the summer water level.
- The coverage variation of *C. vesicaria* and *P. arundinacea* throughout the study period significantly followed third-degree distribution trends at most elevations above the summer water level.
- Study plot species richness decreased strongly the first years after the regulation but was stabilized after approx. 15 years.
- The secondary vegetation succession became stabilized after approx. 15 years. Because of a more stable summer water level, vegetation did not develop in accordance with the original summer water level.
- In most years after the regulation, *C. vesicaria* produced more flowers than *C. rostrata*. The number of flowering shoots in both species was dynamic but significantly correlated during the study period. Both species increased the production of flowering shoots with decreasing elevation.
- The dynamics of *C. vesicaria* flowering shoots increased significantly with high summer temperature and decreased with high water level.
- The production of flowering shoots decreased during years with high water levels, while their coverages were not affected.
- Dense *S. nigricans* forest and thickets dominated by *C. phragmitoides*, *P. arundinacea*, and *D. cespitosa* today dominate most areas between the cultivated fields and the lake water's edge.
- The present results show that the regulation has been negative because a valuable delta area has lost many of its original qualities.

Author details

Arvid Odland
Department of Natural Sciences and Environmental Health, University of South-Eastern Norway, Norway

*Address all correspondence to: arvid.odland@usn.no

References

[1] Burandt P, Grzybowski M, Glińska-Lewczuk K, Gotkiewicz W, Szymańska-Walkiewicz M, Obolewski K. Hydrology as a determinant of riparian habitat structure in lowland river floodplains. Water. 2024;**16**:164. DOI: 10.3390/w16010164

[2] Jing L, Zeng Q, He K, Liu P, Fan R, Lu W, et al. Vegetation dynamic in a large floodplain wetland: The effects of hydroclimatic regime. Remote Sensing. 2023;**15**:2614. DOI: 10.3390/rs15102614

[3] Chapin DM, Paige DK. Response of delta vegetation to water level changes in a regulated mountain lake, Washington State, USA. Wetlands. 2013;**33**:431-444. DOI: 10.1007/s13157-013-0401-5

[4] Krolová M, Cížková H, Hejzlar J, Poláková S. Response of littoral macrophytes to water level fluctuations in a storage reservoir. Knowledge and Management of Aquatic Ecosystems. 2013;**408**:07. DOI: 10.1051/kmae/2013042

[5] Bagchi S, Singh NJ, Briske DD, Bestelmeyer BT, McClaran MP, Murthy K. Quantifying long-term plant community dynamics with movement models: Implications for ecological resilience. Ecological Applications. 2017;**27**:1514-1528. DOI: 10.1002/eap

[6] Bornkamm R. Rates of change in vegetation during secondary succession. In: Poissonet P, Romane F, Austin MA, van der Maarel E, Schmidt W, editors. Vegetation Dynamics in Grasslans, Healthlands and Mediterranean Ligneous Formations, Advances in Vegetation Science. Vol. 4. Dordrecht: Springer; 1981. DOI: 10.1007/978-94-009-7991-8_21

[7] Anderson KJ. Temporal patterns in rates of community change during succession. The American Naturalist. 2007;**169**:780-793. DOI: 10.1086/516653

[8] Odland A, del Moral R. Thirteen years of wetland vegetation succession following a permanent drawdown, Myrkdalen Lake, Norway. Plant Ecology. 2002;**162**:185-198. DOI: 10.1023/A:1020388910724

[9] Odland A. Vascular plant succession on an exposed fluvial delta plain after a permanent lake water level drawdown. A 33-year time course. Polish Journal of Ecology. 2022;**70**:79-88. DOI: 10.3161/15052249PJE2020.70.2.002

[10] Odland A. Patterns in the secondary succession of a *Carex vesicaria* L. wetland following a permanent drawdown. Aquatic Botany. 2002;**74**:233-244. DOI: 10.1016/S0304-3770(02)00103-1

[11] Odland A. Vege-tasjonsøkologiske langtidseffekter etter den permanente reguleringen av Myrkdalsvatnet, Voss. Blyttia. 2024;**82**:187-203. [In Norwegian with English abstract] (In press)

[12] Odland A. Variation in plant cover and flower production of *Carex rostrata* and *C. Vesicaria* within a freshwater littoral zone. A 14-year time course. Phytocoenologia. 2024;**52**:43-53. DOI: 10.1127/phyto/2024/0414

[13] Odland A. Elvesnelle *Equisetum fluviatile* på vandring. Endringer i en elvesnellepopulasjon gjennom 29 år etter en vannstandssenkning. Blyttia. 2017;**75**:113-122. [In Norwegian with English abstract]

[14] Odland A. Endringer i flora og vegetasjon på Bygddeltaet etter

senkingen av Myrkedalsvatnet—utviklingen fra 1987 til 1991. NINA Oppdragsmelding. 1992;**113**:1-36. [In Norwegian with English abstract]

[15] Skoglund J. Seed banks, seed dispersal and regeneration processes in wetland areas [thesis]. Uppsala: Faculty of Science, Uppsala University; 1990

[16] Odland A. A long-term study of *Carex vesicaria* and *Phalaris arundinacea* population dynamics in permanent plots. Polish Journal of Ecology. 2023;**71**:49-57. DOI: 10.3161/15052249PJE2023.71.2.001

[17] ter Braak CJF, Šmilauer P. CANOCO Reference Manual and CanoDraw for Windows User's Guide: Software for Canonical Community Ordination (Version 4.5). New York: Microcomputer Power; 2002

[18] Saarinen T. Demography of *Carex rostrata* in a boreal mesotrophic fen: Shoot dynamics and biomass development. Annales Botanici fennici. 1998;**35**:203-209. Available from: https://www.jstor.org/stable/23726563

[19] De Cáceres M, Legendre P, Valencia R, Cao M, Chang LW, Chuyong G, et al. The variation of tree beta diversity across a global network of forest plots. Global Ecology and Biogeography. 2012;**21**:1191-1202. DOI: 10.1111/j.1466-8238.2012.00770.x

[20] Likens GE, editor. Lake Ecosystem Ecology: A Global Perspective. San Diego: Academic Press; 2010. eBook ISBN: 9780123820037

[21] Wetzel RG. Limnology: Lake and River Ecosystems. 3rd ed. San Diego: Academic Press; 2001. DOI: 10.1016/C2009-0-02112-6

[22] De Kroon H. Competition between shoots in stands of clonal plants. Plant Species Biology. 1993;**8**:85-94. DOI: 10.1111/j.1442-1984.1993.tb00060.x

[23] Klimesová J, Čižková H. Limitations of establishment and growth of *Phalaris arundinacea* in the floodplain. In: Prach K, Jenik J, Large ARG, editors. Floodplain Ecology and Management. Amsterdam: SPB Academic Publishing; 1996. pp. 131-145

[24] Lavergne S, Molofsky J. Reed canary grass (*Phalaris arundinacea*) as a biological model in the study of plant invasions. Critical Reviews in Plant Sciences. 2004;**23**:415-429. DOI: 10.1080/07352680490505934

[25] Danais M. The influence of some environmental factors on the production of *Carex vesicaria* and *Phalaris arundinacea*. Vegetatio. 1986;**67**:45-56. DOI: 10.1007/BF00040317

[26] Bernard JM, Gorham E. Life history aspects of primary production in sedge wetlands. In: Good R, Whigham DF, Simpson RL, editors. Freshwater Wetlands. Ecological Processes and Management Potential. New York: Academic Press; 1978. pp. 39-51

[27] Solander D. Biomass and shoot production of *Carex rostrata* and *Equisetum fluviatile* in unfertilized and fertilized subarctic lakes. Aquatic Botany. 1983;**15**:349-366. DOI: 10.1016/0304-3770(83)90003-7

[28] Bernard JM. Life history and vegetative reproduction in *Carex*. Canadian Journal of Botany. 1990;**68**:1441-1448. DOI: 10.1139/b90-182

[29] Soukupová L. Short life-cycles in two wetland sedges. Aquatic Botany. 1988;**30**:49-62. DOI: 10.1016/0304-3770(88)90006-X

[30] Keddy PA. Wetland Ecology: Principles and Conservation. 2nd ed. New York: Cambridge University Press; 2010. DOI: 10.1017/CBO9780511778179

[31] Sorrell BK, Tanner CC, Brix H. Regression analysis of growth responses to water depth in three wetland plant species. AoB Plants. 2012;**2012**:pls043. DOI: 10.1093/aobpla/pls043

[32] Van Eck WHJM, Van De Steeg HM, Blom CWPM, De Kroon H. Is tolerance to summer flooding correlated with distribution patterns in river floodplains? A comparative study of 20 terrestrial grassland species. Oikos. 2004;**107**:393-405. DOI: 10.1111/j.0030-1299.2004.13083.x

[33] Vervuren PJA, Blom CWPM, de Kroon H. Extreme flooding events on the rhine and the survival and distribution of riparian plant species. Journal of Ecology. 2003;**91**:135-146. DOI: 10.1046/j.1365-2745.2003.00749.x

[34] Pannell JR. The ecology of plant populations: Their dynamics, interactions and evolution. Annals of Botany. 2012;**110**:1351-1355. DOI: 10.1093/aob/mcs224

[35] Blom CW, Voesenek LA. Flooding: The survival strategies of plants. Trends in Ecology and Evolution. 1996;**11**:290-295. DOI: 10.1016/0169-5347(96)10034-3

[36] Zohary T, Ostrovsky I. Ecological impacts of excessive water level fluctuations in stratified freshwater lakes. Inland Waters. 2011;**1**:47-59. DOI: 10.5268/IW-1.1.406

[37] Banach K, Banach AM, Lamers LPM, De Kroon H, Bennicelli RP, Smits AJM, et al. Differences in flooding tolerance between species from two wetland habitats with contrasting hydrology: Implications for vegetation development in future floodwater retention areas. Annals of Botany. 2009;**103**:341-351. DOI: 10.1093/aob/mcn183

[38] Bornette G, Puijalon S. Macrophytes: Ecology of aquatic plants. In: Encyclopedia of Life Sciences. Chichester: John Wiley and Sons, Ltd; 2009. DOI: 10.1002/9780470015902.a0020475

[39] van Eck WHJM, Lenssen JPM, van de Steeg HM, Blom CWPM, de Kroon H. Seasonal dependent effects of flooding on plant species survival and zonation: A comparative study of 10 terrestrial grassland species. Hydrobiologia. 2006;**565**:59-69. DOI: 10.1007/s10750-005-1905-7

[40] Prach K. Vegetation, microtopography and water table in the Lužnice River floodplain, South Bohemia. Czechosolvakia. Preslia Praha. 1992;**64**:357-367

[41] Klimesová J. The effects of timing and duration of floods on growth of young plants of *Phalaris arundinacea* L and *Urtica dioica* L: An experimental study. Aquatic Botany. 1994;**48**:21-29. DOI: 10.1016/0304-3770(94)90071-

[42] Kopecký K. Einfluss langanderder Überflutungen auf die Stoffproduktion von Glanzgraswiesen. Folia Geobotanica and Phytotaxonomica. 1967;**2**:347-382. DOI: 10.1007/BF02854722

[43] Conchou O, Pautou G. Modes of colonization of a heterogeneous alluvial area on the edge of the Garonne River by *Phalaris arundinacea* L. Regulated Rivers. 1987;**1**:37-48. DOI: 10.1002/rrr.3450010105

[44] Barnes WJ. The rapid growth of a population of reed canarygrass (*Phalaris arundinacea* L.) and its impact on some river bottom herbs. The Journal of the Torrey Botanical Society. 1999;**126**:133-138. DOI: 10.2307/2997289

[45] Keddy PA. Shoreline vegetation in Axe Lake. Ontario: Effects of exposure on zonation patterns. Ecology. 1983;**64**:331-344. DOI: 10.2307/1937080

[46] Coops H, Boeters R, Smit H. Direct and indirect effects of wave attack on helophytes. Aquatic Botany. 1991;**41**:333-352. DOI: 10.1016/0304-3770(91)90052-7

[47] Andersson L. Macrophyte development and habitat characteristics in Sweden's large lakes. Ambio. 2001;**30**:503-513. DOI: 10.1579/0044-7447-30.8.503

[48] Vilmundardóttir OK, Magnússon B, Gísladóttir G, Thorsteinsson T. Shoreline erosion and aeolian deposition along a recently formed hydro-electric reservoir, Blöndulón, Iceland. Geomorphology. 2010;**114**:542-555. DOI: 10.1016/j.geomorph.2009.08.012

[49] Odland A. Development of vegetation in created wetlands in western Norway. Aquatic Botany. 1997;**59**:45-63. DOI: 10.1016/S0304-3770(97)00030-2

[50] Bakker JP, Olff H, Willems JH, Zobel M. Why do we need permanent plots in the study of long-term vegetation dynamics. Journal of Vegetation Science. 1996;7:147-155. DOI: 10.2307/3236314

[51] White ER. Minimum time required to detect population trends: The need for long-term monitoring programs. Bioscience. 2019;**69**:40-46. DOI: 10.1093/biosci/biy144

[52] Weiher E, Keddy P. The assembly of experimental wetland plant communities. Oikos. 1995;**73**:323-335. DOI: 10.2307/3545956

[53] Roozen AJM, Westhoff V. A study on long-term salt marsh succession using permanent plots. Vegetatio. 1985;**61**:23-32. DOI: 10.1007/BF00039807

[54] Magurran AE, Baillie SR, Buckland ST, Dick JM, Elston DA, Scott EM, et al. Long-term datasets in biodiversity research and monitoring: Assessing change in ecological communities through time. Trends in Ecology and Evolution. 2010;**25**:574-582. DOI: 10.1016/j.tree.2010.06.016

[55] McCain CM, Szewczyk T, Knight KB. Population variability complicates the accurate detection of climate change responses. Global Change Biology. 2016;**22**:2081-2093. DOI: 10.1111/gcb.13211

[56] Guo QF. Temporal species richness-biomass relationships along successional gradients. Journal of Vegetation Science. 2003;**14**:121-128. DOI: 10.1111/j.1654-1103.2003.tb02134.x

[57] Furey PC, Nordin RN, Mazumder A. Water level drawdown affects physical and biogeochemical properties of littoral sediments of a reservoir and a natural lake. Lake and Reservoir Management. 2004;**20**:280-295. DOI: 10.1080/07438140409354158

[58] Stadler J, Klotz S, Brandl R, Knapp S. Species richness and phylogenetic structure in plant communities: 20 years of succession. Web Ecology. 2017;**17**:37-46. DOI: 10.5194/we-17-37-2017

[59] Carmignani JR, Roy AH. Ecological impacts of winter water level drawdowns on lake littoral zones: A review. Aquatic Sciences. 2017;**79**:803-824. DOI: 10.1007/s00027-017-0549-9

[60] Toivonen H, Lappalainen T. Ecology and production of aquatic macrophytes in the oligotrophic, mesohumic lake Suomunjärvi, eastern Finland. Annales

Botanici Fennici. 1980;**17**:69-85. Available from: http://www.jstor.org/stable/23726094

[61] Spence DHN. The zonation of plants in fresh-water lakes. Advances in Ecological Research. 1982;**12**:37-125. DOI: 10.1016/S0065-2504(08)60077-X

Chapter 4

The Role of Pre-Desert Vegetation in the Rehabilitation of Degraded Soil

Mallem Hamida

Abstract

In arid and Saharan zones, drought and human activities accentuate the risk of degradation of pre-desert rangelands. Some plants disappear, while others persist and adapt. In this study, we aimed to list the plants that can grow in a sandy environment and investigate their distribution levels and effect on sand dunes. We conducted vegetation inventories in Laghouat province in Algeria, and we calculated the ecological parameters using the transect method. The height of fixed windblown deposits was measured under all inventoried plants. Under three perennial plants (*Retama raetam, Aristida pungens*, and *Astragalus armatus*), soil texture, and physical and microbiological characteristics were analyzed. We found that the area studied is species-poor but has certain diversity, so the plant biological type influenced sand fixation. Fixed Aeolian deposits stabilized by plants ranged in height from 28 to 63 cm. Micro-dunes stabilized by *R. raetam* exhibited higher nitrogen, carbon, and bacterial richness compared to those stabilized by *A. pungens*. In contrast, *A. pungens* micro-dunes had greater calcium carbonate, electrical conductivity, and fungal richness. Our findings highlight the distinct contributions of each species to soil rehabilitation. Moreover, the complementary effects of these plants suggest their potential for synergistic wind erosion control.

Keywords: vegetation, soil, Saharan, dune, restoration

1. Introduction

North Africa is one of the regions most vulnerable to the consequences of climatic aridity and the impact of human activities on the natural environment. The steppe areas of North Africa are particularly affected by the problems of desertification [1]. In Algeria, Morocco, and Tunisia, the proportion of national territory affected by desertification was estimated at over 80% in the early 1980s [2]. Similarly, the desertification sensitivity map drawn up by the Algerian National Center for Space Technology reveals that 53% of the total Algerian steppe area is classified as very sensitive to desertification [3]. In Algeria, outside the Saharan zone, it is above all the high plains that are most affected by the phenomenon of silting-up, with 500,000 ha of aeolian formations extending to the north of the Saharan Atlas [4]. Wind erosion is a major physical factor in the depletion of agricultural land [5], and causes severe

 IntechOpen

environmental degradation by impoverishing soils and reducing their production [5, 6]. Nearly 600,000 hectares of land in the steppic zone are totally desertified, with no possibility of biological recovery, and almost 6 million hectares are threatened by the effects of wind erosion [7]. The Algerian steppe is subject to ecologically unsustainable exploitation. Desertification is gaining ground due to recurrent drought, overgrazing, and the extension of rain-fed and sometimes irrigated agriculture, which is unsuited to the conditions of the natural environment [8]. The extension of plowing and the introduction of mechanization are just as important parameters of degradation as overgrazing. The plowing techniques used by agropastoralists have an erosive action, destroying the superficial horizon and sterilizing the soil, often irreversibly. The woody species that hold the soil in place are destroyed and replaced by adventitious species that promote wind erosion [9]. Psammophyte steppes are linked to the sandy texture of the surface horizons and to aeolian inputs. They follow the silting corridors and are also distributed in the depressions formed by the chotts, they are more frequent in arid and pre-Saharan zones, forming steppes with *Stipagrostis pungens* and *Thymelaea microphylla* or shrub steppes with *Retama raetam*, and their pastoral values vary from 200 to 250 FU/ha [10]. According to the results of the 2003 ROSELT program, psammophytic vegetation, which was non-existent in 1978 in the west, has expanded due to greater silting during the drought period of 1981–1987. The cover of psammophytic vegetation often exceeds 30%, thanks to the proliferation of annual species [11]. Sand dunes are characterized by the presence of plant cover, of variable density, which fixes their sand to a greater or lesser extent. Dune soils are generally poor in nutrients and water [12]. Houyou et al. [13] have demonstrated that sandstorms cause (64.32 t/ha/year) of land loss in the Mokrane area of Laghouat, accentuated by plowing and wind speed (7.3 m/s). In view of the silting-up of this cereal-growing and sheep-breeding site, we set out to draw up an inventory of the plant species colonizing the sandy accumulations and assess the pastoral value of these rangelands. In arid areas, plowing and overgrazing increase the risk of degradation of steppe rangelands, leading to sand encroachment and wind erosion. Some plants disappear, while others persist and adapt. What are soil characteristics of the plants canopy that can grow on sandy accumulations? In this study, we aimed to identify steppe plants that can grow in a sandy environment and to investigate their distribution levels on sand dunes and their effects on dune soil.

2. Material and methods

2.1 Study area

The studied site is named Mokrane, situated 4 km from Laghouat, the capital of Laghouat Province (N 33°48′–E 2°48′) (**Figure 1**).

Mokrane is an alluvial plain characterized by steppe vegetation. Frequent sandstorms, coupled with agricultural activities, and extensive grazing, exacerbate wind erosion and sedimentation (**Figure 2**).

2.2 Floristic study

A vegetation survey was conducted in March 2009 using the Braun-Blanquet method [14, 15]. Twelve linear transects were established based on vegetation

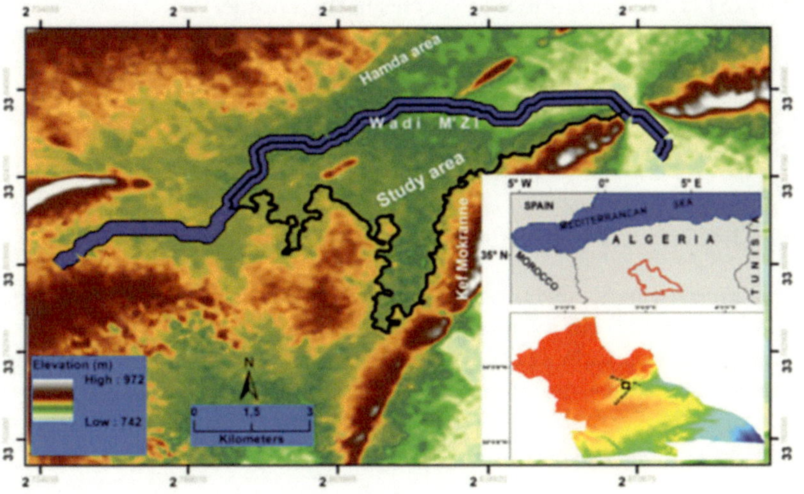

Figure 1.
Location of the study area in Algeria.

Figure 2.
Appearance of vegetation in the study site.

physiognomy and the Djebaili method [15], covering areas ranging from 128 to 420 m^2. Each 20-meter transect involved recording data at 20 cm intervals, totaling 100 data points per transect.

Floristic richness, Shannon diversity index (H′), equitability (J), sand cover, and vegetation cover were assessed. A higher H′ value indicates greater diversity. Equitability helps detect changes in community structure, especially those of anthropogenic origin. These indices are sensitive to frequent species rather than overall richness [15]. Plant identification followed the flora of Quezel and Santa [16]. Simpson's index (1-D) and others indexes were calculated according to Krebs [17] as below:

a. *Richness* (*S*) is just the number of species in the sample.

b. *Dominance* index (D): Dominance index values range from 0 to 1.

D = $\sum(niN)2$, With: ni = individual number of a species, N = individual number all species.

c. *Equitabilty or Pielou's measure of species evenness*, $J = H'/\ln(S)$ where H' is Shannon Weiner diversity and S is the total number of species in a sample.

d. *Simpson diversity index* (1-D) represents the probability that two individuals randomly selected from a sample will belong to different species. It is a measure of diversity that considers both Richness (the number of species per sample) and Evenness (the relative abundance of the different species making up the richness of an area). Formula: $1 - D = 1 - \sum(n/N)2$. The results can be from 0 (less diverse) to 1 (most diverse).

e. *Shannon diversity index* (*H'*) is a measure of the average degree of uncertainty in predicting to what species and individual chosen at random from a collection of species and individuals will belong. Formula: $H' = -\sum(Pi * \ln Pi)$. The values can range from 0 (less diverse) to ln (1/S), but usually they will fall between 1.5 and 3.5 and rarely above 5.

f. *Berger-Parker dominance index (1/d)*, is the number of individuals in the dominant taxon relative to n. Its value does not take into account the number of classes but it is highly influenced by equity. It is a simple measure of the numerical importance of the most abundant species. Formula: 1/d = Nmax/N. An increase in the value of the index accompanies an increase in diversity and a reduction in dominance.

To evaluate rangeland degradation, a disturbance index (DI) was computed using the ratio described by Hébrard et al. [18]:

$$DI = \frac{\text{Chamaephytes number} + \text{Therophytes number}}{\text{Total number of species}} \quad (1)$$

2.3 Pastoral productivity of Mokrane's grazing land

Pastoral productivity represents the vegetation's energy output per unit area and time, expressed as Forage unit per hectare and year (FU ha^{-1} t^{-1}). The formula used in numerous Algerian steppe studies [19] was adopted (Eq. (2)). It correlates pastoral value with forage productivity.

$$Pp = 6.74\,Pv + 14.77 \quad (2)$$

Where:

- Pp: Pastoral productivity of the facies in FU/ha
- Pv: Pastoral value

According to Aidoud [19], pastoral value is an allometric assessment suitable for estimating rangeland forage production. It is calculated by multiplying each species' contribution by its corresponding index and summing the results. To determine pastoral value, we employed Eq. (3), previously used by Hirche et al. [20] for Algerian steppe rangelands. This equation originates from Daget and Poissonet in [21].

$$\mathrm{Pv}\,(\%) = 0.1 \times \mathrm{VC} \times \Sigma\,\mathrm{SCi} \times \mathrm{SI} \tag{3}$$

Where:

- SCI: Specific contribution of plant species
- SI: Specific index of plant species
- VC: Overall vegetation cover

The specific vegetation indices for the Algerian steppe are empirically derived using a numerical scale developed by Aidoud [19]. This scale considers factors such as bromatological quality, digestibility, and assessments by steppe breeders. Each species is assigned a score or grade, which we adopted for this study.

Animal load refers to the livestock carrying capacity of a grazing area, often expressed as the number of animals per unit area.

2.4 Sediment height measurement under and outside the canopy

Sediment height measurement was conducted concurrently with vegetation surveys. Using a modified differential leveling method [22], sediment accumulation at the study site's ground surface was measured. Altitude differences were determined every 20 cm along the transect using a leveling instrument aimed at a graduated rod placed on the accumulated sediment.

2.5 Soil analysis

Soil samples were collected from eight randomly selected sediment accumulation points associated with each host species (fixed soil) and eight points without vegetation (unfixed soil). Samples were taken in March 2014, at the end of the rainy season, using a hand auger to collect the top 30 cm of sediment, approximately 10–15 cm from the host plant center [23]. The following soil parameters were measured: moisture content (determined by drying at 105°C for 24 hours), particle size distribution, total nitrogen, limestone content, electrical conductivity (EC), pH, and organic matter content. Soil particle size distribution was determined by sieving and sedimentation to obtain percentages of sand, silt, and clay [24]. Organic matter content was estimated through loss on ignition at 600°C for 5 hours. Soil pH and electrical conductivity (EC) were measured in a 1:5 soil-to-water suspension after shaking for 30 minutes using a digital conductivity meter (YSI Inc., OH, USA). Total nitrogen (N) was quantified using the Kjeldahl method. Calcium carbonate ($CaCO_3$) content was determined according to the procedures of the United States Salinity Laboratory (Jackson, 1962). For microbial enumeration, PDA (Potato Dextrose Agar) and LPGA (Yeast Peptone Glucose Agar) media were used for fungi and bacteria, respectively [25].

2.6 Statistical data

Floristic diversity was assessed using Paleontological Statistics software (version 2.17c). Soil parameters were compared using ANOVA in XLSTAT software (version 3.5) at a significance level of $P \leq 0.05$. Principal component analysis (PCA) was conducted using XLSTAT to examine relationships between soil parameters and vegetation.

3. Results and discussion

3.1 Vegetation

All the plant species inventoried in the Mokrane area have already been reported by other authors interested in the plant biodiversity of North African steppes [26, 27]. The floristic analysis of the sandy agricultural zone of Mokrane allowed us to identify the dominance of seven steppe plants: *Bromus rubens*, *Schismus barbatus*, *Thymelaea microphylla*, *Echinops spinosus*, *Retama raetam*, *Aristida pungens*, and *Astragalus armatus* (**Figure 3**).

We recorded the dominance of three families: Asteraceae, Fabaceae, and Poaceae (**Table 1**). Aidoud [19] has already confirmed the dominance of the families Asteraceae, Poaceae, Fabaceae and Chenopodiaceae in the desert flora. Similar observations have been noted in the floristic description of the North Algerian Sahara [28]. Asteraceae, Poaceae, and Fabaceae are three families that represent 35–40% of the flora in each Saharan sector [29, 30]. This predominance is justified since these are cosmopolitan families that are widespread throughout the steppe and the Saharan Atlas [31].

On a physionomic and floristic level, Djebaili [15] defines the variation in the floristic composition of the grouping according to the nature of the lithological substrate. When it develops on heavily sanded erosion glacis, this grouping forms facies with psammophyte species, such as *Retama raetam* and *Thymelaea microphylla*. According to Benaradj et al. [31], sandblasting favors the installation of psammophyte species, such as *Retama raetam* and *Malcolmia aegyptiaca*. Pouget [32] indicated

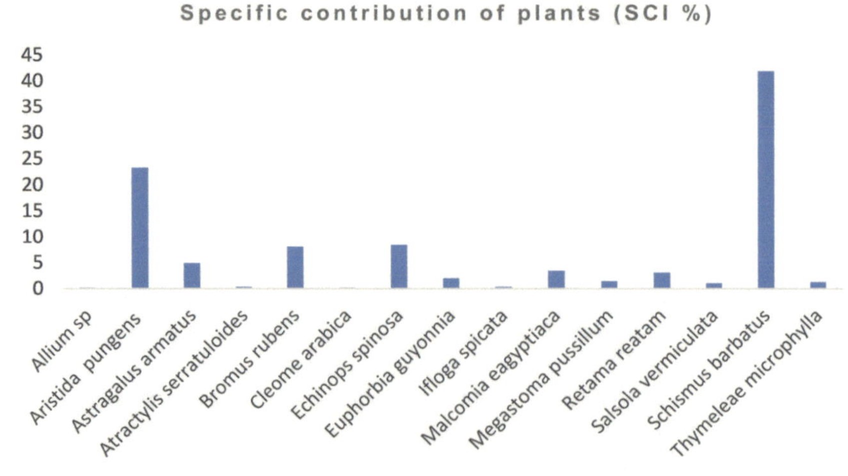

Figure 3.
Specific contribution of plants.

Species	Familly	Abbreviation	Cycle	Biological types	Biogeography
Aristida pungens Desf.	Poaceae	*A. pungens*	P	Hemicryptophyte	Saharo South Africa
Astragalus armatus Willd.	Fabaceae	*A. armatus*	P	Chamaephyte	Mediterranean
Atractylis serratuloides Sieber ex Cass	Asteraceae	*A. serratuloides*	p	Chamaephyte	Saharan
Bromus rubens L.	Poaceae	*B. rubens*	A	Therophyte	Mediterranean
Cleome arabica L.	Capparidaceae	*C. arabica*	A	Therophyte	Saharo Sindian
Echinops spinosa L.	Asteraceae	*E. spinosa*	P	Chamaephyte	Mediterranean
Erodium glaucophyllum L.	Geraniaceae	*E. glaucophyllum*	P	Hemicryptophyte	Endemic Mediterranean
Eruca vesicaria (L) Thell.	Brassicaceae	*E. vesicaria*	A	Therophyte	Mediterranean
Euphorbia guyoniana Boiss & Rent.	Euphorbiaceae	*E. guyoniana*	P	Hemicryptophyte	Endemic N. Africa
Evax desertorum Pomel.	Asteraceae	*E. desertorum*	A	Therophyte	Mediterranean
Helianthemum getulum Pomel.	Cistaceae	*H. getulum*	P	Chamaephyte	Mediterranean Saharan
Ifloga spicata Vahl.	Asteraceae	*I. spicata*	A	Therophyte	Saharo Sindian
Launaea resedifolia	Asteraceae	*L. resedifolia*	P	Hemicryptophyte	Mediterranean Saharo Arab
Malcomia aegyptiaca Spr.	Brassicaceae	*M. aegyptiaca*	A	Therophyte	Saharo Arabo Subtropical
mediterraneanicago laciniata L.	Boraginaceae	*M. laciniata*	A	Therophyte	Mediterranean. Saharo Sindian
Megastoma pusillum Coss. & Durieu	Boraginaceae	*M. pusillum*	A	Therophyte	Mediterranean
ononis pusilla L.	Fabaceae	*O. pusilla*	P	Hemicryptophyte	Mediterranean
Peganum harmala L.	Zygophyllaceae	*P. harmala*	p	Hemicryptophyte	Irano-Turanian
Plantago albicans L.	Plantaginaceae	*P. albicans*	P	Hemicryptophyte	Mediterranean
Retama raetam Forssk.	Fabaceae	*R. raetam*	P	Nanophanerophyte	Saharo Sindian
Salsola vermiculata L.	Chenopodiaceae	*S. vermiculata*	P	Chamaephyte	Saharo Mediterranean
Schismus barbatus L. Thel.	Poaceae	*S. barbatus*	A	Therophyte	Mediterranean
Silene arenarioides Desf.	Caryophyllaceae	*S. arenarioides*	A	Therophyte	Endemic N. Africa

Species	Familly	Abbreviation	Cycle	Biological types	Biogeography
Stipa tinassicima L.	Poaceae	*S. tinassicima*	P	Hemicryptophyte	Endemic
Thymelaea microphylla Coss et Dur.	Thymelaeaceae	*T. microphylla*	P	Chamaephyte	Endemic N. Africa

Table 1.
List of steppe plants in the Mokrane area (P: perennial, A: annual).

that *Aristida pungens* and *Malcomia eagyptiaca* are found on fixed micro dunes. On the other hand, non-fixed dunes are colonized by *saccocalyx saturoides*; *Euphorbia guyoniana*, and *Ononis natrix*; these species are resistant to water shortage and are best adapted to this type of ecosystem. According to QUEZEL [33], on mobile dunes a few species can survive under 50 mm of rainfall, such as *Euphorbia guyoniana* and *Aristida pungens*.

Killiain [34] has already indicated that *Euphorbia guyoniana* releases roots 15 cm from the wet horizon. Therophytes like *Bromus rubens* and *Lolium rigidum* require a certain percentage of water. Pouget [35] indicated the presence of *Malcomia aegyptiaca, Aristida pungens,* and *Retama reatam* in the coarse texture of the soils. In our case, we investigated the existence of plants linked to fixed micro dunes, the species linked to sandy sails and unbound dunes.

The results of the analysis of the biological type (**Table 1**) showed the dominance of therophytes in the Mokrane pasture. The observed therophytization is linked on the one hand to the harshness of the climate and on the other hand to human activities which increasingly degrade the conditions for the establishment of new species.

Emberger [36] and Daget [37] state that the rate of therophytes increases with the aridity of the environment. This richness is due to the process of the biological rise, of the reconstitution, regeneration, and reappearance of species threatened with destruction by degradation factors. About 30% of psammophyte vegetation is due to the proliferation of ephemeral plants [11]. We have thus observed the impact of climatic drought by the increase in the rate of therophytes and by the presence of Saharan affinity species such as: *Euphorbia guyoniana* and *Launaea residifolia*. Chamaephytes can also develop forms of adaptation to drought. The chamaephytization recorded in the Mokrane area has its origin in the phenomenon of aridity, according to Floret and Pontanier [26], chamaephytes adapt better to summer drought. Grazing also generally favors chamaephytes rejected by herds, such as *Thymelaea microphylla, Hammada scoparia, Anvillea radiata,* and *Gymnocarpos decander* [38], and *Astragalus armatus* according to Floret and Pontanier [26]. According to Kadi Hanifi [38], grazing generally favors chamaephytes refused by herds. Kadi Hanifi [39] confirmed that the regression of steppe formations generally results in chamaephytization by thorny species without economic interest abandoned and rejected by livestock. Le Houerou [27] also reported that the increase in woody chamaephytes in Poaceae formations is due to overgrazing by sheep and cattle. In our study area, the rangelands are mainly exploited by sheep and goats which allowed the appearance of the *A. armatus* group. Phanerophytes are less important, these decreasing gradually with the aridification of the climate. The proliferation of hemicryptophytes can also be explained by the poverty of the soil in organic matter, which has been demonstrated by soil analysis. The analysis of the phytogeographic spectrum (**Table 1**) showed the dominance of the Mediterranean, endemic, and Saharan tendencies. However, these results confirm the trends observed on the scale of the North African sub-region by Le Houerou [27] who showed that among the 2630 vascular plant species present in the Maghreb steppes, 60% are Mediterranean affinity species and 30% tropical affinity species. The high Saharan spectrum rate can be explained by the aridity of the environment.

From 2009 to 2015, the ecological indices (total richness, equitability, and species diversity) have not shown a significant difference (**Table 2**), The environment exhibited low biodiversity, with a species richness ranging from 15 to 21 taxa and Shannon index ranging from 1.8 to 2.13. Species evenness, as measured by the equitability index, fluctuated between 0.66 and 0.74.

Between 2009 and 2015, vegetation cover on the site increased significantly from 46.16% to 61.99%, coinciding with a substantial decrease in sand cover from 53.84% to 31.5% (**Figure 4**). This suggests that the fixation of sand by perennial plants contributed to the observed vegetation expansion, El-Bana et al. [40], Guerrache et al. [41], and Akkouche et al. [42] recorded the improvement of the vegetation cover on the sand dunes fixed by *Retama raetam*. *Astragalus armatus*, a chamaephyte with low pastoral value, has shown remarkable proliferation on the site, it is a species that appears in formerly cultivated areas and overgrazed areas [26]. Chaieb [43] explained this proliferation by the root architecture of this plant, favoring water absorption and also its high germination power.

Floral modifications of pastoral ecosystems in arid and desert regions, under the effect of animal pressure and water deficit, first affect palatable grasses and chamaephytes [44]. Maintaining the biological productivity of the environment remains, under such conditions, linked to the appearance of a new type of vegetation that is not very productive but in balance with this new environment [26]. However, although *Astragalus armatus* is considered to be a species marking degradation, it can contribute, even if only partially, to the process of restoring ecological balance in these degraded environments [43]. The trapping of sand and the reconstitution of the wind veil by the highly developed clumps lead to an improvement in the soil water balance and favor the germination of species hitherto rare. Finally, its ability to fix atmospheric nitrogen contributes to improving soil fertility.

3.2 Pastoral value

The Mokrane area in Laghouat rangeland (Drinn rangeland: *A. pungens* Steppe) presented an average pastoral value of 11–14%, we estimated this rate during a period

Ecological indices	Year 2009	Year 2010	Year 2011	Year 2012	Year 2014	Year 2015
Taxa_S	15.00	16.00	19.00	19.00	21.00	18.00
Dominance_D	0.25	0.21	0.20	0.16	0.09	0.16
Simpson_1-D	0.75	0.79	0.80	0.84	0.91	0.84
Shannon_H'	1.80	1.99	2.13	2.13	2.64	2.13
Equitability_J	0.66	0.72	0.72	0.72	0.87	0.74
Berger-Parker	0.42	0.40	0.38	0.26	0.15	0.27

Table 2.
The ecological indices measured through 6 years.

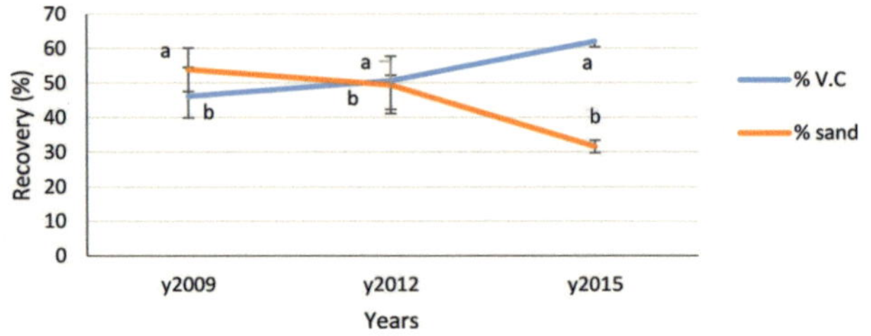

Figure 4.
Change of vegetation cover (VC) and sand cover over time.

when the vegetation was supposed to be optimal. Compared to Hirche et al. [20], they gave a value of 7% for a pasture of the South-West Algerian steppe in the Naâma region. In the same region in 2007, Benaradj et al. [31] calculated a pastoral value of 4% for a Sparte pasture (*Lygeum spartum*). On the other hand, the results established by OSS [45] on the potential of West Algerian rangelands report a higher pastoral value of around 11%, a value close to our results. The different types of vegetation and the richness in these rangelands explain the gaps recorded between the pastoral values. As well as the vegetation cover itself is always dependent on local edaphoclimatic conditions which differ from one region to another and from one year to the next.

The pastoral value is a synthetic index used to characterize the capacities offered by the rangelands. The Mokrane rangeland presented a pastoral value close to the alfa grass steppe but better than that of sparte and remt (**Table 3**). In our case, the low average pastoral value of Mokrane can also be explained by the low rate of soil surface cover by vegetation, by the low richness of the flora that characterizes the rangeland and by the nature of the plant species present which are not all palatable.

Among the inventoried plants, more than 68% of the species have poor specific pastoral indices (≤ 2) indicative of low energy values. About 16% of the species have average energy qualities with specific pastoral indices ($3 < Is \leq 5$). However, some species have good forage qualities with specific indices greater than 5 (*Medicago laciniata, megastoma pussillum,* and *Plantago albicans*) but weakly present in the place.

The low pastoral value can also be due to the dominance of psammophytes in the rangeland, including *Thymelaea microphylla, Aristida pungens, Retama raetam*, and *Astragalus armatus*, which also have average and poor forage qualities (specific indices). In the psammophyte steppes of southern Oranian city in Algeria, Nedjraoui [46] estimated maximum pastoral productivities between 200 and 250 FU/ha. Our values for (Mokrane) Laghouat vary between 89 and 108 FU/ha, like those reported by Salemkour et al. [47] for other pastures in the Laghouat steppe. However, the productivity of the Laghouat rangeland reveals a low nutritional value of the vegetation present and the mediocrity of the grazing area. The differences in productivity between the Laghouat rangelands and those of the southern Oranais steppe can be explained by the type of vegetation that dominates, the area it covers, the nature of the rangeland soil, and especially the rainfall in the area [46]. Indeed, the amount of rainfall received is a major factor in the development of plant communities in the steppes [45, 48, 49].

The estimated pastoral values and productivities result in average pastoral loads for the rangeland varying from 3 to 5 ha/Sheep Unit. This value is close to those estimated during the same period in other places in the Algerian steppe [46, 47, 50, 51]. The estimated load for the Mokrane rangeland clearly reflects a situation of pastoral imbalance and a poor state of health of this pasture, reflecting the situation of the Algerian steppe. The calculation of the disturbance index (DI) fluctuated between 0.74 and 0.66

Year	Year 2009	Year 2012	Year 2015
Vegetation cover (%)	46.24	62.13	63.17
Pastoral value (%)	13.85	10.99	13.75
Pastoral productivity (FU/an)	108.12	88.85	107.40
Pastoral charge (hectare)	3.70	4.50	3.72

Table 3.
Pastoral value of Mokrane area (2009–2015).

during the study years, suggesting that the Mokrane area is disturbed. Anthropogenic disturbances are largely responsible for the current state of vegetation structures in the Maghreb [52]. The natural vegetation cover is permanently subjected to a double impact, on the one hand from the soils (too dry and light) and the climate (low rainfall) and on the other hand from the actions of humans and their animals [53].

3.3 Height of accumulated sediments under and outside the canopy

The results of the statistical analysis concerning the preference of certain plants for certain elevations of accumulated sediments, allowed us to conclude that there are significant differences in this distribution (P = 0.001, **Figure 5**).

The three perennial species *A. pungens*, *R. raetam,* and *A. armatus* were established at a lower soil level than some annual plants that developed on the highest peaks of the sandy accumulations. The differences between the distribution levels of the inventoried plants could be due to seed dispersal modes, seed weight, wind direction, wind speed, and the phenomenon of runoff by rainwater. Bochet [54] explained that the seed dissemination capacity varies greatly between species and is mainly linked to the characteristics of the seed. In our case, seeds on and in the soil can be moved horizontally to new locations by different biotic factors (animals) or abiotic factors (wind, runoff, and gravity). Seed transport by runoff depends on soil characteristics, slope, and rainfall. Seeds can also be moved vertically by animals, in the opposite direction of the soil seed bank. Sediment movement can also explain the differences in the position levels of plant species on the accumulated sediments and their deposits in the study area. The appearance of the surface of Mokrane has been attributed to the Quaternary colluvial and alluvial deposits, and the phenomenon seems to be continuing to this day. Houyou et al. [13] observed two phenomena in Laghouat—sand burial and erosion depending on the wind speed and direction in the Mokrane area. Sandstorms in Laghouat caused the transport of seeds and their deposits with the sand. Once the seeds are deposited in the soil, they germinate, and then the plants grow if conditions are favorable. If there is a lack of water, the annual plants (ephemeral) disappear, otherwise the perennials persist. Our findings agreed with those of Cusseddu et al. [55], who reported that on coastal dunes in Sardinia, plant communities are organized in a hierarchical manner with the elevations of the

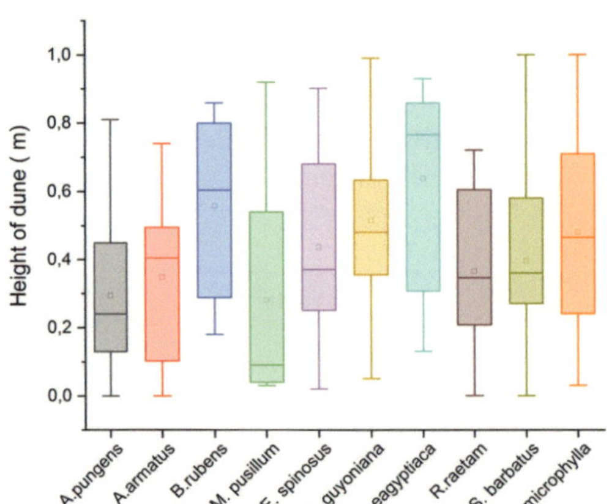

Figure 5.
Sediment depth under canopy.

dune. These authors added that the high summer temperatures led to a competitive dominance of woody plants on the back of the dune. In our case, although *A. pungens*, *A. armatus,* and *R. raetam* have different root architectures; they fixed the sand at similar heights, not exceeding 50 cm. The dispersal of their seeds is successively due to different processes (wind, water, and animals). The annual plants spend the bad season in the form of seeds and will be moved with the sand sediments by the wind, knowing that in the Mokrane area the wind blows all year round [13]. The new sand deposits (50–97 cm) would contain seeds of annual plants, which will then appear with the first spring rains. This behavior is different from the seeds of perennial plants that have been studied; these require specific germination conditions due to their dormancy [56]. This original method for calculating the elevations of fixed and non-fixed windblown deposits allowed us to conclude that the Mokrane area is invaded by deposits that do not exceed 1 meter in height, so we call the windblown deposits in this area micro dunes and not nebkas, given the succession observed.

3.4 Soil characteristics under and outside the canopy

The results of the soil analysis of the Mokrane area showed that the Mokrane soil generally had a sandy texture (**Table 4**), an alkaline pH, a low level of moisture, nitrogen, and organic matter, was slightly to moderately calcareous, had a low salinity, and a low microbial richness. Our results showed that soils under plant cover had higher moisture, organic matter, carbon, nitrogen, and microbial richness than those of soil outside the canopy. This could be due to the higher moisture content, which stimulates the growth and activity of soil microorganisms according to the statements of Hesp and Mclachlan [57].

The high soil moisture content measured under the canopy in Laghouat could be influenced by the lower temperature, which also stimulates the decomposition of organic matter. In fact, Lopez-Pintor et al. [58] found that woody vegetation provides a less stressful microclimate on the environment below the cover, reducing

Soil parameters	*R. raetam*	*A. armatus*	*A. pungens*	Not fixed soil	*P*-value
Humidité (%)	2.39a ± 0.48	1.89a ± 0.86	2.24a ± 0.48	1.06b ± 0.64	0.01
pH	8.30b ± 0.46	8.87a ± 0.07	8.76a ± 0.07	8.72a ± 0.12	0.009
CE (ms/cm)	0.40b ± 0.07	0.39b ± 0.08	0.92a ± 0.06	0.50b ± 0.06	<0.001
C (%)	0.82a ± 0.27	0.70a ± 0.26	0.68a ± 0.29	0.25b ± 0.18	0.01
N (%)	0.08a ± 0.01	0.06ab ± 0.01	0.04bc ± 0.01	0.02c ± 0.01	0.005
$CaCO_3$ (%)	1.75 ± 0.89	1.55 ± 1.11	2.37 ± 0.89	1.96 ± 0.47	0.07
M.O (%)	1.42a ± 0.48	1.21a ± 0.45	1.16a ± 0.51	0.43b ± 0.32	0.001
C/N	8.57 ± 2.06	11.25 ± 4.17	14.54 ± 11.04	13.97 ± 4.93	0.66
Coarse sands (%)	34	30	36	46	—
Fine sands (%)	62	51	64	50	—
Silt (%)	2	9	0	2	—
Clay (%)	2	10	0	2	—

Table 4.
Soil characteristics under und out of the canopy.

the direct effects of strong solar radiation and temperature, providing greater water availability, and causing an accumulation of organic matter and nutrients. Pyke and Archer [59] indicated that the soil under the tree canopy could maintain a higher population of bacteria, conservation and accumulation of carbon and nitrogen. The statements of these latter authors agree with our results under perennials (**Figure 5**).

The nitrogen content in the soil under the cover of perennial plants can be explained by the presence of shrubs that fix atmospheric nitrogen such as *R. raetam* and *A. armatus*, the same findings were made by Muñozvalles et al. [60] in Spain who recorded high carbon and nitrogen contents under the canopy of *Retama monosperma*. Similarly, Halvorson et al. [61] observed a high N content in the soil under desert shrubs in the United States. Vegetation also contributes to a reduction in groundwater runoff [62] and nutrient uptake in the soil under the canopy [40, 63, 64].

Between 2009 and 2015, we observed the formation of a microbial crust on the surface of the soil in the study area. This is an important factor for conserving soil moisture, several studies have indicated that the existence of the microbial crust is necessary for the development of plant communities [65], it affects the water regime by altering runoff [62, 66] and reducing evaporation [67]. Rhizosphere microbial communities can be stimulated or inhibited by root exudate components [68]. The roots of some plants, as they develop, secrete rhizodeposits, the soluble fraction of which is rich in polysaccharides that increase aggregate cohesion [69]. Our results recorded that the three perennial species reacted differently for microbial richness at the soil level. We found that the soil under *A. pungens* contains more fungi than bacteria compared to the soils under *R. raetam* and *A. armatus* (**Figure 6**). Fungi are the most effective in improving aggregate stability due to their secretions with high adhesive power [69]. The N_2 fixation process of the two Fabaceae plants, involving a variable number of atmospheric nitrogen-fixing bacteria on the roots [70, 71].

Our results showed that the soil texture, more precisely the accumulated sediments under the plants studied, was different from that of the non-fixed sedimentary accumulations, the soil under the canopy showed higher rates of fine particles (silt and clay). El-Bana et al. [40] reported that there were significant differences in the soil texture of the nebkas fixed by *Retama raetam*, the proportion of the finest soil particles was higher under the cover than that of the soil between nebkas. The principal component analysis (PCA, **Figure 7**) allowed a better visualization of the behavior of the soil in the presence of vegetation and a study of the relationship between soil variables. The first axis F1 was positively correlated with N, C,

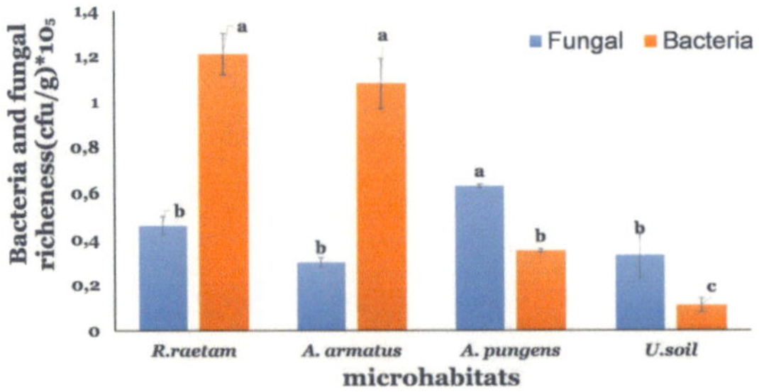

Figure 6.
Canopy influence on sediment microbial richness.

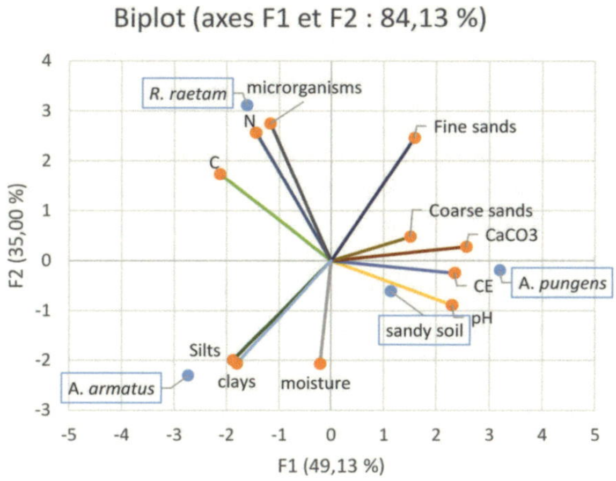

Figure 7.
The principal component analysis demonstrating the relationship between soil variables and vegetation.

bacterial and moisture contents, this variability was linked to the presence of *R. raetam* (r = 0.66). Axis F1 was negatively correlated with coarse sand content, C/N ratio and accumulated sediment height, this variability was linked to the absence of vegetation on non-fixed soil (r = 0.87). The second axis (F2) was positively correlated with fungal content, fine sand percentage, $CaCO_3$ content, and soil EC, this variability was linked to the presence of *A. pungens* (r = 0.86). The negative side of axis F2 was correlated with silt and clay contents, this variability was linked to the presence of *A. armatus* (r = 0.56).

Each plant reacted differently:

- *R. raetam* provided more information on nitrogen, carbon, and bacterial richness. According to El-Bana et al. [72], the *R. raetam* nebka can improve floristic richness in areas where the soil is subjected to severe deflation. *R. raetam* branches are effective in capturing and retaining soil [40]. Danin [63] reported that this *R. raetam* shrub could survive in sand deflation cycles. Our results were similar to those of Sarig et al. [71] who explained that N2-fixing Fabaceae trees, such as *R. raetam*, could improve nutrient content and soil structure by adding nitrogen and organic matter to the soil due to falling litter and thanks to symbiotic associations at the root level with rhizobial bacteria and mycorrhizal fungi. Through our study, we have added similar effects of *A. armatus*, which presented characteristics close to those of *R. raetam*.

- *A. armatus* had higher silt and clay values under its canopy, which could be due to the alluvial and colluvial deposits in the Mokrane area. *A. armatus* is also known as a degradation indicator plant [73].

- *A. pungens* can be considered a "facilitator" (sensu CONNELL and SLATYER [74]) rather than an initiator of the nebka. This role has already been described in Tunisia [18] for *A. pungens*, which regresses and disappears once sand fixation is complete. Our results showed that *A. pungens* had a marked effect on capturing fine sand, and the root exudates of this plant attract fungi more than the two other plants studied.

4. Conclusion

Arid and semi-arid ecosystems are particularly vulnerable to desertification due to climatic constraints and anthropogenic pressures. This study aimed to assess the role of vegetation in mitigating wind erosion and improving soil conditions in the Mokrane region of Algeria. Our findings reveal that while the plant species richness is limited, the existing vegetation plays a crucial role in stabilizing sand dunes and enhancing soil properties. The formation of fixed Aeolian deposits by selected plant species demonstrates their effectiveness in controlling wind erosion. Moreover, the observed variations in soil characteristics under different plant species highlight the importance of species diversity in ecosystem restoration. The results of this study emphasize the potential of *Retama raetam* and *Aristida pungens* for rehabilitating degraded sandy soils in arid environments. A combined approach utilizing these species could offer a promising strategy for sustainable land management and combating desertification in the region. Further research is recommended to explore the long-term impacts of these plant species on soil fertility, biodiversity, and ecosystem services.

Author details

Mallem Hamida
Mechanics Laboratory, Natural Resources and Environment Team, Faculty of Science, Amar Telidji University, Laghouat, Algeria

*Address all correspondence to: h.mallem@lagh-univ.dz

References

[1] Réquier-Desjardins M, Jauffret S, Khatra NB. Lutter contre la désertification. In: MediTERRA 2009. Paris: Presses de Sciences Po; 2009. pp. 137-182. Available from: https://www.cairn.info/mediterra-2009--978272461109-page-137.htm

[2] Dregne HE. Combating desertification: evaluation of progress. Environmental Conservation. 1984;**11**(2):115-121

[3] Oussedik A, Iftene T, Zegrar A. Réalisation par télédétection de la carte d 'Algérie de sensibilité à la désertification. Science et changements planétaires/Sécheresse. 2003;**14**(3):195-201

[4] Bleu P, Antipolis S. Les menaces sur les sols dans les pays méditerranéens méditerranéens. 2003. Available from: https://planbleu.org/wp-content/uploads/2003/05/cahiers2_sols_fr.pdf

[5] Bielders CL, Rajot JL, Amadou M. Transport of soil and nutrients by wind in bush fallow land and traditionally managed cultivated fields in the Sahel. Geoderma. 2002;**109**(1-2):19-39

[6] Ikazaki K, Shinjo H, Tanaka U, Tobita S, Funakawa S, Iwai K, et al. Soil and nutrient loss from a cultivated field during wind erosion events in the Sahel, West Africa. Pedologist. 2012;**55**(3):355-363

[7] Ghazi A, Lahouati R. Algérie 2010. Sols et ressources biologiques. Alger: Institut national des études de stratégie globale (INESG); 1997

[8] Khaldi A. La gestion non–durable de la steppe algérienne. In: VertigO-la revue électronique en sciences de l'environnement. 2014. Available from: https://journals.openedition.org/vertigo/15152

[9] Dalila N, Slimane B. La désertification dans les steppes algériennes: Causes, impacts et actions de lutte. VertigO - la revue électronique en sciences de l'environnement. 2008;**8**(1):1-20. DOI: 10.4000/vertigo.5375. Available from: https://journals.openedition.org/vertigo/5375

[10] Nedjraoui D. Evaluation des ressources pastorales des régions steppiques algériennes et définition des indicateurs de dégradation. Cahiers Options Méditerranéennes. 2004;**62**:239-243

[11] Id A. Les potentialités agro-pastorales de la steppe algérienne, rapport Cellule SIG et télédétection. MADR: HCDS; 2011

[12] Hatimi A, Tahrouch S. Caractérisations chimique, botanique et microbiologique du sol des dunes littorales du Souss-Massa. Biomatec Echo. 2007;**2**(5):85-97

[13] Houyou Z, Bielders CL, Benhorma HA, Dellal A, Boutemdjet A. Evidence of Strong Land Degradation by Wind Erosion as a Result of Rainfed Cropping in the Algerian Steppe: A Case Study at Laghouat. Land Degradation and Development. 2016;**27**(8):1788-1796

[14] Braun-Blanquet J. Prodrome des groupements végétaux. In: Comité international du Prodome Phytosociologique. CBNPMP, Cedex; 1933

[15] Djebaili S. Recherches phytosociologiques et écologiques sur la végétation des hautes plaines steppiques et de l'Atlas Saharien Algériens.

Doctoral dissertation. USTHB Alger; 1978. 203 p. Available from: https://www.ccdz.cerist.dz/admin/notice.php?id=00000000000000041669000000

[16] Quezel P, Santa S. Nouvelle flore de l'Algérie et des régions désertiques méridionales. Eds. du Centre Nat. Paris: de la Recherche Scientifique; 1963

[17] Krebs CJ. Ecological Methodology. New York: Harper Collins Publishers; 1989

[18] Hébrard JP, Loisel R, Roux C, Gomila H, Bonin G. Incidence of clearing on phanerogamic and cryptogamic vegetation in South-Eastern France: disturbance indices. In: Functioning and dynamics of natural and perturbed ecosystems. Vol. 1995. Paris: Lavoisier; 1995. pp. 747-758

[19] Aidoud A. Contribution à l'étude des écosystèmes steppiques du Sud oranais: phytomasse, productivité primaire et applications pastorales. Algeria: USTHB; 1983

[20] Hirche A, Boughani A, Nedjraoui D. About grassland quality assessment in arid areas. 1999. Available from: https://www.cabidigitallibrary.org/doi/full/10.5555/19990706153

[21] Daget P. Les prairies du Cantal. Revue de la Haute-Auvergne. 1974. pp. 4-35

[22] Mallem H, Houyou Z, Benrima A, Wesselingh R, Kouidri M, Ameur D, et al. Particular effects of steppic plants in the rehabilitation of degraded soils, ability to control wind erosion. Journal of Biodiversity and Environmental Sciences. 2018;**13**:347-355

[23] El-Bana MI, Li ZQ, Nijs I. Role of host identity in effects of phytogenic mounds on plant assemblages and species richness on coastal arid dunes. Journal of Vegetation Science. 2007;**18**(5):635-644

[24] Sanglerat G, Costet J. Cours pratique de mécanique des sols: Plasticité et calcul des tassements. Paris: Dunod; 1981

[25] Klement Z, Rudolph K, Sands DC. Methods in phytobacteriology [Internet]. 1990. Available from: https://www.cabidigitallibrary.org/doi/full/10.5555/19912310531

[26] Floret C, Pontanier R. L'aridité en Tunisie présaharienne: climat, sol, végétation et aménagement. Paris: Travaux et documents de l'O.R.S.T.O.M.; 1982. 544 p

[27] Le Houérou HN. Bioclimatologie et biogéographie des steppes arides du Nord de l'Afrique: Diversité biologique, développement durable et désertisation. In: Options Mediterraneennes Serie B: Etudes et Recherches (CIHEAM) no 10 [Internet]. Montpellier; 1995. Available from: https://agris.fao.org/search/en/providers/123819/records/6473602e53aa8c89630a9025

[28] Chehma A, Youcef F. Variations saisonnières des caractéristiques floristiques et de la composition chimique des parcours sahariens du Sud-Est algérien. Sécheresse (Montrouge). 2009;**20**(4):373-381

[29] Ozenda P. Flore du Sahara. 2e éd ed. Paris. P: CNRS; 1983

[30] Monod T. About deserts. 1992. Available from: https://www.cabidigitallibrary.org/doi/full/10.5555/19931981119

[31] Abdelkrim B, Hafidha B, Okkacha H, Khalladi M, Abdelkrim S. Rehabilitation of the Steppe *Lygeum spartum* in the Region of Naama (Western Algeria). Energy Procedia. 2013;**1**(36):349-357

[32] Pouget M. Etude agropédologique du bassin du Zahrez Rharbi (feuille du rocher de sel). Alger: Secretaria d'etat à l'hydraulique; 1971. 220 p. Available from: https://horizon.documentation.ird.fr/exl-doc/pleins_textes/divers16-01/06527.pdf

[33] Quézel P (Pierre). La végétation du Sahara : du Tchad à la Mauritanie. Annales de géographie. 1967;**415**:357-359. Available from: https://www.persee.fr/doc/geo_0003-4010_1967_num_76_415_15022

[34] Ch K. Conditions édaphiques et réactions des plantes indicatrices de la région alfatière. Ann Agron. 1948;**1**

[35] Pouget M. Les relations sol-végétation dans les steppes sud-algéroises. Paris: Travaux et documents de l'O.R.S.T.O.M.; 1980. 570 p

[36] Emberger L. Aperçu général sur la végétation du Maroc. Mémoire hors série de la Soc. Berne. 156p: Sci Nat Maroc Hans Huber; 1939

[37] Daget P. Le nombre de diversité de Hill, un concept unificateur dans la théorie de la diversité écologique. 1980. Available from: https://pascalfrancis.inist.fr/vibad/index.php?action=getRecordDetail&idt=PASCAL8110048027

[38] Kadi-Hanifi-Achour H. L'alfa en Algérie. Syntaxonomie, relation milieu-végétation, dynamique et perspectives d'avenir [thcse Doct]. Es science University of Science and Technology Houari Boumediene Algeria. 1998. 270 p

[39] Kadi-Hanifi H. Diversité biologique et phytogéographique des formations à Stipa tenacissima L. de l 'Algérie. Science et changements planétaires/Sécheresse. 2003;**14**(3):169-179

[40] El-Bana MI, Nijs I, Khedr AA. The Importance of Phytogenic Mounds (Nebkhas) for Restoration of Arid Degraded Rangelands in Northern Sinai. Restoration Ecology. 2003;**11**(3):317-324

[41] Guerrache N, Akkouche S, Kadik L. Evaluation of the biodiversity and stabilization of the soil after the fixating of the dunes by *Retama retam* Webb., Tamarix gallica L. and Tamarix aphylla (L.) Karst in the dunes cordon of El-Mesrane (W. Djelfa) in Algeria. African Journal of Agricultural Research. 2014;**9**(19):1467-1479

[42] Akkouche S, Kadik L, Guerrache N, Bouderbala R. Characterizing the fixation of dunes in the region of Djelfa in Algeria. International Journal of Ecology & Development. 2017;**32**(3):1-11

[43] Chaïeb M. Comportement biologique comparé d'Astragalus armatus Willd. subsp. tragacanthoïdes (Desf.) M. et de Rhanterium suaveolens Desf. sur la steppe sableuse dégradée de la zone aride tunisienne. Ecologia mediterranea. 1997;**23**(3):45-52

[44] Shmida A. Plant growth form strategies and vegetation types in arid environment. Plant form Structure; 1988. pp. 211-241

[45] OSS. The North Western Sahara aquifer system (Algeria, Tunisia, Libya), joint management of transborder basin water, Synthesis collection. N°1. Tunisia: Observatoire du sahara et du sahel; 2008. 48 p

[46] Nedjraoui D. Evaluation des ressources pastorales des régions steppiques algériennes et définition des indicateurs de dégradation. In: Ferchichi A. (comp.), Ferchichi A. (collab.). Réhabilitation des pâturages et des parcours en milieux méditerranéens. Zaragoza: CIHEAM. Cahiers Options Méditerranéennes. 2004;**62**:239-240.

Available from: https://om.ciheam.org/om/pdf/c62/04600165.pdf

[47] Salemkour N, Benchouk K, Nouasria D, Kherief N, Belhamra M. Effets de la mise en repos sur les caractéristiques floristiques et pastorale des parcours steppiques de la région de Laghouat (Algérie). Journal Algérien des Régions Arides. 2013;**12**(1):103-114. Available from: https://www.asjp.cerist.dz/en/downArticle/102/12/1/76909

[48] Daget P, Poissonet J. Biodiversité et végétation pastorale. Revue Élev. Méd. vét. Pays trop. 1997;**50**(2):141-144. Available from: https://agritrop.cirad.fr/389287

[49] Djebaili S, Djellouli Y, Daget P. Les steppes pâturées des Hauts Plateaux algériens. Fourrages. 1989;**120**:393-400. Available from: https://agris.fao.org/search/en/providers/123819/records/64735a1953aa8c896307db7c

[50] Le Houérou HN. Man-Made Deserts: Desertization Processes and Threats. Arid Land Research and Management. 2002;**16**(1):1-36

[51] Rekik F, Bentouati A, Aidoud A. Evaluation des potentialités fourragères d'un parcours steppique à dominance de Salsola vermiculata L. dans l'est de l'Algérie. Livestock Research for Rural Development. 2014;**26**:231. Available from: http://lrrd.cipav.org.co/lrrd26/12/reki26231.html

[52] Quézel P, Barbero M. Variations climatiques au Sahara et en Afrique sèche depuis le Pliocène: enseignements de la flore et de la végétation actuelles. Bulletin d'écologie. 1993;**24**(2-4):191-202

[53] Floret C, Le Floc'h E, Romane F. Dynamique de systèmes écologiques de la zone aride. Application à l'aménagement sur des bases écologiques d'une zone de la Tunisie présaharienne. cta Oecologica. Oecologia Applicata. 1981;**2**(3):195-214. ISSN 0243-7678. Available from: https://horizon.documentation.ird.fr/exl-doc/pleins_textes/pleins_textes_5/b_fdi_20-21/28728.pdf

[54] Bochet E. The fate of seeds in the soil: A review of the influence of overland flow on seed removal and its consequences for the vegetation of arid and semiarid patchy ecosystems. The Soil. 2015;**1**(1):131-146

[55] Cusseddu V, Ceccherelli G, Bertness M. Hierarchical organization of a Sardinian sand dune plant community. PeerJ. 2016;**4**:e2199. DOI: 10.7717/peerj.2199

[56] Neffati M, Akrimi N, Le Floc'h. L'étude des conditions de germination des espèces: une nécessité pour le succès des travaux de réhabilitation. Ecologia mediterranea. 1993;**19**(1-2):39-47

[57] Hesp P, McLachlan A. Morphology, dynamics, ecology and fauna of Arctotheca populifolia and Gazania Rigens Nabkha dunes. Journal of Arid Environments. 2000;**44**(2):155-172

[58] López-Pintor A, Gómez Sal A, Rey Benayas JM. Shrubs as a source of spatial heterogeneity—The case of *Retama sphaerocarpa* in Mediterranean pastures of central Spain. Acta Oecologica. 2006;**29**(3):247-255

[59] Archer S, Pyke D, DA. Plant–animal interactions affecting plant establishment and persistence on revegetated rangeland. Journal of Range Management. 1991;**44**:558-565

[60] Muñoz Vallés S, Gallego B, Fernández J, Dellafiore C, Cambrollé J. Effects on soil, microclimate and vegetation of the native-invasive *Retama monosperma* (L.) in coastal dunes. Plant Ecology. 2011;**212**(2):169-179

[61] Halvorson J, Bolton H, Smith J, Rossi R. Geostatistical analysis of resource islands under *Artemisia tridentata* in the shrub-steppe. Great Basin Naturalist [Internet]. 1994;**54**(4):313-328. Available from: https://scholarsarchive.byu.edu/gbn/vol54/iss4/3

[62] Yair A, Lavee H, Greitser N. Spatial and Temporal Variability of Water Percolation and Movement in a System of Longitudinal Dunes, Western Negev, Isreal. Hydrological Processes. 1997;**11**(1):43-58

[63] Danin A. Plant adaptations in desert dunes. Journal of Arid Environments. 1991;**21**(2):193-212

[64] Alpert P, Mooney HA. Resource heterogeneity generated by shrubs and topography on coastal sand dunes. Vegetatio. 1996;**122**(1):83-93

[65] A Danin, Y Bar-Or, I Dor, T Yisraeli. The role of cyanobacteria in stabilization of sand dunes in Southern Israel. Ecologia Mediterranea. 1989;**15**(1-2):55-64. Available from: https://www.persee.fr/doc/ecmed_0153-8756_1989_num_15_1_1633

[66] Kidron GJ. Differential water distribution over dune slopes as affected by slope position and microbiotic crust, Negev Desert, Isreal. Hydrological Processes. 1999;**13**(11):1665-1682

[67] Verrecchia E, Yair A, Kidron GJ, Verrecchia K. Physical properties of the psammophile cryptogamic crust and their consequences to the water regime of sandy soils, North-Western Negev Desert, Isreal. Journal of Arid Environments. 1995;**29**(4):427-437

[68] Hartmann A, Schmid M, Tuinen D v, Berg G. Plant-driven selection of microbes. Plant and Soil. 2009;**321**(1):235-257

[69] Annabi M, Haithem B. Influence des racines sur la stabilite structurale d'un sol du Nord de la Tunisie/Influence of roots on soil aggregate stability. Etude et Gestion des Sols. 2011;**18**(2):63-73

[70] Leyla B, Ferjani R, Mellal H, Mimouni W, Tir R, Gaci M, et al. Diversity of Bacteria Associated with Root Nodules of Hedysarum Pallidum Desf. Growing in Different Regions of Algeria. Asian Journal of Microbiology, Biotechnology and Environmental Sciences. 2023;**22**:217-227

[71] Sarig S, Fliessbach A, Steinberger Y. Soil Microbial Biomass Under the Canopy of Coastal Sand Dune Shrubs. Arid Soil Research and Rehabilitation. 1999;**13**(1):75-80

[72] El-Bana MI, Nijs I, Kockelbergh F. Micro environmental and vegetation heterogeneity induced by phytogenic nebkhas in an arid coastal ecosystem. Plant and Soil. 2002;**247**(2):283-293

[73] Le Floc'h E. Biodiversité et gestion pastorale en zones arides et semi-arides méditerranéennes du Nord de l'Afrique. Bocconea. 2001;**13**:223-237

[74] Connell JH, Slatyer RO. Mechanisms of Succession in Natural Communities and Their Role in Community Stability and Organization. The American Naturalist. 1977;**111**(982):1119-1144

Section 2

Vegetation Monitoring and Management

Chapter 5

Floristic Morpho-Ecological Strategies: Methodological Approach to Characterize Robustness of Vegetation

João Rocha, Margarida L.R. Liberato, Teresa Fonseca, Raul Morais, Nuno Silva and António L. Crespí

Abstract

Resistance and resilience are usually applied to discuss the functionality of systems, once both describe the responses of biological units to environmental alterations. With this purpose, an analytical methodology focused on the description of functional of vegetation, and their potential correlation with the environmental matrices, was elaborated in the present work. Comparison between amplitudes and averages of morpho-ecological variables, here analyzed by Fluctuation Asymmetries (FAs), was the fundament of this methodology, applied on three different areas of Portugal with similar extensions from north, center, and south Portugal (designed as MB, CP, and FS). The results obtained described very similar functional responses for these study areas. The discussion of these results explained them by high resistance for the three study areas, with lower resilience for one of them (CP), recently exposed to intensive forest fires. So, the functional robustness for these areas will be sustained by their relevant resistance, where environmental variability (geomorphology and climatology) will be the cause of the resilience associated. In this context, increases of the forest production pressions and higher frequencies of fires and climate changes will trigger significant changes in the functionality of their vegetation. Changes in the resilience will be necessary in order to maintain the robustness of vegetation.

Keywords: resistance, resilience, functional matrices, ecological variables, morphological variables, fluctuating asymmetries

1. Introduction

Fire has profoundly shaped both ecological and evolutionary processes and human societies in most Mediterranean-type climate regions [1]. It also acts as a major disturbance to forest ecosystems, as seen with the wildfires and its devastating effects in southern Europe [2]. Wildfires are increasingly impacting the Mediterranean Basin due to more extreme fire seasons with faster and more intense fires that often overwhelm suppression efforts [3]. In recent decades, climate and land use changes have led to a rising prevalence of megafires in Mediterranean-type climate regions [1, 4]. Like other Mediterranean climate regions, mainland Portugal has faced major fires for several decades, being the most severely impacted country by large fires in southern Europe [5]. The occurrence of wildfires is heavily influenced by weather and climate conditions [6–8]. Over the past 20 years, Portugal has seen a continuous decrease in ignitions and significant annual variation in burnt areas, driven by interannual weather variability. From 1984 to 2018, the variability in burnt areas has increased due to climate change, forest continuity, and a focus on fire response rather than mitigation [9]. The same source states that between 2008 and 2017, even though the number of ignitions decreased, the likelihood and size of fires ≥1 ha, as well as the chances of large fires (≥100 ha) and reignitions, increased or remained stable.

Despite considerable uncertainty about projected trends, due to global warming, fire danger and burned areas are anticipated to increase in Mediterranean-type climate regions and will be further exacerbated by ongoing changes in land use and management that increase fuel loads and continuity [1]. The review by Dupuy et al. [2] indicates that fire danger and fire season length are expected to rise in southern Europe over the century. Projection studies estimate that future fire danger and burnt areas will increase, on average, by 2 to 4% and 5 to 50% per decade, respectively.

Models by Sousa et al. [10] project about 2–3 more times mean burnt area in Iberian Peninsula around 2075. Climate change is expected to worsen fire behavior, extending the fire season and increasing the number of days with weather conditions that promote high intensity wildfires [3]. Results on the impact of future climate change on fire activity across Portugal [11] indicated that temperature is predicted to increase and precipitation to decrease. This will lead to a higher Forest Fire Weather Index (FWI), along with more severe and prolonged fire seasons. Consequently, future fire activity is expected to increase dramatically across the entire country, with substantial rises in both the area burned and fire occurrences.

Wildfire management policies have a significant impact on fire system behavior [1]. A focus on suppression rather than prevention (which can be exacerbated by public pressure, especially after severe fire events) can lead to a reduction in preventive measures, creating a 'firefighting trap' in which increased fuel accumulation leads to more severe and larger fires and larger burned areas [1, 12]. While fire suppression efforts can sometimes limit fire size, they are largely ineffective under Extreme Wildfire Events (EWE) conditions, even with substantial resource deployment. Given current and projected trends in climate, ecology, socioeconomic factors, and land use, the frequency of extreme wildfire events is expected to rise, even with increasing investments in fire suppression [1].

A functional approach to the study of the vegetation cover of ecosystems provides a perspective between the biological diversity and their characteristics in the systems under analysis [13]. In this sense, the difficulty arises in the choice of variables used to describe this functionality. Bibliography distinguishes two basic types of functional descriptive variables: those referring to general typified behaviors, or those related to attributes of each taxon of systems studied [14]. Among the first, there are several possible functional variables to use: physiognomic, structural, or floristic variables. In relation to those about functional attributes of the observed taxa, we face very different variables, since morphological [14, 15], phenological, reproductive, germinative [16], to metabolic, physiological, or germinative characteristics [17–19] are involved. These functional variables describe quantitatively or qualitatively the functioning of the analyzed diversity [20].

The matrices obtained from these functional variables are correlated with environmental information, in order to describe the response of flora to environmental dynamic of systems analyzed [21]. Functional types (plant functional types or PFTs [13]) are obtained by these analyses, which in some cases respond more to the classic vision of ecological indicators [22, 23]. Restrictive applications like these, where species are correlated with very limited numbers of variables under restricted environmental matrices, certainly limit the information provided by functional matrices. In fact, functional diversity of vegetation provides systemic responses of vegetation to environmental changes [24–26], and the focus on some variables distorts the spatial scale and field of study [27]. In this context, the application of functional groups to obtain species indicators of ecological trends involves complex hierarchical methodologies [28], and the use of functional groups seems more appropriate [29].

The parameters used to characterize the responses of any functional group fall within resistance and resilience to ecological alterations [30, 31]. Both parameters describe the strategies of existing diversity to variations in the environmental matrix [24–26], by their functional attributes [32]. In this context, functional diversity reaches special relevance, as the range of functional responses provided the capacity of reaction by the diversity of each system studied [33, 34].

Forest fires are often examples of this type of analysis [35–37] and, in fact, this work was elaborated from those occurred along 2017 in Portugal, which caused significant human, economic, and environmental damages [38, 39]. Among the decisions adopted for the socioeconomic and environmental recovering process, the possibility of evaluating the resistance and resilience of natural vegetation of this area was established. With this purpose, the functional study of the potential flora of this region was carried out, and, in order to compare with vegetation and environmental matrices of very different areas, two other areas with similar sizes but clearly different thermopluviometric and geomorphological conditions were also analyzed. In this work, functional resistance was understood as the functional response of every species, and its resilience would be associated to the variability observed. In this way, resistance was established by the functional profiles obtained. The more resilience associated to those functional profiles obtained for every system, the more functional variabilities they should contain. So, the resistance of systems analyzed could be more or less resilient, based on the functional variability observed for each case.

Resolution 49/2020 of the Council of Ministers in Portugal [40] establishes the Landscape Transformation Program, targeting forest areas with high fire risk. The objective is to promote a transformation that ensures the resilience, sustainability and enhancement of the territory, in line with the National Forest Strategy 2030 and the National Strategy for Nature Conservation and Biodiversity 2030. This resolution continues earlier regulations that established working groups to promote, monitor, and drive an integrated plan to transform burned areas and promote agroforestry management nationwide. The study presented here builds upon these efforts for the municipalities in the Pinhal Interior Sul region, guided by Order No. 7417/2019 [41], which have been severely affected by recurrent fires in recent decades. Based on these previous requirements, the methodology here suggested was sustained in the difference between functional resistance and resilience, as parameters to describe the robustness of systems under characterization. The most generalized definition for resilience is considered as "the ability of a system to resist disturbances or absorb them and rapidly self-reorganize to maintain its main structure and functions" [42]. We understand resistance as "the ability of a community to remain unchanged when challenged by disturbances" [43]. So, while resilience describes the ability of systems to maintain their functionality face to environmental alterations by changing their structure and functioning, resistance guarantees them without any change. The research question arises on how to quantify both, as apparently one depends on the other [44]. Even more, the relationship between them is not necessarily parallel, as it could be inverse [45]: resistant systems could maintain lower resilience to guarantee its functional strategy. For this reason, each species keeps the capacity to resist until being forced into a response to assure its functionality [46]. In this sense, the existing biological diversity and its functionality will be involved in the functional response of systems to environmental alterations [25, 47, 48].

The purpose of this study will be focused in the functional characterization, based on the robustness of vegetation, provided by its resistance and resilience. In this sense, the main objectives will be the establishment of one methodology to describe the functionality of the study areas selected, in this case based on their potential vascular flora, and its robustness in accordance with their environmental (altitudinal and climatic) and functional (morphological) variabilities.

2. Methodology

2.1 Study areas and floristic catalogs

The election of study areas was based on those biophysic variabilities, as well as the frequency of forest fires along the last decades [49]. Based on this information, two more areas, with similar areas, were contrasted with the case study area, CP (Castelo Branco-Proença-a-Nova): one on the north, designated as MB (Montalegre-Boticas); and finally, one on the south designated as FS (Faro-Loulé-São Brás de Alportel).

For these areas, equidistantly located along a north-south longitudinal gradient (**Figure 1**), the potential floristic catalog was elaborated. The catalogs were prepared from the information stored in the Flora Digital database (https://jb.utad.pt/flora), where the floristic information of Portugal is organized by biogeographic regions and municipalities. The database includes a MySQL table that links bioregions with species, containing entries for the species identifier and the bioregion identifier. This allows the association of each species' presence or absence in the 25 defined georegions of mainland Portugal, and each species can be recorded in one or more of these regions. This structure facilitates the creation of biogeographic distribution maps, which reflect not only the ecological and biogeographic characteristics of the species but also their occurrence at the municipal level, providing a detailed and accurate view of species distribution across the mainland territory. Although the development of floristic catalogs at the municipal level is not yet fully implemented, the current structure of the database already supports this functionality, ensuring the possibility of future updates and improvements.

2.2 Functional matrices

Functional matrices were elaborated for each floristic catalog. These matrices were performed on two different types of variables: ecological and functional variables. The ecological variables describe general characteristics for each taxon, and 38 variables were here selected: general distribution (nine variables), physiognomic types (10 variables), flowering months (12 variables, one per month of the year), preferential

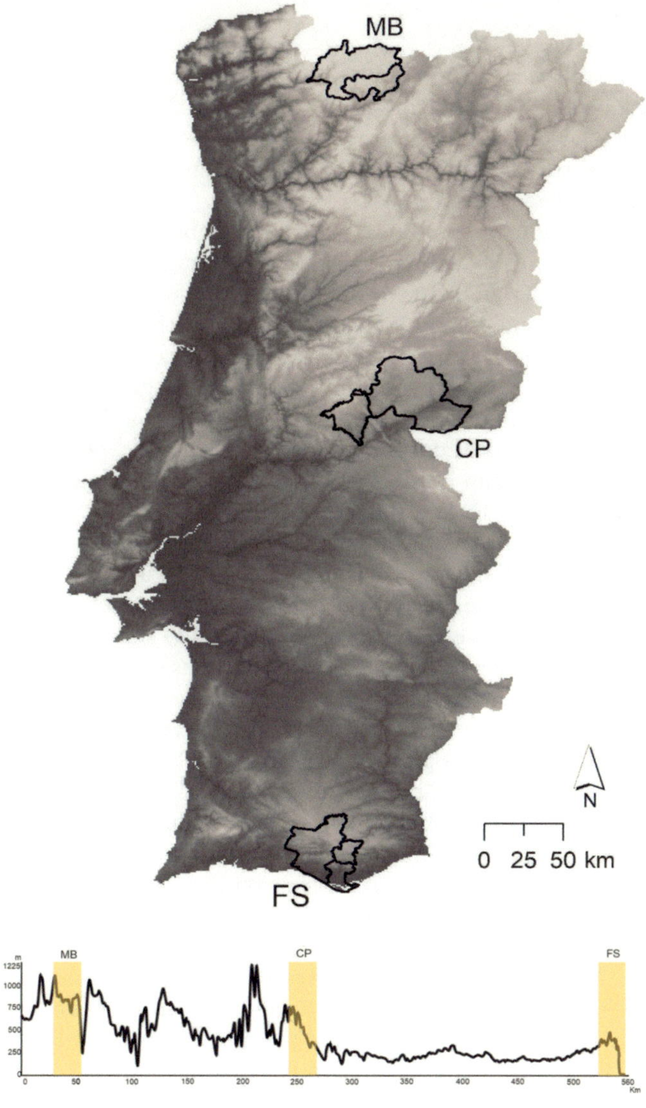

Figure 1.
N-S geomorphologic profile of Portugal, with the location of the study areas (MB, Montalegre-Boticas; CP, Castelo Branco-Proença-a-Nova; and FS, Faro-Loulé-São Brás de Alportel).

habitat (six variables), and systematic group (11 variables). Each of these variables are described and characterized in **Table A1(a)**.

About the functional variables, based on morphological characteristics of each taxon per catalog, 90 were chosen. The relationship of these variables is also presented in the annexed documentation (**Table A1(b)**).

The morphological matrix was standardized, and the average of each morphological variable shortened by the ecological variables was calculated, in order to obtain the contingency matrix, now organized by eco-morphological variables. This methodology is shown in **Figure A1a**.

2.3 Climatic analysis

The first methodological approach in this sense was based on the thermopluviometric variability for the three analyzed areas. This climatological analysis is performed for distinct periods, one representing the recent past (1981–2010) and the other representing future climate change scenarios: for the mid-twenty-first century (2040–2059) and the end of the twenty-first century (2080–2099). In all cases, temperature and precipitation observational data from public datasets were obtained and observed. Historical monthly temperature and precipitation data from synoptic weather stations are available from the Portuguese Met office (IPMA) and are used to compute the corresponding climatological normals and characterize the climate for the meteorological data of the largest council of each area: Montalegre (for MB), Castelo Branco (for CP), and Faro (for FS). Additionally, gridded observational data products were analyzed, such as the European Centre for Medium-Range Weather Forecasts, ERA5 reanalysis product [50], here used as a reference data set for validation of the General Circulation Models (GCMs). Data are used in a 0.25° latitude-longitude grid and 6-hourly time resolution, spanning from 1981 to 2020.

Additionally, future climate projections are obtained from regional climate models' simulations database from the Coupled Model Intercomparison Project Phase 6 (CMIP6) [51] of the World Climate Research Programme, under different projected time periods and Shared Socio-economic Pathways (SSPs) adopted by the Intergovernmental Panel on Climate Change (IPCC) on the Sixth Assessment Report. SSPs represent a range of plausible trends in the evolution of society over the twenty-first century. The SSPs' quantitative projections of socio-economic drivers include population, gross domestic product (GDP), and urbanization [52]. The SSP2 is a 'middle-of-the-road' path and CO_2 emissions remaining around current levels until the middle of the century. It assumes intermediate levels of technological progress, moderate population growth, and a balanced approach to economic and environmental goals. The SSP5 is the 'fossil fuel-intensive' development pathway, which projects rapid economic growth and high population, with strong reliance on fossil fuels [52, 53]. In this latter assessment, environmental concerns are typically disregarded in favor of economic development. SSP5–8.5 do not represent a typical 'business-as-usual' projection but are only useful as high-end, high-risk scenarios [54]. In this study, a multi-model ensemble monthly dataset is considered for the historical reference period (1995–2014) and for the SSP2–4.5 and the SSP5–8.5 scenarios for mid-twenty-first century (2040–2059) and the end of the twenty-first century (2080–2099), at a 0.25° $\times$ 0.25° resolution.

On the other hand, the description of the climatic and altitudinal variability for the three study areas was also performed by worldclim.org, version WorldClim 2.1 [55], with a spatial resolution of 30 seconds (approximately 1 km^2) in TIFF format, was used. Climate variables used are as follows: Elevation (derived from the SRTM elevation data), 19 bioclimatic variables (version 2; **Table A2**), maximum temperature (° C), minimum temperature (°C) and precipitation (mm). Administrative borders used were obtained from CAOP 2023–Official Administrative Boundaries (available on https://www.dgterritorio.gov.pt/).

2.4 Functional analysis

The functional analysis was elaborated from the contingency matrix, and the methodological scheme was sustained in the comparison between eco-morphological variables, per study area, and the analysis of their asymmetries. The comparison of the eco-morphological variables was performed by Pearson correlations obtained from the contingency matrix per study area and the Principal Component Analysis (PCA) to describe the information included in those correlation matrices. Fluctuating Asymmetries (FA) [56] of means or amplitude of the eco-morphological information per study area was the analytical approach used to characterize functional differences between those areas (MB, CP, and FS). This analytical tool is usually calculated from the variance between extremes or their means [57]. In our case, FAs were represented by the range of variability (amplitudes) or its mean, standardized by base 10 logarithm. Once FAs are usually employed to describe asymmetries between bilateral traits [58], we used it here to compare those fluctuating asymmetries between two descriptive parameters for the eco-morphological variables. The results were exposed in bar diagrams, to better compare those asymmetries for variables or study areas. The description of these diagrams for FAs is graphically represented in **Figure A1b**.

The analysis of distribution of the morpho-ecological variables, across the PCA for the contingency matrix of the study areas, was performed by the occurrence of the eco-morphological variables by the eight correlation areas, here designed as "correlation units" (Us). Those morpho-ecological variables of each study area (MB, CP, and FS) included in the same U were distinguished from variables distributed in different Us. The hypothesis here applied described resistant functional responses in the first case (Rs), in contrast with those in different Us describing resilient functional ones (Rl). In order to compare de Rs and Rl morpho-ecological variables, the FA analysis was elaborated for the average values for the amplitude of the morpho-ecological variables included in each group (Rs and Rl) per study area (MB, CP, and FS). This methodology is graphically explained in **Figure A1b**.

3. Results

3.1 Study areas and floristic catalogs

The potential floristic catalogs for each area showed differences in terms of number of species: while 1032 species were registered for CP (58.4% of taxonomic diversity for three study areas), 719 were obtained for MB (40.7%), and 1068 for FS (60.5%). The catalogs were classified by family and order in **Table A3**. **Table 1** shows the percentages species, families and orders by clade (m to s9) per study area (CP, MB, and FS). No significant differences were observed about the distribution of taxonomic diversity by systematic groups. In fact, s7, s8, and s9 gather 77.6% of species and subspecies diversity for MB, 80.5% for CP, and 81.2% for FS.

3.2 Functional matrices

The correlation between the percentages of species per ecological variable showed very similar responses for study area, and all the variables (**Figure 2a**). Fluctuating asymmetries (FAs) for the average percentages of ecological variables (Ln{Ecol}) and their amplitudes of variability (Ln[Ecol]) per study area are confirmed this conservative response for the taxonomical distribution per ecological variable and study area (**Figure 2b**). No significance differences were observed in terms of taxonomic distribution of taxa gathered by ecological variables. These percentages of taxa per ecological variable are exposed in Annex. Based on these values, higher fluctuations for phenology and preferential habitats were confirmed for these areas, as well as for

	MB		CP		FS	
S	Fam	Spp	Fam	Spp	Fam	Spp
m	8	28	12	26	8	19
g	1	2	2	3	2	4
s2	5	14	7	23	7	19
s3	2	15	3	19	3	18
s4	1	5	2	10	1	13
s5	3	14	6	18	5	15
s6	8	83	14	102	15	113
s7	**12**	**158**	**12**	**238**	**15**	**222**
s8	**20**	**150**	**32**	**288**	**31**	**274**
s9	**20**	**250**	**18**	**305**	**18**	**371**

Table 1.
Number of families (Fam), and species and subspecies (Spp) per systematic group (S) for the study areas analyzed (1765 species and subspecies for the three study areas).

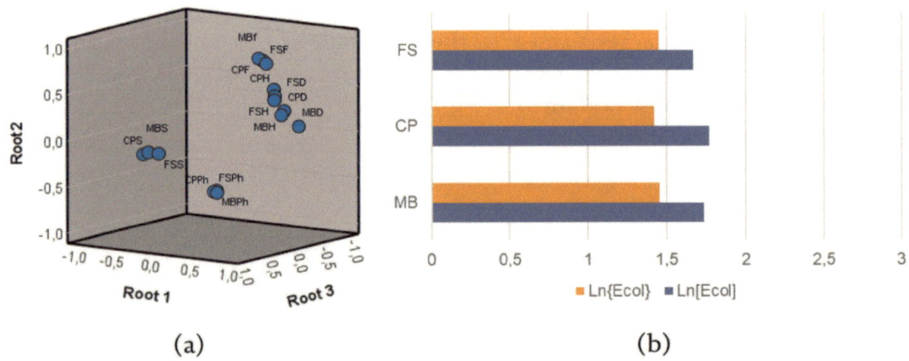

Figure 2.
Analysis of percentage of taxa per ecological variable: a) PCA for the correlation matrix obtained from the distribution of percentages per ecological variable; and b) FA diagram for average percentages (Ln{Ecol}) and amplitudes (Ln[Ecol]) per study area.

vegetative and floristic morphological characters. So, the functional variability of the flora of these regions seems to be concentrated around flowering periods and distribution along habitats, in terms of ecological variables and vegetative and floristic morphological responses.

3.3 Climatic analysis

The climate of the three areas of mainland Portugal considered in this study is classified as being temperate climate according to the Köppen-Geiger climate classification system, based on seasonal precipitation and temperature patterns (Csa and Csb according to Köppen-Geigen classification [59]), with a warm summer in the northern area (Csb in Montalegre, MB) and a dry summer in the central and southern areas ((Csa in Castelo Branco, CP) and Faro, FS). The climate in the considered areas is predominantly influenced by latitude, orography and its proximity to the Atlantic Ocean. Climate variables (**Figure 3a**) precipitation, and temperature present a north-south gradient as well as a very evident monthly, seasonal, and inter-annual variability with average temperatures between 5 and 25°C. Average total annual precipitation in these areas of mainland Portugal shows a strong spatial variability, with the highest values observed (exceeding 1450 mm) in the mountainous regions of Montalegre (1005 m), values below 800 mm in Castelo Branco synoptic weather station (386 m), and the lowest values, below 520 mm, in the southern region of Faro (7 m). On average, most annual precipitation occurs during winter (December to February) and precipitation of the driest months in summer (June to August) are well below 40 mm and much below the third of the precipitation of the respective wettest month in winter. Transition seasons – spring (March to May) and autumn (September to November) – show a very variable inter-annual distribution. The southern area in this study is classified as being on semiarid climate conditions. Semiarid regions are characterized by a mean annual precipitation between 200 and 700 mm [60].

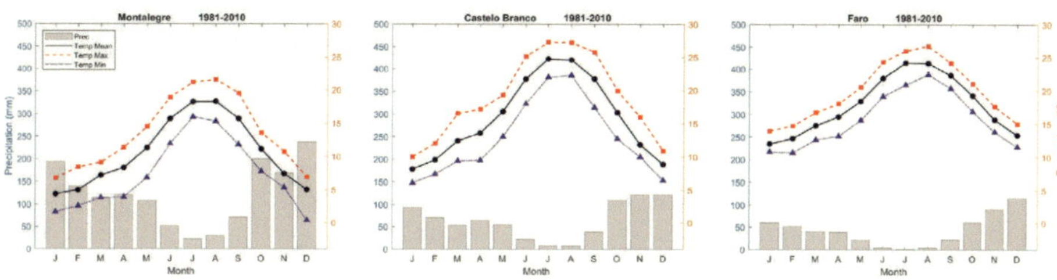

(a)

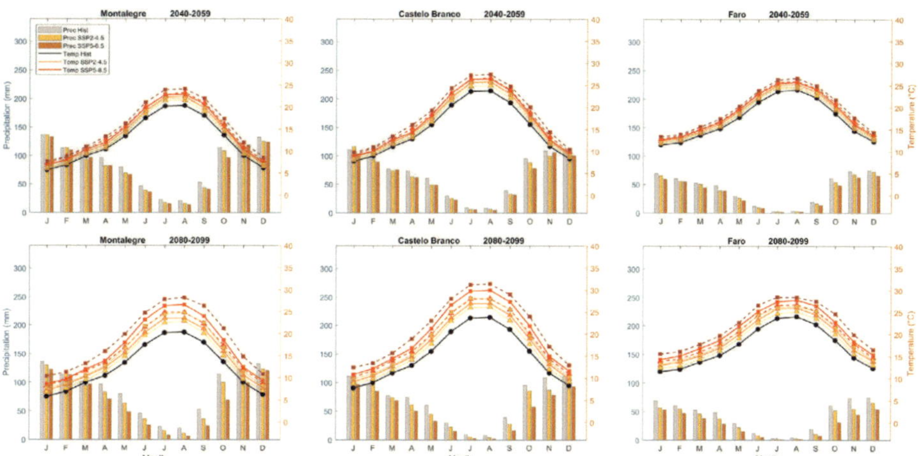

(b)

Figure 3.
Monthly thermopluviometric graph for the three areas (MB, CP and FS): a) for observational data for the normal period 1981–2010 -the bars represent average accumulated precipitation; the lines represent average temperature (black: mean; red: maximum; blue: minimum)-; b) for the historical period (reference) and for the two periods – for the mid-twenty-first century (2040–2059) and the end of the twenty-first century (2080–2099) – and for the two climate change scenarios (the intermediate scenario SSP2–4.5 and the very high GHG emissions scenario SSP5–8.5). The bars and the solid lines represent the 50th percentiles; the dotted and dashed lines represent the 10th and 90th percentiles for the SSP2–4.5 and the SSP5–8.5 scenarios, respectively.

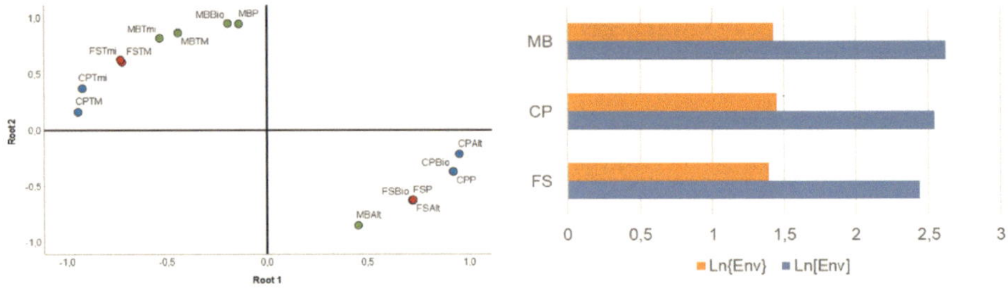

(a) (b)

Figure 4.
Analysis of correlation and fluctuation for the climatic matrix: a) PCA for the first two roots (more than 90% of correlation described) for the average values of the environmental variables (Alt, elevation; P, precipitation; TM, maximum temperature; Tmi, minimal temperature), per study area; b) fluctuation of average and amplitude values for the environmental matrix (logarithm for averages -Ln{Env}- and for amplitudes -Ln[Env]- of the environmental variables).

The results for the mid-twenty-first century (2040–2059) and the end of the twenty-first century (2080–2099) (**Figure 3b**) are in agreement with previously published studies [61–64]. They also suggest the CP area might become semi-arid throughout the twenty-first century, resulting from the projected changes in precipitation and potential evapotranspiration [65]. Temperature and precipitation variability is also increasing, which may lead not only to longer, more intense, and widespread droughts but also to greater exposure to heavy precipitation and floods [66, 67].

The climatic matrix was performed by 6137 locations across the study areas: 1754 for MB, 2777 for CP, and 1606 for FS. The analysis of the amplitude for the environmental variables (altitudinal and climatic) standardized is exposed in the **Figure 4**.

The correlation analysis showed two opposite environmental responses, MB and CP-FS (**Figure 4a**): while MB is characterized by wider altitudinal variations, higher average precipitation, and lower temperature (minimal and maximum); more restricted altitudinal variations and higher temperature were recorded for CP and FS (statistical values in **Table A4**). So, the highest FA between the average values and the amplitudes for the environmental values was for MB and the lowest for FS (**Figure 4b**).

3.4 Functional analysis

The PCA on the Pearson correlation matrix calculated from the functional contingency matrix described the distribution of eco-morpho functional variables as shown in **Table 2**.

The analysis of the correlation for the functional contingency matrix, for Rl and Rs eco-morphological variables, sustain the functionality on the Rs variables, once their variability is described by the first root – Root 1 – in contrast with the Rl variables, described by the second root – Root2 (**Figure 5a**). The average variation of amplitudes for the eco-morphological variables reflects FA differences between regions (**Figure 5b**). Very short resilient FAs (Rl), in contrast with clearly larger resistant FAs (Rs): lower values of Rl for CP and higher for MB and FS.

Var	R1	R2	R3	U	Var	R1	R2	R3	U	Var	R1	R2	R3	U
Mg	0,535	−0,223	−0,121	U5	Cg	0,477	−0,244	−0,164	U5	Fg	0,517	−0,259	−0,068	U5
Mm	0,280	−0,155	−0,163	U5	Cm	0,305	−0,179	−0,149	U5	Fm	0,347	−0,173	−0,136	U5
Ms2	0,925	−0,236	−0,078	U5	Cs2	0,899	−0,318	−0,091	U5	Fs2	0,901	−0,241	−0,128	U5
Ms3	0,147	0,415	0,001	U2	Cs3	0,106	0,311	−0,022	U1	Fs3	0,157	0,317	0,068	U2
Ms4	0,200	0,085	0,266	U2	Cs4	0,130	0,031	0,343	U2	Fs4	0,135	−0,006	0,291	U6
Ms5	0,177	0,375	0,662	U2	Cs5	0,165	0,331	0,634	U2	Fs5	0,231	0,296	0,643	U2
Ms6	−0,095	0,311	0,908	U3	Cs6	−0,121	0,184	0,949	U3	Fs6	−0,078	−0,011	0,935	U7
Ms7	0,882	−0,338	−0,283	U5	Cs7	0,864	−0,367	−0,300	U5	Fs7	0,879	−0,344	−0,285	U5
Ms8	−0,248	0,890	0,219	U3	Cs8	−0,350	0,862	0,234	U3	Fs8	−0,232	0,899	0,231	U3
Ms9	−0,892	−0,361	−0,114	U8	Cs9	−0,870	−0,411	−0,107	U8	Fs9	−0,875	−0,405	−0,097	U8
MA13	0,920	−0,100	−0,033	U5	CA13	0,932	−0,284	0,018	U6	FA13	0,915	−0,253	−0,007	U5
MA46	−0,945	−0,128	0,010	U7	CA46	−0,916	0,103	−0,272	U4	FA46	−0,780	0,505	−0,207	U4
MA78	0,025	0,683	0,039	U2	CA78	−0,504	0,696	0,240	U3	FA78	−0,770	0,335	0,286	U3
MA9	−0,845	−0,463	0,050	U7	CA9	0,006	−0,857	0,005	U6	FA9	0,259	−0,887	−0,114	U5
MTr	0,492	−0,190	0,208	U6	CTr	0,193	0,175	0,327	U2	FHb	−0,577	−0,322	−0,417	U8
MSr	−0,215	0,830	0,261	U3	CSr	−0,256	0,697	0,439	U3	Faq	0,900	−0,311	0,086	U6
Maq	0,345	0,203	0,721	U2	Caq	0,891	−0,237	0,129	U6	FSr	−0,089	0,627	0,468	U3
MHb	−0,219	−0,702	−0,406	U8	CHb	−0,169	−0,535	−0,440	U8	FTr	0,392	−0,074	0,403	U6
Mwi	0,661	−0,039	−0,013	U5	Cwi	−0,273	−0,257	0,438	U7	Fwi	0,518	−0,115	0,455	U6
Msp	0,587	0,759	0,204	U2	Csp	0,069	0,947	0,242	U2	Fsp	−0,096	0,943	0,275	U3
Msu	−0,888	−0,304	−0,026	U8	Csu	−0,803	−0,493	0,093	U7	Fsu	−0,811	−0,537	−0,020	U8
Mou	−0,624	−0,572	−0,110	U8	Cou	−0,357	−0,832	0,124	U7	Fou	0,017	−0,944	0,045	U6
Mfo	0,744	0,182	0,469	U2	Cfo	0,863	−0,143	0,310	U6	Ffo	0,896	−0,278	0,164	U6

Var	R1	R2	R3	U	Var	R1	R2	R3	U	Var	R1	R2	R3	U
Msh	−0,096	0,872	−0,134	U6	Csh	−0,602	0,669	−0,071	U4	Fsh	−0,529	0,720	0,007	U3
MHr	−0,648	−0,338	−0,380	U8	CHr	−0,880	0,063	−0,216	U4	FHr	−0,699	−0,243	−0,521	U8

Table 2.
Distribution of eco-morphological variables along the first three roots of the PCA on the functional contingency matrix (Var, eco-morphological variables; R1–3, roots of the PCA; U1–8, correlation units).

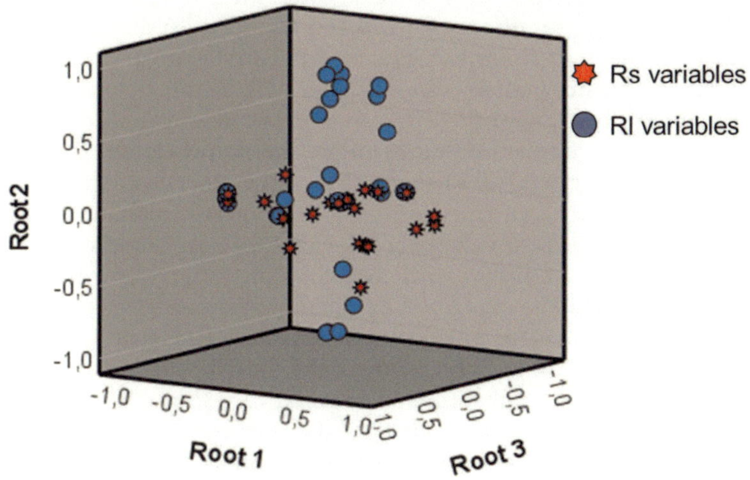

(a)

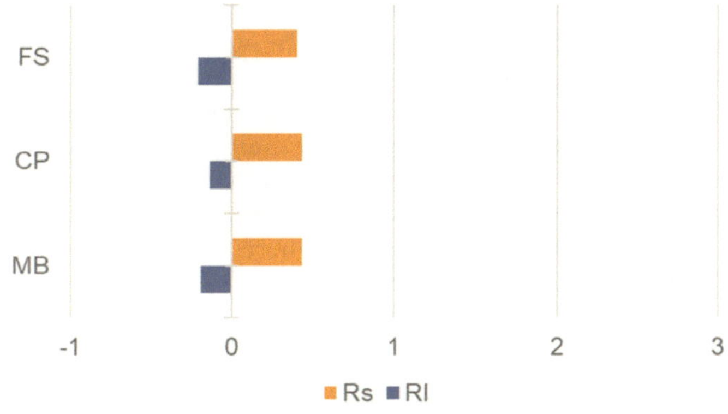

(b)

Figure 5.
Analysis of the contingency functional matrix, after the distribution of the eco-morphological variables per correlation unit (U): a) PCA from the correlation of the contingency functional matrix, with the variables divided into resilient (Rl) and resistant (Rs); b) FA of the average amplitudes for the Rl and Rs variables per study area.

4. Discussion

The results obtained along the functional characterization of the potential vascular flora for the areas analyzed (MB, CP, and FS) showed significant differences in terms of number of taxa, especially between MB (719 species and subspecies) and CP-FS (1032 and 1068, respectively). On contrary, the functional characterization described remarkable similarities for all the study areas. A result like this could be contradictory, but once again is confirmed the "highly complex and context-dependent" [68] relationship between taxonomic diversity and functional variability. In fact, the analysis of the contingency functional matrix revealed differences in resilient terms, not in resistance. This was more significant for MB and FS, once CP showed more restricted in this sense. More diverse gemorphologies for MB could explain those resilient values, as highly contrasting reliefs trigger higher diverse microclimatic variations [69, 70]. The climatic analysis of the three areas exposed notorious differences between MB and CP-FS. Cooler and more humid thermopluviometric regimes, with shorter dry periods, described MB in contrast with the other two study areas. At the same time, the FA analysis confirmed wider fluctuations for climatic variability in MB, in contrast with the lowest values for FS. However, and based on this geomorphologic justification, lower resilience could be expected for FS. In this case, functional diversities could be explained by the morphogenetic flow along glacial crisis in this region [71, 72]. This biogeographic dynamic has been decisive to increase taxonomic diversity in this area. Nevertheless, the functional similarity based on the percentage of taxa per ecological variable described very similar areas. Something like this confirmed the same ecological strategy of the flora for these study areas. However, the morpho-ecological approach (based on the analysis of the contingency functional matrix) reflected higher functional variability for the same ecological strategies. In this case, as FS as MB described higher resilience (represented by the Rl fluctuations) than CP, as well as similar resistances for all the study areas. In other words, similar ecological profiles, but different functional profiles: larger functionality profiles for FS and MB, in contrast with CP. The analysis of correlations (in this case by the correlations units – U) revealed decisive to characterize and distinguish differences in those functional profiles. This analytical methodology, sustained by the functional characterization of flora through their morphologies, was here confirmed as a reliable methodology to monitor the vegetation than the floristic diversity approach, as suggested in other contributions [73, 74].

This combined functional characterization, where a contingency matrix was elaborated by eco-morphological variables, was decisive to describe the functional differences between vegetation of areas with very similar ecological strategies. Environmental variability is decisive to explain these functional differences, as previous bibliography shows in this context [75–78]. At the same time, evolutionary changes of climatic niches seem to be the way to explore functional abilities of spaces [79], what reveals correlations between geomorphology and climatic variability. The presence of refugees, as a result of that climate dynamism involved here explained by the glacial-interglacial process, has been suggested to explain the unexpected functional diversity in other areas [80], and the southwestern of the Iberian Peninsula has been pointed as a significant congregation area of germplasm [81–83].

If we assumed robustness as a resistance and resilient combination to guarantee functionality of vegetation face to environmental changes or stress [84, 85], the study areas could be described as resistant scenarios, with low resilient functionalities (clearly lower for CP). Intermediate geomorphology or climatic variabilities could be suggested to explain that difference when compared with FS or MB. Taxonomic richness close to FS was not enough to describe high resilience, which confirms the lack of correlation between taxonomical diversity and functional diversity [86, 87]. High resistance and similar ecological variabilities must perform low capacity to change the functional strategy of vegetation, based on the restricted resilience observed for the three study areas. In this sense, it could be described as a stable scenario of vegetation because of the constancy of the functionality observed across the three areas [84]. This circumstance is even more evident for CP, with lower functional resilience for its taxonomic diversity.

Robustness is described as the ability to maintain the system under certain conditions [88], face to perturbations to change it. In this sense, the three systems showed similar robustness, here sustained in low resilience and higher resistance, as expected taking into consideration the negative correlation between them: the bigger the resistance, the lower resilience [89]. That prevalence of resistance over resilience of the study areas described the functional mechanism to guarantee stability, remaining unchanged under disturbances avoiding reorganizing to maintain the same functionality [43]. But, at the same time, that stability (functional constancy along time) became the guarantee to avoid vulnerability (damages caused by alterations) of the systems [84].

If we assume the taxonomic diversity and their functional variability as a response of vegetation to environmental dynamics [73, 90], the robustness of these systems, and their functional homogenization sustained by high resistance and low resilience, must be understood as functional responses to specific ranges of alterations. On contrary, systems with higher resilience and lower resistance should be expected for regimes with higher perturbations. This phenomenon has been observed when predetermined environmental conditions were not adapted to substantial changes of environmental profiles [91]. Based on the evolution of forest fires, in terms of frequencies and surfaces affected, the environmental matrices for the areas analyzed will be exposed to significant perturbations. So, robustness with high resistance and low resilience will not provide efficient functional responses of vegetation.

The thermopluviometric changes expected for these areas provided substantial alterations, more especially for CP and FS, where dry periods by concentration of precipitations were very relevant. It is clear than frequent forest fires will introduce environmental heterogeneity in order to increase functional resilience for the study areas (and for Portugal, once these areas extended along the country). More homogeneous environmental variability reduces the resilience of systems [92]. The distribution of the taxonomic diversity per systematic group was very similar for the study areas (more than 75% was concentrated in s7, s8, and s9). This could explain the functional similarity across the areas analyzed, and the adaptation to their environmental matrices. Alterations on the environmental matrix will trigger the resilient functional response of vegetation and the decreasing of its resistance [40]. That response will be associated with changes in the ecological variables, by changes in the taxonomic diversity involved. So, the robustness of vegetation to climatic changes or forest fires will be preserved by this functional mechanism promoted by opposite responses between resistance and resilience. The effects on the taxonomic diversity involved in those ecological alterations of systems (here analyzed by the ecological variables) remain as the main question for future contributions.

5. Conclusions

The functional diversity approach by the elaboration of eco-morphological matrices, and its analysis by units of correlation, revealed a useful tool to describe the functional robustness of flora. Adaptative functional responses of flora were discussed here as a robustness strategy based on higher resistance of vegetation. Alterations of the environmental matrices, previewed in modeling scenarios of climate change and higher forest pressions, will trigger alterations in the ecological strategies of flora, in order to increase resilience in the robustness of vegetation. Interferences caused for higher frequencies of forest fires could interfere in this functional process, in order to introduce alterations in the ecological functionalities of vegetation.

Acknowledgements

This work is supported by National Funds by FCT – Portuguese Foundation for Science and Technology, under the projects UI/04033, and LA/P/0126/2020 (https://doi.org/10.54499/LA/P/0126/2020).

Conflict of interest

Authors declare no conflict of interest.

Appendix

(a) Ecological variables

Variable	Abb.	Description
Distribution	A1	Cosmopolitan (both hemispheres)
	A2	Subcosmopolitan (north hemisphere)
	A3	Eurasiatic
	A4	Western Eurasia (from Urals mountains, western Asia and northern Africa)
	A5	Mediterranean region
	A6	Atlantic region (Europe and western Morocco)
	A7	Western Mediterranean region
	A8	Southern France, Iberian Peninsula and Morocco and western Argel)
	A9	Allochthonous (introduced since XV century)
Physiognomy types	Ter	Terophyte
	Hem	Hemicriptophyte
	Geo	Geophyte
	Cam	Camephyte
	Hel	Helophyte
	Hid	Hidrophyte
	Nan	Nanophanerophyte
	Mic	Microphanerophyte
	Mes	Mesophanerophyte
	Meg	Megaphanerophyte

Variable	Abb.	Description
Phenology	Jan	January
	Feb	February
	Mar	March
	Apr	April
	My	May
	Jun	June
	Jul	July
	Aug	August
	Set	Setember
	Oct	October
	Nov	November
	Dec	December
Preferential habitat	For	Forests
	Rip	Riparian
	Shr	Shrubs (dominated by nano and microphanerophytes)
	Inc	Shrubs (dominated by herbaceous and nanophanerophytes)
	Anth	Anthropic areas (cultivated, ruderal)
	Wet	Wetlands and natural prairies
Systematic	m	See description in Annex
	g	See description in Annex
	s2	See description in Annex
	s3	See description in Annex
	s4	See description in Annex
	s5	See description in Annex
	s6	See description in Annex
	s7	See description in Annex
	s8	See description in Annex
	s9	See description in Annex

(b) Functional (morphological) variables.

Vegetative		Reproductive			
		Flowers		Fruits/Spores	
Cotiledons	Monocot	Inflorescence	Isolated	Reproduction	Spores
	Dicot		Racemose		Seeds
Root	Allhorhizic	Cymose	Fruit dehiscence	Dehiscence	
	Homeorhizic	Vegetative whorls		Perigonial	Not deshiscence
	Parasitic		Perianth	Type of fruit	Achene

Vegetative			Reproductive		
			Flowers		Fruits/Spores
Leaves	Insertion	Alternate		Without vegetatitve whorls	Silicua
		Opossite	Epicalyx	Without	Silicula
		Whorled		With	Capsule
	Petioles	Petiolate	Number of sepals	0	Folicle
		Not petiolate		≤ 3	Legume
	Composition	Simple		4	Berry
		Composite		5	Drupaceous
	Tendrils	Without		>5	Pomoid
		Whit	Number of petals/tepals	0	Pepponeid
	Lobulation	No lobulation		≤ 3	
		On the base		4	
		Lobed		5	
		Cloven		>5	
		Sect	Fusion of sepals	Free	
		Divided		Not free	
	Venation	Single-nerve	Fusion of petals/tepals	Free	
		Penninervia		Not free	
		Parallel	Simetry of vegetative whorls	Radiated	
		Palminervia		Not radiated	
	Stipules	With	Sex of flowers	Unissex	
		Without		Hermaphrodite	
			Number of stamens	≤ 3	
				4–10	
				>10	
			Fusion of stamens	Not	
				Yes	
			Insertion of stamens	At the receptacle	
				At the vegetative whorl	
			Number of carpels	≤ 2	
				3–5	
				>5	

Vegetative	Reproductive		
	Flowers		**Fruits/Spores**
	Fusion of carpels	Free	
		Partial	
		Total	
	Insertion of ovary	Inferior	
		Semi-inferior	
		Upper	
	Color of petals/ tepals (frequency of visible)	740–625	
		625–590	
		590–565	
		565–520	
		520–500	
		500–435	
		435–380	
		White	

Table A1.
Relation of variables used for the elaboration of the functional analysis of vegetation.

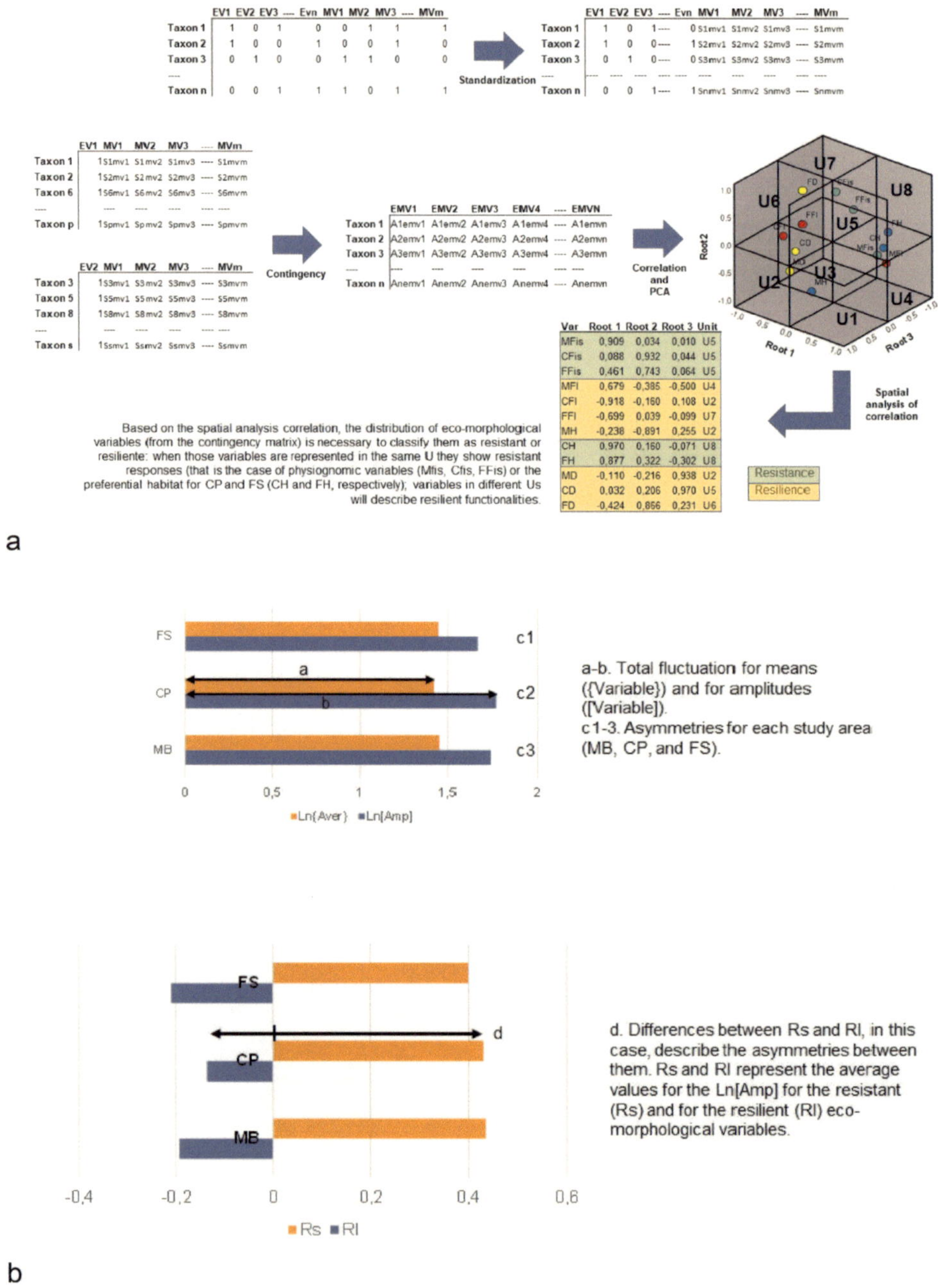

Figure A1.
Elaboration of contingency matrix, and its analysis: (a) methodology to perform the contingency matrix, from the basic functional matrix organized by the ecological (EV), and morphological variables (MV) -S, standardized values; EMV, eco-morphological variables-, and application of the correlation units (U) in the Principal Component Analysis (PCA), to distinguish resistant and resilient eco-morphological variables (Rs and Rl, respectively); (b) description of the FAs diagrams, to describe the information represented.

Abbr	Description
Elev	Elevation
P	Precipitation per month
Tmin	Minimum Temperature per month
Tmax	Maximum Temperature per month
b1	Annual Mean Temperature
b2	Mean Diurnal Range (Mean of monthly (max temp - min temp))
b3	Isothermality ((b2/b7)*100)
b4	Temperature Seasonality (standard deviation*100)
b5	Max Temperature of Warmest Month
b6	Min Temperature of Coldest Month
b7	Temperature Annual Range (b5-b6)
b8	Mean Temperature of Wettest Quarter
b9	Mean Temperature of Driest Quarter
b10	Mean Temperature of Warmest Quarter
b11	Mean Temperature of Codest Quarter
b12	Annual Precipitation
b13	Precipitation of Wettest Month
b14	Precipitation of Driest Month
b15	Precipitation Seasonality (Coefficient of Variation)
b16	Precipitation of Wettest Quarter
b17	Precipitation of Driest Quarter
b18	Precipitation of Warmest Quarter
b19	Precipitation of Coldest Quarter

Table A2.
Altitudinal and climatic variables (climatic variables available in https://www.worldclim.org/).

S	Species	Family	MB	CP	FS	S	Species	Family	MB	CP	FS
g	*Juniperus oxycedrus* L. subsp. *oxycedrus*	*Cupressaceae*	1	1	1	s8	*Genista florida* L.	*Fabaceae*	1	0	0
g	*Juniperus turbinata Guss.*	*Cupressaceae*	1	1	1	s8	*Genista hirsuta* Vahl subsp. hirsuta	*Fabaceae*	0	0	1
g	*Pinus halepensis* Mill.	*Pinaceae*	1	0	0	s8	*Genista hystrix* Lange	*Fabaceae*	0	1	0
g	*Pinus pinaster* Aiton	*Pinaceae*	1	1	1	s8	*Genista micrantha* Ortega	*Fabaceae*	1	0	0
g	*Pinus pinea* L.	*Pinaceae*	1	1	1	s8	*Genista polyanthos* R. Roem. ex Willk.	*Fabaceae*	0	1	1
g	*Pinus sylvestris* L.	*Pinaceae*	1	1	1	s8	*Genista triacanthos* Brot.	*Fabaceae*	0	1	1
m	*Adiantum capillus-veneris* L.	*Pteridaceae*	0	0	1	s8	*Geranium columbinum* L.	*Geraniaceae*	1	1	1
m	*Anogramma leptophylla* (L.) Link	*Pteridaceae*	1	0	0	s8	*Geranium dissectum* L.	*Geraniaceae*	1	1	1
m	*Asplenium adiantum-nigrum* L. var. adiantum-nigrum	*Aspleniaceae*	1	0	1	s8	*Geranium lucidum* L.	*Geraniaceae*	1	1	0
m	*Asplenium ceterach* L.	*Aspleniaceae*	1	1	0	s8	*Geranium molle* L. subsp. molle	*Geraniaceae*	1	1	1
m	*Asplenium obovatum* F.W.Schultz subsp. billotii	*Aspleniaceae*	0	1	0	s8	*Geranium pyrenaicum* Burm. f.	*Geraniaceae*	1	0	0
m	*Asplenium onopteris* L.	*Aspleniaceae*	1	1	1	s8	*Geranium robertianum* L. subsp. purpureum (Vill.) Nyman	*Geraniaceae*	1	1	1
m	*Asplenium petrarchae* (Guérin) DC. in Lam et DC.	*Aspleniaceae*	0	0	1	s8	*Geranium robertianum* L. subsp. robertianum	*Geraniaceae*	1	1	0
m	*Asplenium ruta-muraria* L.	*Aspleniaceae*	0	1	0	s8	*Geranium rotundifolium* L.	*Geraniaceae*	1	1	1
m	*Asplenium trichomanes* L. subsp. quadrivalens	*Aspleniaceae*	0	1	0	s8	*Geum sylvaticum* Pourr.	*Rosaceae*	0	1	0
m	*Athyrium filix-femina* (L.) Roth	*Woodsiaceae*	1	1	0	s8	*Geum urbanum* L.	*Rosaceae*	1	0	0
m	*Azolla filiculoides* Lam.	*Marsiliaceae*	0	0	1	s8	*Halimium calycinum* (L.) K. Koch	*Cistaceae*	0	0	1
m	*Blechnum spicant* (L.) Roth	*Blechnaceae*	1	0	0	s8	*Halimium halimifolium* (L.) Willk. subsp. *halimifolium*	*Cistaceae*	0	0	1
m	*Cheilanthes acrostica* (Balb.) Tod.	*Pteridaceae*	1	1	0	s8	*Halimium halimifolium* (L.) Willk. subsp. *multiflorum* (Sazmani ex Dunal) Maire	*Cistaceae*	0	0	1

DOI: http://dx.doi.org/10.5772/intechopen.1008900

S	Species	Family	MB	CP	FS	S	Species	Family	MB	CP	FS
m	*Cheilanthes hispanica* Mett.	*Pteridaceae*	1	1	0	s8	*Halimium lasianthum* (Lam.) Spach subsp. lasianthum	*Cistaceae*	1	0	1
m	*Cheilanthes maderensis* Lowe	*Pteridaceae*	1	0	0	s8	*Halimium lasianthum* (Lam.) Spach subsp. alyssoides	*Cistaceae*	1	1	0
m	*Cheilanthes tinaei* Tod.	*Pteridaceae*	1	0	0	s8	*Halimium ocymoides* (Lam.) Willk.	*Cistaceae*	1	1	1
m	*Cosentinia vellea* (Aiton) Tod.	*Pteridaceae*	1	1	0	s8	*Halimium umbellatum* (L.) Spach subsp. umbellatum	*Cistaceae*	1	0	0
m	*Cystopteris viridula* (Desv.) Desv.	*Woodsiaceae*	1	0	0	s8	*Halimium umbellatum* (L.) Spach subsp. viscosum	*Cistaceae*	1	1	1
m	*Dryopteris affinis* (Lowe) Fraser-Jenkins subsp. affinis	*Dryopteridaceae*	1	0	0	s8	*Hedysarum glomeratum* F.Dietr.	*Fabaceae*	0	0	1
m	*Dryopteris affinis* (Lowe) Fraser-Jenkins subsp. borreri (Newman) Fraser-Jenkins	*Dryopteridaceae*	1	0	0	s8	*Helianthemum aegyptiacum* (L.) Mill.	*Cistaceae*	0	1	1
m	*Dryopteris dilatata* (Hoffm.) A. Gray	*Dryopteridaceae*	1	1	1	s8	*Helianthemum ledifolium* (L.) Mill.	*Cistaceae*	0	1	1
m	*Dryopteris expansa* (C. Presl) Fraser-Jenk. et Jermy	*Dryopteridaceae*	0	0	1	s8	*Helianthemum nummularium* (L.) Mill.	*Cistaceae*	1	0	0
m	*Dryopteris filix-mas* (L.) Schott	*Dryopteridaceae*	0	1	0	s8	*Helianthemum salicifolium* (L.) Mill.	*Cistaceae*	0	0	1
m	*Dryopteris guanchica* Gibby et Jermy	*Dryopteridaceae*	0	1	1	s8	*Hippocrepis biflora* Spreng.	*Fabaceae*	0	0	1
m	*Dryopteris oreades* Fomin	*Dryopteridaceae*	0	0	1	s8	*Hippocrepis ciliata* Willd.	*Fabaceae*	0	0	1
m	*Equisetum arvense* L.	*Equisetaceae*	0	1	1	s8	*Hippocrepis multisiliquosa* L.	*Fabaceae*	0	0	1
m	*Equisetum ramosissimum* Desf.	*Equisetaceae*	0	1	0	s8	*Hirschfeldia incana* (L.) Lagr.-Foss.	*Brassicaceae*	0	1	0
m	*Equisetum telmateia* Ehrh.	*Equisetaceae*	0	1	1	s8	*Hymenocarpos hamosus* (Desf.) Vis.	*Fabaceae*	0	0	1
m	*Isoetes durieui* Bory	*Isoetaceae*	1	1	1	s8	*Hymenocarpos lotoides* (L.) Vis.	*Fabaceae*	0	1	1
m	*Isoetes histrix* Bory	*Isoetaceae*	1	0	0	s8	*Hypericum androsaemum* L.	*Hypericaceae*	1	0	0
m	*Isoetes setaceum* Lam.	*Isoetaceae*	1	1	1	s8	*Hypericum elodes* L.	*Hypericaceae*	1	1	1

S	Species	Family	MB	CP	FS	S	Species	Family	MB	CP	FS
m	*Isoetes velatum* A. Braun	*Isoetaceae*	1	1	0	s8	*Hypericum humifusum* L.	*Hypericaceae*	1	1	1
m	*Marsilea batardae* Launert	*Marsiliaceae*	1	0	0	s8	*Hypericum linariifolium* Vahl	*Hypericaceae*	1	1	0
m	*Ophioglossum lusitanicum* L.	*Ophioglossaceae*	1	1	0	s8	*Hypericum perfoliatum* L.	*Hypericaceae*	0	1	0
m	*Osmunda regalis* L.	*Osmundaceae*	1	1	1	s8	*Hypericum perfoliatum* L. subsp. angustifolium (DC.) A. Fröhl.	*Hypericaceae*	0	1	1
m	*Phyllitis scolopendrium* (L.) Newman subsp. scolopendrium	*Aspleniaceae*	0	1	0	s8	*Hypericum perforatum* L. subsp. perforatum	*Hypericaceae*	1	1	1
m	*Polypodium cambricum* L. subsp. cambricum	*Polypodiaceae*	0	1	0	s8	*Hypericum pulchrum* L.	*Hypericaceae*	1	0	0
m	*Polypodium interjectum* Shivas	*Polypodiaceae*	0	0	1	s8	*Hypericum tomentosum* L.	*Hypericaceae*	0	0	1
m	*Polypodium vulgare* L.	*Polypodiaceae*	0	0	1	s8	*Hypericum undulatum* Schousb. ex Willd.	*Hypericaceae*	1	1	1
m	*Polystichum setiferum* (Forssk.) Woynar	*Dryopteridaceae*	1	1	1	s8	*Lathyrus amphicarpos* L.	*Fabaceae*	0	0	1
m	*Pteridium aquilinum* (L.) Kuhn	*Dennstaedtiaceae*	0	1	1	s8	*Lathyrus angulatus* L.	*Fabaceae*	0	1	1
m	*Selaginella denticulata* (L.) Spring	*Selaginellaceae*	1	0	0	s8	*Lathyrus annuus* L.	*Fabaceae*	0	0	1
s2	*Alisma lanceolatum* With.	*Alismataceae*	1	1	1	s8	*Lathyrus aphaca* L.	*Fabaceae*	0	0	1
s2	Alisma plantago-aquatica L.	Alismataceae	0	1	1	s8	Lathyrus cicera L.	Fabaceae	0	1	1
s2	Arisarum simorrhinum Durieu	Araceae	0	1	1	s8	Lathyrus clymenum L.	Fabaceae	0	1	1
s2	*Arum italicum* Miller subsp. Italicum	Araceae	1	1	1	s8	*Lathyrus hirsutus* L.	Fabaceae	0	1	0
s2	*Arum italicum* Miller subsp. neglectum (Townsend) Prime	Araceae	0	1	0	s8	*Lathyrus linifolius* (Reichard) Bässler	Fabaceae	1	0	0
s2	Baldellia alpestris (Coss.) Vasc.	Alismataceae	1	1	0	s8	*Lathyrus ochrus* (L.) DC.	Fabaceae	0	0	1
s2	*Baldellia ranunculoides* (L.) Parl. Subsp. ranunculoides	Alismataceae	1	0	1	s8	*Lathyrus odoratus* L.	Fabaceae	0	1	0
s2	Biarum tenuifolium (L.) Schott subsp. arundanum (Boiss. & Reut.) Nyman	Araceae	0	0	1	s8	*Lathyrus sphaericus* Retz.	Fabaceae	0	1	1

S	Species	Family	MB	CP	FS	S	Species	Family	MB	CP	FS
s2	*Butomus umbellatus* L.	Butomaceae	0	1	0	s8	*Lathyrus sylvestris* L.	Fabaceae	0	1	0
s2	Colchicum multiflorum Brot.	Liliaceae	0	1	0	s8	*Lathyrus tingitanus* L.	Fabaceae	0	1	1
s2	Cymodocea nodosa (Ucria) Asch.	Potamogetonaceae	0	0	1	s8	*Lavatera arborea* L.	Malvaceae	0	0	1
s2	Damasonium bourgaei Coss.	Alismataceae	0	0	1	s8	*Lavatera cretica* L.	Malvaceae	1	1	1
s2	Dioscorea communis (L.) Caddick & Wilkin	Dioscoreaceae	1	1	1	s8	Lavatera triloba L.	Malvaceae	0	0	1
s2	Erythronium dens-canis L.	Liliaceae	1	0	0	s8	*Lavatera trimestris* L.	Malvaceae	0	0	1
s2	Fritillaria lusitanica Wikstr. Var. lusitanica	Liliaceae	0	1	0	s8	*Lepidium coronopus* (L.) Al-Shehbaz	Brassicaceae	0	1	0
s2	Gagea soleirolii F. W. Schultz	Liliaceae	1	0	0	s8	*Lepidium heterophyllum* Benth.	Brassicaceae	1	1	1
s2	Gagea lusitanica A.Terracc.	Liliaceae	0	1	0	s8	*Lepidium latifolium* L.	Brassicaceae	0	1	0
s2	*Lemna gibba* L.	Araceae	0	1	1	s8	*Lindernia dubia* (L.) Pennell	Linderniaceae	0	1	0
s2	*Lemna minor* L.	Araceae	1	1	1	s8	*Linum bienne* Mill.	Linaceae	1	1	1
s2	*Lemna minuta* Kunth	Araceae	0	1	0	s8	Linum setaceum Brot.	Linaceae	0	0	1
s2	*Najas marina* L. subsp. marina	Hydrocharitaceae	0	0	1	s8	Linum strictum L.	Linaceae	0	1	1
s2	*Najas minor* All.	Hydrocharitaceae	0	1	1	s8	Linum tenue Desf.	Linaceae	0	0	1
s2	Narthecium ossifragum (L.) Huds.	Liliaceae	1	0	0	s8	*Linum trigynum* L. subsp. trigynum	Linaceae	0	1	0
s2	Paradisea lusitanica (P. Cout.) Samp.	Liliaceae	1	1	0	s8	Lotus castellanus Boiss. & Reut.	Fabaceae	0	0	1
s2	*Potamogeton crispus* L.	Potamogetonaceae	0	1	0	s8	*Lotus conimbricensis* Brot.	Fabaceae	0	1	1
s2	*Potamogeton natans* L.	Potamogetonaceae	1	0	0	s8	*Lotus corniculatus* L. subsp. carpetanus (Lacaita) Rivas Mart.	Fabaceae	1	1	0
s2	*Potamogeton nodosus* Poir. in Lam. Subsp. Nodosus	Potamogetonaceae	1	0	0	s8	*Lotus corniculatus* L. subsp. corniculatus	Fabaceae	1	1	0
s2	*Potamogeton pectinatus* L.	Potamogetonaceae	0	0	1	s8	*Lotus creticus* L.	Fabaceae	0	0	1
s2	*Potamogeton polygonifolius* Pourr.	Potamogetonaceae	1	0	0	s8	*Lotus edulis* L.	Fabaceae	0	0	1
s2	*Potamogeton pusillus* L.	Potamogetonaceae	0	1	0	s8	*Lotus hispidus* Desf. ex DC.	Fabaceae	0	1	1

S	Species	Family	MB	CP	FS	S	Species	Family	MB	CP	FS
s2	Potamogeton trichoides Cham. & Schltdl.	Potamogetonaceae	0	1	0	s8	*Lotus ornithopodioides* L.	Fabaceae	0	0	1
s2	*Smilax aspera* L.	Smilaceae	0	1	1	s8	*Lotus parviflorus* Desf.	Fabaceae	0	1	1
s2	Triglochin barrelieri Loisel.	Juncaginaceae	0	1	0	s8	*Lotus pedunculatus* Cav.	Fabaceae	1	1	1
s2	*Tulipa sylvestris* L. subsp. australis (Link) Pamp.	Liliaceae	1	1	0	s8	*Lotus subbiflorus* Lag.	Fabaceae	0	1	0
s2	*Zannichellia palustris* L.	Potamogetonaceae	0	0	1	s8	*Ludwigia palustris* (L.) Elliott	Onagraceae	0	1	1
s2	*Zantedeschia aethiopica* (L.) Spreng.	Araceae	0	1	1	s8	*Lunaria annua* L. subsp. annua	Brassicaceae	0	1	0
s2	*Zostera marina* L.	Zosteraceae	0	0	1	s8	*Lupinus angustifolius* L.	Fabaceae	1	1	1
s2	*Zostera noltii* Hornem.	Zosteraceae	0	0	1	s8	Lupinus cosentinii Guss.	Fabaceae	0	0	1
s3	Adonis microcarpa DC.	Ranunculaceae	0	0	1	s8	Lupinus gredensis Gand.	Fabaceae	1	1	0
s3	Anemone palmata L.	Ranunculaceae	0	1	1	s8	*Lupinus luteus* L.	Fabaceae	1	1	1
s3	*Anemone trifolia* L. subsp. albida (Mariz) Ulbr.	Ranunculaceae	1	0	0	s8	*Lupinus micranthus* Guss.	Fabaceae	0	0	1
s3	*Aquilegia vulgaris* L. subsp. dichroa (Freyn) T. E. Díaz	Ranunculaceae	1	0	0	s8	Lythrum borysthenicum (Schrank) Litv.	Lythraceae	0	1	1
s3	Aristolochia baetica L.	Aristolochiaceae	0	0	1	s8	*Lythrum hyssopifolia* L.	Lythraceae	0	1	1
s3	Aristolochia paucinervis Pomel	Aristolochiaceae	1	1	1	s8	Lythrum junceum Banks et Sol. ex Russell	Lythraceae	0	1	1
s3	Aristolochia pistolochia L.	Aristolochiaceae	0	1	0	s8	*Lythrum portula* (L.) D. A. Webb	Lythraceae	1	1	1
s3	*Caltha palustris* L.	Ranunculaceae	1	0	0	s8	*Lythrum salicaria* L.	Lythraceae	1	1	1
s3	*Clematis campaniflora* Brot.	Ranunculaceae	0	1	0	s8	*Lythrum thymifolia* L.	Lythraceae	0	1	0
s3	*Clematis cirrhosa* L.	Ranunculaceae	0	0	1	s8	*Malus sylvestris* Mill.	Rosaceae	1	0	0
s3	*Clematis flammula* L.	Ranunculaceae	0	0	1	s8	Malva hispanica L.	Malvaceae	0	1	1
s3	Delphinium gracile DC.	Ranunculaceae	0	1	1	s8	*Malva neglecta* Wallr.	Malvaceae	1	1	1

S	Species	Family	MB	CP	FS	S	Species	Family	MB	CP	FS
s3	Delphinium pentagynum Lam.	Ranunculaceae	0	0	1	s8	*Malva nicaeensis* All.	Malvaceae	0	1	0
s3	*Laurus nobilis* L.	Lauraceae	0	0	1	s8	Malva olbia (L.) Alef.	Malvaceae	0	0	1
s3	*Nigella damascena* L.	Ranunculaceae	0	0	1	s8	*Malva parviflora* L.	Malvaceae	0	1	1
s3	Platanus hispanica Mill. ex Münchh.	Platanaceae	0	1	0	s8	*Malva sylvestris* L.	Malvaceae	1	1	1
s3	*Ranunculus bulbosus* L. subsp. aleae (Willk.) Rouy & Foucaud	Ranunculaceae	1	1	1	s8	Malva tournefortiana L.	Malvaceae	1	1	0
s3	Ranunculus bullatus L.	Ranunculaceae	0	0	1	s8	Matthiola fruticulosa (Loefl. ex L.) Maire	Brassicaceae	0	1	0
s3	Ranunculus bupleuroides Brot.	Ranunculaceae	1	0	0	s8	*Medicago arabica* (L.) Huds.	Fabaceae	0	1	0
s3	*Ranunculus ficaria* L.	Ranunculaceae	1	1	1	s8	*Medicago doliata* Carmign.	Fabaceae	0	1	1
s3	*Ranunculus flammula* L.	Ranunculaceae	1	0	0	s8	*Medicago intertexta* (L.) Mill.	Fabaceae	0	0	1
s3	*Ranunculus gramineus* L.	Ranunculaceae	0	0	1	s8	*Medicago italica* (Mill.) Fiori in Fiori et Paol.	Fabaceae	0	1	0
s3	Ranunculus macrophyllus Desf.	Ranunculaceae	0	0	1	s8	*Medicago littoralis* Rohde ex Loisel.	Fabaceae	0	1	1
s3	*Ranunculus muricatus* L.	Ranunculaceae	1	1	0	s8	*Medicago lupulina* L.	Fabaceae	0	1	1
s3	Ranunculus nigrescens Freyn in Willk. et Lange	Ranunculaceae	1	0	0	s8	*Medicago marina* L.	Fabaceae	0	0	1
s3	Ranunculus ollissiponensis Pers. Subsp. Olissiponensis	Ranunculaceae	1	1	0	s8	*Medicago minima* (L.) L.	Fabaceae	0	1	1
s3	Ranunculus omiophyllus Ten.	Ranunculaceae	1	0	0	s8	*Medicago monspeliaca* (L.) Trautv.	Fabaceae	0	1	1
s3	Ranunculus paludosus Poir.	Ranunculaceae	0	1	1	s8	*Medicago orbicularis* (L.) Bartal.	Fabaceae	0	1	1
s3	*Ranunculus parviflorus* L.	Ranunculaceae	0	1	0	s8	*Medicago polymorpha* L.	Fabaceae	1	1	1
s3	Ranunculus peltatus Schrank subsp. Peltatus	Ranunculaceae	1	1	1	s8	*Medicago sativa* L.	Fabaceae	0	0	1
s3	Ranunculus peltatus Schrank subsp. saniculifolius (Viv.) C. D. K. Cook	Ranunculaceae	0	1	0	s8	*Medicago truncatula* Gaertn.	Fabaceae	0	0	1

S	Species	Family	MB	CP	FS	S	Species	Family	MB	CP	FS
s3	Ranunculus penicillatus (Dumort.) Bab.	Ranunculaceae	0	1	0	s8	*Melilotus albus* Medik.	Fabaceae	1	0	0
s3	*Ranunculus repens* L.	Ranunculaceae	1	1	0	s8	*Melilotus indicus* (L.) All.	Fabaceae	0	1	0
s3	*Ranunculus trilobus* Desf.	Ranunculaceae	0	1	1	s8	*Melilotus segetalis* (Brot.) Ser. in DC.	Fabaceae	0	0	1
s3	Ranunculus tripartitus DC.	Ranunculaceae	0	1	0	s8	*Melilotus siculus* (Turra) B. D. Jacks.	Fabaceae	0	0	1
s3	Thalictrum speciosissimum L. in Loefl.	Ranunculaceae	1	1	0	s8	Melilotus spicatus (Sm.) Breistr.	Fabaceae	0	1	0
s4	Ceratocapnos claviculata (L.) Lidén	Papaveraceae	1	0	0	s8	*Mercurialis annua* L.	Euphorbiaceae	1	0	1
s4	*Chelidonium majus* L.	Papaveraceae	1	1	1	s8	Mercurialis elliptica Poir. in Lam.	Euphorbiaceae	0	0	1
s4	Fumaria agraria Lag.	Papaveraceae	0	0	1	s8	Mercurialis tomentosa L.	Euphorbiaceae	0	0	1
s4	*Fumaria capreolata* L.	Papaveraceae	0	1	1	s8	*Myrtus communis* L.	Myrtaceae	0	1	1
s4	Fumaria muralis Sond. ex W. D. J. Koch	Papaveraceae	1	1	1	s8	*Oenothera glazioviana* Micheli in Mart.	Onagraceae	0	1	0
s4	*Fumaria officinalis* L. subsp. officinalis	Papaveraceae	0	0	1	s8	*Oenothera rosea* L*Hér. ex Aiton	Onagraceae	0	0	1
s4	*Fumaria parviflora* Lam.	Papaveraceae	0	0	1	s8	Ononis baetica Clemente	Fabaceae	0	0	1
s4	*Fumaria reuteri* Boiss.	Papaveraceae	0	1	0	s8	*Ononis mitissima* L.	Fabaceae	0	0	1
s4	Fumaria sepium Boiss. et Reut. in Boiss.	Papaveraceae	0	0	1	s8	Ononis pinnata Brot.	Fabaceae	0	1	0
s4	*Glaucium flavum* Crantz	Papaveraceae	0	0	1	s8	*Ononis pubescens* L.	Fabaceae	0	1	1
s4	Hakea sericea Schrad.	Proteaceae	0	1	0	s8	Ononis ramosissima Desf.	Fabaceae	0	1	1
s4	*Hypecoum imberbe* Sm.	Papaveraceae	0	1	0	s8	*Ononis reclinata* L. subsp. reclinata	Fabaceae	0	1	1
s4	Hypecoum littorale Wulfen in Jacq.	Papaveraceae	0	0	1	s8	*Ononis spinosa* L. subsp. australis (Širj.) Greuter et Burdet	Fabaceae	0	1	1
s4	*Papaver argemone* L.	Papaveraceae	0	1	0	s8	*Ononis spinosa* L. subsp. spinosa	Fabaceae	0	1	0
s4	*Papaver dubium* L.	Papaveraceae	1	1	1	s8	Ononis variegata L.	Fabaceae	0	0	1
s4	*Papaver hybridum* L.	Papaveraceae	0	1	1	s8	*Ononis viscosa* L. subsp. brachycarpa (DC.) Batt. in Batt. et Trab.	Fabaceae	0	0	1
s4	*Papaver rhoeas* L.	Papaveraceae	1	1	1	s8	*Ononis viscosa* L. subsp.breviflora (DC.) Nyman	Fabaceae	0	0	1

DOI: http://dx.doi.org/10.5772/intechopen.1008900

S	Species	Family	MB	CP	FS	S	Species	Family	MB	CP	FS
s4	*Papaver somniferum* L. subsp. setigerum (DC.) Arcang.	Papaveraceae	0	0	1	s8	*Ornithopus compressus* L.	Fabaceae	1	1	1
s5	*Bryonia dioica* Jacq.	Cucurbitaceae	1	1	1	s8	*Ornithopus perpusillus* L.	Fabaceae	1	1	0
s5	Chrysosplenium oppositifolium L.	Saxifragaceae	1	0	1	s8	*Ornithopus pinnatus* (Mill.) Druce	Fabaceae	0	1	1
s5	*Crassula tillaea* Lest.-Garl.	Crassulaceae	1	1	1	s8	*Ornithopus sativus* Brot. subsp. isthmocarpus (Coss.) Dostál	Fabaceae	0	1	1
s5	Crassula vaillantii (Willd.) Roth	Crassulaceae	0	1	0	s8	*Ornithopus sativus* Brot. subsp. sativus	Fabaceae	0	1	0
s5	Cynomorium coccineum L.	Cynomoriaceae	0	0	1	s8	*Oxalis corniculata* L.	Oxalidaceae	1	0	1
s5	*Ecballium elaterium* (L.) A. Rich.	Cucurbitaceae	0	0	1	s8	*Oxalis pes-caprae* L.	Oxalidaceae	0	1	1
s5	Micranthes clusii (Gouan) Fern.Prieto, Vázquez, Vallines & Cires subsp. lepismigena (Planellas) Gornall	Saxifragaceae	1	0	1	s8	*Oxalis purpurea* L.	Oxalidaceae	0	1	0
s5	*Myriophyllum alterniflorum* DC. in Lam. et DC.	Haloragaceae	0	1	0	s8	*Parietaria judaica* L.	Urticaceae	0	1	1
s5	Paeonia broteri Boiss. et Reut.	Paeoniaceae	0	1	1	s8	Parietaria mauritanica Durieu	Urticaceae	0	1	0
s5	Pistorinia hispanica (L.) DC.	Crassulaceae	0	1	0	s8	*Pistacia lentiscus* L.	Anacardiaceae	0	1	1
s5	Saxifraga granulata L.	Saxifragaceae	1	1	1	s8	*Pistacia terebinthus* L.	Anacardiaceae	0	1	1
s5	Saxifraga spathularis Brot.	Saxifragaceae	1	0	1	s8	*Pisum sativum* L. subsp. elatius (M. Bieb.) Asch. et Graebn.	Fabaceae	0	1	0
s5	*Sedum album* L.	Crassulaceae	1	1	1	s8	Polygala microphylla L.	Polygalaceae	0	1	0
s5	Sedum amplexicaule DC.	Crassulaceae	0	1	0	s8	*Polygala vulgaris* L.	Polygalaceae	0	1	0
s5	Sedum andegavense (DC.) Desv.	Crassulaceae	0	1	0	s8	*Populus alba* L.	Salicaceae	0	1	0
s5	Sedum anglicum Huds.	Crassulaceae	1	0	0	s8	*Populus nigra* L. var. nigra	Salicaceae	0	1	0
s5	Sedum arenarium Brot.	Crassulaceae	1	1	0	s8	*Potentilla erecta* (L.) Raeusch.	Rosaceae	1	1	0
s5	Sedum brevifolium DC.	Crassulaceae	1	1	0	s8	*Potentilla reptans* L.	Rosaceae	1	1	1

S	Species	Family	MB	CP	FS	S	Species	Family	MB	CP	FS
s5	*Sedum forsterianum* Sm. in Sm.	Crassulaceae	1	1	1	s8	*Potentilla sterilis* (L.) Garcke	Rosaceae	1	0	0
s5	Sedum hirsutum All.	Crassulaceae	1	1	0	s8	*Prunus avium* L.	Rosaceae	1	1	0
s5	Sedum mucizonia (Ortega) Raym.-Hamet	Crassulaceae	0	0	1	s8	*Prunus dulcis* (Mill.) D. A. Webb	Rosaceae	0	0	1
s5	Sedum pruinatum Brot.	Crassulaceae	1	0	0	s8	*Prunus insititia* L.	Rosaceae	1	0	0
s5	Sedum rubens L.	Crassulaceae	0	1	1	s8	*Prunus lusitanica* L. subsp. lusitanica	Rosaceae	0	1	0
s5	Sedum sediforme (Jacq.) Pau	Crassulaceae	0	1	1	s8	*Prunus spinosa* L.	Rosaceae	0	1	0
s5	Umbilicus rupestris (Salisb.) Dandy in Ridd.	Crassulaceae	1	1	1	s8	Pterospartum tridentatum (L.) Willk. subsp. cantabricum (Spach) Talavera et P. E. Gibbs	Fabaceae	1	0	0
s5	*Vitis vinifera* L.	Vitaceae	0	1	0	s8	Pterospartum tridentatum (L.) Willk. subsp. lasianthum (Spach) Talavera et P. E. Gibbs	Fabaceae	1	1	0
s6	*Agrostemma githago* L.	Caryophyllaceae	1	1	0	s8	Pterospartum tridentatum (L.) Willk. subsp. tridentatum	Fabaceae	0	1	0
s6	Aizoon hispanicum L.	Aizoaceae	0	0	1	s8	Pyrus bourgaeana Decne.	Rosaceae	0	1	1
s6	*Amaranthus albus* L.	Amaranthaceae	1	1	0	s8	*Pyrus cordata* Desv.	Rosaceae	1	0	0
s6	*Amaranthus blitoides S. Watson*	Amaranthaceae	0	0	1	s8	Quercus coccifera L.	Fagaceae	0	1	1
s6	*Amaranthus blitum* L. subsp.. Blitum	Amaranthaceae	0	0	1	s8	Quercus faginea Lam. subsp. broteroi (P. Cout.) *A. camus*	Fagaceae	0	1	0
s6	*Amaranthus blitum* L. subsp.. emarginatus (Moq. ex Uline & W.L.Bray)	Amaranthaceae	0	0	1	s8	Quercus faginea Lam. subsp. faginea	Fagaceae	0	1	1
s6	*Amaranthus deflexus* L.	Amaranthaceae	1	1	0	s8	Quercus pyrenaica Willd.	Fagaceae	1	1	0
s6	*Amaranthus hybridus* L.	Amaranthaceae	1	1	1	s8	*Quercus robur* L.	Fagaceae	1	1	0
s6	Amaranthus powelii *S. Watson*	Amaranthaceae	0	1	0	s8	Quercus rotundifolia Lam.	Fagaceae	0	1	1
s6	*Amaranthus retroflexus* L.	Amaranthaceae	0	1	1	s8	*Quercus suber* L.	Fagaceae	1	1	1

S	Species	Family	MB	CP	FS	S	Species	Family	MB	CP	FS
s6	*Anagallis arvensis* L. subsp. Arvensis	Primulaceae	0	1	0	s8	*Radiola linoides* Roth	Linaceae	0	1	1
s6	*Anagallis foemina* Mill.	Primulaceae	0	1	0	s8	*Raphanus raphanistrum* L. subsp. raphanistrum	Brassicaceae	1	1	1
s6	*Anagallis monelli* L.	Primulaceae	0	1	0	s8	*Reseda lutea* L. subps. Lutea	Resedaceae	0	0	1
s6	*Anagallis tenella* (L.) L.	Primulaceae	0	1	0	s8	*Reseda luteola* L.	Resedaceae	1	1	1
s6	*Anredera cordifolia* (Ten.) Steenis	Basellaceae	0	0	1	s8	Reseda media Lag.	Resedaceae	1	1	1
s6	*Arbutus unedo* L.	Ericaceae	0	1	0	s8	*Retama monosperma* (L.) Boiss.	Fabaceae	0	0	1
s6	Arenaria algarbiensis Welw. ex Willk.	Caryophyllaceae	0	0	1	s8	Retama sphaerocarpa (L.) Boiss.	Fabaceae	0	1	0
s6	*Arenaria emarginata* Brot.	Caryophyllaceae	0	0	1	s8	*Rhamnus alaternus* L.	Rhamnaceae	0	1	1
s6	*Arenaria montana* L. subsp. montana	Caryophyllaceae	1	1	0	s8	*Rhamnus lycioides* L. subsp. oleoides (L.) Jahand. & Maire	Rhamnaceae	0	1	1
s6	Arenaria querioides Pourr. ex Willk. Subsp. querioides	Caryophyllaceae	1	0	0	s8	*Rhus coriaria* L.	Anacardiaceae	0	0	1
s6	Armeria beirana Franco	Plumbaginaceae	1	1	0	s8	*Ricinus communis* L.	Euphorbiaceae	0	0	1
s6	Armeria macrophylla Boiss. et Reut.	Plumbaginaceae	0	0	1	s8	*Robinia pseudoacacia* L.	Fabaceae	0	1	0
s6	*Armeria pungens* (Link) Hoffmanns. et Link	Plumbaginaceae	0	0	1	s8	*Rorippa nasturtium-aquaticum* (L.) Hayek	Brassicaceae	1	1	1
s6	Armeria transmontana (Samp.) G. H. M. Lawr.	Plumbaginaceae	1	1	0	s8	*Rorippa palustris* (L.) Besser	Brassicaceae	1	0	1
s6	Armeria velutina Welw. ex Boiss. et Reut.	Plumbaginaceae	0	0	1	s8	*Rorippa sylvestris* (L.) Besser subsp. sylvestris	Brassicaceae	0	1	0
s6	Arthrocnemum macrostachyum (Moric.) Moris in Moris et Delponte	Amaranthaceae	0	1	1	s8	*Rosa canina* L.	Rosaceae	1	1	0
s6	Asterolinon linum-stellatum (L.) Duby	Primulaceae	0	1	0	s8	*Rosa corymbifera* Borkh.	Rosaceae	1	0	0
s6	*Atriplex halimus* L.	Amaranthaceae	0	0	1	s8	*Rosa micrantha* Borrer ex Sm.	Rosaceae	0	1	0

S	Species	Family	MB	CP	FS	S	Species	Family	MB	CP	FS
s6	*Atriplex portulacoides* L.	Amaranthaceae	0	0	1	s8	*Rosa pouzinii* Tratt.	Rosaceae	0	1	0
s6	*Atriplex prostrata* Boucher ex DC. in Lam. t DC.	Amaranthaceae	0	0	1	s8	Rosa squarrosa (A. Rau) Boreau	Rosaceae	1	0	0
s6	*Beta macrocarpa* Guss.	Amaranthaceae	0	0	1	s8	*Rosa vosagiaca* N.H.F. Desp.	Rosaceae	1	1	0
s6	*Beta maritima* L.	Amaranthaceae	0	0	1	s8	Rubus brigantinus Samp.	Rosaceae	1	0	0
s6	Bufonia macropetala Willk.	Caryophyllaceae	0	1	0	s8	Rubus henriquesii Samp.	Rosaceae	1	0	0
s6	*Calluna vulgaris* (L.) Hull	Ericaceae	0	1	0	s8	Rubus lainzii H.E. Weber	Rosaceae	1	0	0
s6	*Carpobrotus edulis* (L.) N. E. Br. in E. Phillips	Aizoaceae	0	0	1	s8	*Rubus ulmifolius* Schott var. ulmifolius	Rosaceae	1	1	1
s6	*Cerastium brachypetalum* Desp. ex Pers. Var. brachypetalum	Caryophyllaceae	1	0	0	s8	Ruta angustifolia Pers.	Rutaceae	0	1	0
s6	*Cerastium diffusum* Pers.	Caryophyllaceae	1	1	0	s8	*Ruta chalepensis* L.	Rutaceae	0	1	0
s6	*Cerastium fontanum* Baumg. subsp. vulgare (Hartm.) Greuter et Burdet	Caryophyllaceae	1	1	0	s8	Ruta montana (L.) L.	Rutaceae	0	1	0
s6	*Cerastium glomeratum* Thuill.	Caryophyllaceae	1	1	1	s8	*Salix alba* L.	Salicaceae	0	1	0
s6	Cerastium ramosissimum Boiss.	Caryophyllaceae	1	0	0	s8	*Salix atrocinerea* Brot.	Salicaceae	1	1	1
s6	*Cerastium semidecandrum* L.	Caryophyllaceae	1	0	0	s8	*Salix fragilis* L.	Salicaceae	0	1	0
s6	Chaetonychia cymosa (L.) Sweet	Caryophyllaceae	1	1	1	s8	*Salix repens* L	Salicaceae	1	0	1
s6	*Chenopodium album* L. var. album	Amaranthaceae	1	1	1	s8	Salix salviifolia Brot. subsp. salviifolia	Salicaceae	1	1	1
s6	*Chenopodium ambrosioides* L.	Amaranthaceae	1	1	1	s8	Salix salviifolia Brot. Subsp. australis Franco	Salicaceae	0	1	0
s6	*Chenopodium bonus-henricus* L.	Amaranthaceae	1	0	0	s8	*Sanguisorba hybrida* (L.) Font Quer	Rosaceae	0	1	1
s6	*Chenopodium chenopodioides* (L.) Aellen	Amaranthaceae	0	0	1	s8	Sanguisorba verrucosa (Link ex G. Don) Ces.	Rosaceae	1	1	1
s6	*Chenopodium multifidum* L.	Amaranthaceae	0	0	1	s8	*Scorpiurus muricatus* L.	Fabaceae	0	1	0
s6	*Chenopodium murale* L.	Amaranthaceae	1	1	1	s8	*Scorpiurus sulcatus* L.	Fabaceae	0	1	1

S	Species	Family	MB	CP	FS	S	Species	Family	MB	CP	FS
s6	*Chenopodium opulifolium* Schrad. ex W.D.J. Koch & Ziz	Amaranthaceae	0	1	0	s8	*Scorpiurus vermiculatus* L.	Fabaceae	0	1	1
s6	*Chenopodium vulvaria* L.	Amaranthaceae	0	0	1	s8	Sesamoides purpurascens (L.) G. López	Resedaceae	1	1	1
s6	*Corrigiola litoralis* subsp. perez-larae Chaudhri, Muñoz Garm. & Pedrol	Caryophyllaceae	0	0	1	s8	Sesamoides suffruticosa (Lange) Kuntze	Resedaceae	1	1	0
s6	Corrigiola littoralis L. subsp. Littoralis	Caryophyllaceae	1	1	0	s8	*Sinapis alba* L. subsp. mairei (H. Lindb.) Maire	Brassicaceae	0	1	0
s6	Corrigiola telephiifolia Pourr.	Caryophyllaceae	1	1	1	s8	*Sinapis arvensis* L.	Brassicaceae	0	1	0
s6	Dianthus broteri Boiss. et Reut.	Caryophyllaceae	0	0	1	s8	Sisymbrella aspera (L.) Spach subsp. aspera	Brassicaceae	0	1	0
s6	Dianthus laricifolius Boiss. et Reut. Subsp. Laricifolius	Caryophyllaceae	1	1	0	s8	*Sisymbrium officinale* (L.) Scop.	Brassicaceae	1	1	1
s6	Dianthus lusitanus Brot.	Caryophyllaceae	1	1	0	s8	*Sorbus aucuparia* L.	Rosaceae	1	0	0
s6	Dianthus pungens L. subsp. langeanus	Caryophyllaceae	1	0	0	s8	*Spartium junceum* L.	Fabaceae	0	0	1
s6	*Disphyma crassifolium* (L.) L.Bolus	Aizoaceae	0	0	1	s8	Stauracanthus boivinii (Webb) Samp.	Fabaceae	0	0	1
s6	*Drosera rotundifolia* L.	Droseraceae	1	0	0	s8	Stauracanthus genistoides (Brot.) G. Sampaio	Fabaceae	0	1	1
s6	Drosophyllum lusitanicum (L.) Link	Drosophyllaceae	0	1	1	s8	*Teesdalia coronopifolia* (J.P.Bergeret) Thell.	Brassicaceae	1	1	1
s6	*Dysphania pumilio* (R.Br.) Mosyakin & Clemants	Amaranthaceae	0	1	0	s8	*Teesdalia nudicaulis* (L.) R.Br.	Brassicaceae	1	1	1
s6	*Emex spinosa* (L.) Campd.	Polygonaceae	0	0	1	s8	Thymelaea villosa (L.) Endl.	Thymelaeaceae	0	0	1
s6	*Erica arborea* L.	Ericaceae	0	1	0	s8	*Tribulus terrestris* L. subsp. terrestris	Zygophyllaceae	0	1	1
s6	Erica australis L.	Ericaceae	0	1	0	s8	*Trifolium angustifolium* L.	Fabaceae	1	1	1
s6	Erica ciliaris L.	Ericaceae	0	1	0	s8	*Trifolium arvense* L. var. arvense	Fabaceae	1	1	1

S	Species	Family	MB	CP	FS	S	Species	Family	MB	CP	FS
s6	*Erica cinerea* L.	Ericaceae	0	1	0	s8	*Trifolium bocconei* Savi	Fabaceae	0	1	0
s6	*Erica lusitanica* Rudolphi	Ericaceae	0	1	0	s8	*Trifolium campestre* Schreb. in Sturm	Fabaceae	1	1	1
s6	Erica scoparia L. subsp. scoparia	Ericaceae	0	1	0	s8	*Trifolium cernuum* Brot.	Fabaceae	1	1	1
s6	*Erica tetralix* L.	Ericaceae	0	1	0	s8	*Trifolium cherleri* L.	Fabaceae	0	1	1
s6	Erica umbellata Loefl. ex L.	Ericaceae	0	1	0	s8	*Trifolium dubium* Sibth.	Fabaceae	1	1	1
s6	*Fallopia convolvulus* (L.) Á. Löve	Polygonaceae	1	1	0	s8	*Trifolium fragiferum* L.	Fabaceae	0	0	1
s6	Frankenia boissieri Reut. ex Boiss.	Frankeniaceae	0	0	1	s8	*Trifolium gemellum* Pourr. ex Willd.	Fabaceae	0	1	0
s6	Frankenia laevis L.	Frankeniaceae	0	0	1	s8	*Trifolium glomeratum* L.	Fabaceae	1	1	1
s6	*Frankenia pulverulenta* L.	Frankeniaceae	0	0	1	s8	*Trifolium hirtum* All.	Fabaceae	0	1	0
s6	*Glinus lotoides* L.	Molluginaceae	0	1	1	s8	*Trifolium incarnatum* L.	Fabaceae	0	1	0
s6	Halopeplis amplexicaulis (Vahl) Ung.-Sternb. ex Ces., Passer. et Gibelli	Amaranthaceae	0	0	1	s8	*Trifolium isthmocarpum* Brot.	Fabaceae	0	1	1
s6	Herniaria lusitanica Chaudhri subsp. Lusitanica	Caryophyllaceae	1	1	0	s8	*Trifolium lappaceum* L.	Fabaceae	0	1	1
s6	Herniaria scabrida Boiss. subsp. scabrida var. glabrescens Boiss.	Caryophyllaceae	1	1	0	s8	*Trifolium ligusticum* Balb. ex Loisel.	Fabaceae	0	1	0
s6	*Illecebrum verticillatum* L.	Caryophyllaceae	1	1	1	s8	*Trifolium pratense* L. var. villosum	Fabaceae	1	0	0
s6	Limoniastrum monopetalum (L.) Boiss. in DC.	Plumbaginaceae	0	0	1	s8	*Trifolium pratense* L. var. pratense	Fabaceae	1	1	0
s6	Limonium algarvense Erben	Plumbaginaceae	0	0	1	s8	*Trifolium repens var. repens*	Fabaceae	1	1	1
s6	Limonium ferulaceum (L.) Chaz.	Plumbaginaceae	0	0	1	s8	*Trifolium resupinatum* L.	Fabaceae	0	1	1
s6	Limonium ovalifolium (Poir.) Kuntze	Plumbaginaceae	0	0	1	s8	*Trifolium scabrum* L.	Fabaceae	0	1	1
s6	*Limonium vulgare* Mill.	Plumbaginaceae	0	0	1	s8	*Trifolium spumosum* L.	Fabaceae	0	0	1
s6	Loeflingia baetica Lag.	Caryophyllaceae	0	1	1	s8	*Trifolium squamosum* L.	Fabaceae	0	0	1

DOI: http://dx.doi.org/10.5772/intechopen.1008900

S	Species	Family	MB	CP	FS	S	Species	Family	MB	CP	FS
s6	*Lychnis flos-cuculi* L. subsp. Flos-cuculi	Caryophyllaceae	1	0	0	s8	*Trifolium stellatum* L.	Fabaceae	0	1	1
s6	Lysimachia ephemerum L.	Primulaceae	0	1	0	s8	*Trifolium striatum* L. subsp. striatum	Fabaceae	0	1	0
s6	*Mesembryanthemum crystallinum* L.	Aizoaceae	0	0	1	s8	*Trifolium subterraneum* L. subsp. subterraneum	Fabaceae	0	1	1
s6	*Mesembryanthemum nodiflorum* L.	Aizoaceae	0	0	1	s8	*Trifolium suffocatum* L.	Fabaceae	0	1	1
s6	Minuartia recurva (All.) Schinz et Thell.	Caryophyllaceae	1	0	0	s8	*Trifolium tomentosum* L.	Fabaceae	0	1	1
s6	*Mirabilis jalapa* L.	Nyctaginaceae	0	1	0	s8	*Trifolium vesiculosum* Savi	Fabaceae	0	1	0
s6	Moehringia pentandra J. Gay	Caryophyllaceae	1	1	0	s8	*Tripodion tetraphyllum* (L.) Fourr.	Fabaceae	0	0	1
s6	*Moehringia trinervia* (L.) Clairv.	Caryophyllaceae	1	1	0	s8	*Tropaeolum majus* L.	Tropaeolaceae	0	0	1
s6	*Moenchia erecta* (L.) G. Gaertn., B. Mey. et Scherb. Subsp. Erecta	Caryophyllaceae	1	1	1	s8	Tuberaria globulariifolia (Lam.) Willk.	Cistaceae	1	0	1
s6	*Mollugo verticillata* L.	Molluginaceae	0	1	0	s8	*Tuberaria guttata* (L.) Fourr.	Cistaceae	1	0	1
s6	*Montia fontana* L. subsp. amporitana Sennen	Portulacaceae	1	1	0	s8	Ulex argenteus Welw. ex Webb subsp. argenteus	Fabaceae	0	0	1
s6	*Montia perfoliata* (Donn ex Willd.) Howell	Portulacaceae	0	1	0	s8	Ulex argenteus Welw. ex Webb subsp. subsericeus (P. Cout.) Rothm.	Fabaceae	0	0	1
s6	Myriolimon diffusum (Pourr.) Lledó, Erben & M.B.Crespo	Plumbaginaceae	0	0	1	s8	*Ulex europaeus* L. subsp. europaeus	Fabaceae	1	0	0
s6	*Opuntia dillenii* (Ker Gawl.) Haw.	Cactaceae	0	0	1	s8	*Ulex europaeus* L. subsp. latebracteatus (Mariz) Rothm.	Fabaceae	1	1	0
s6	Opuntia maxima Mill.	Cactaceae	0	1	1	s8	*Ulex minor* Roth	Fabaceae	1	1	0
s6	Ortegia hispanica Loefl. ex L.	Caryophyllaceae	0	1	0	s8	*Ulmus minor* Mill.	Cannabaceae	0	1	1
s6	Osyris alba L.	Santalaceae	1	1	1	s8	*Urtica dioica* L.	Urticaceae	0	1	0
s6	Paronychia argentea Lam.	Caryophyllaceae	1	1	1	s8	Urtica membranaceae Poir.	Urticaceae	0	1	1
s6	*Paronychia echinulata* Chater	Caryophyllaceae	0	0	1	s8	*Urtica urens* L.	Urticaceae	0	1	1

S	Species	Family	MB	CP	FS	S	Species	Family	MB	CP	FS
s6	Patellifolia procumbens (C.Sm.) A.J.Scott, Ford-Lloyd & J.T.Williams	Amaranthaceae	0	0	1	s8	*Vicia benghalensis* L. var. benghalensis	Fabaceae	0	1	1
s6	*Persicaria decipiens* (R.Br.) K.L.Wilson	Polygonaceae	0	1	1	s8	*Vicia dasycarpa* Ten.	Fabaceae	1	1	0
s6	*Persicaria maculosa* Gray	Polygonaceae	1	0	0	s8	*Vicia disperma* DC.	Fabaceae	1	1	1
s6	*Petrorhagia nanteuilii* (Burnat) P. W. Ball et Heywood	Caryophyllaceae	1	1	1	s8	*Vicia eriocarpa* (Hausskn.) Halácsy	Fabaceae	0	1	1
s6	*Phytolacca americana* L.	Phytolaccaceae	1	1	1	s8	*Vicia hirsuta* (L.) Gray	Fabaceae	1	0	0
s6	Plumbago europaea L.	Plumbaginaceae	0	0	1	s8	Vicia lens (L.) Coss. & Germ. Subsp. lamottei (Czefr.) H.Schaef., Coulot & Rabaute	Fabaceae	0	0	1
s6	*Polycarpon alsinifolium* (Biv.) DC.	Caryophyllaceae	0	0	1	s8	*Vicia lutea* L. subsp. lutea	Fabaceae	1	1	1
s6	*Polycarpon tetraphyllum* (L.) L. subsp. tetraphyllum	Caryophyllaceae	1	1	1	s8	*Vicia lutea* L. subsp. vestita (Boiss.) Rouy in Rouy et Foucaud	Fabaceae	0	0	1
s6	Polygala microphylla L.	Polygalaceae	1	0	0	s8	Vicia parviflora Cav.	Fabaceae	0	1	1
s6	Polygala monspeliaca L.	Polygalaceae	0	0	1	s8	Vicia pubescens (DC.) Link	Fabaceae	0	1	1
s6	*Polygala serpyllifolia* Hose	Polygalaceae	1	0	0	s8	*Vicia sativa* L. subsp. nigra (L.) Ehrh.	Fabaceae	1	1	1
s6	*Polygala vulgaris* L. subsp. vulgaris	Polygalaceae	1	0	0	s8	*Vicia sativa* L. subsp. sativa	Fabaceae	1	1	1
s6	*Polygonum arenastrum* Boreau	Polygonaceae	1	0	0	s8	*Vicia villosa* Roth subsp. villosa	Fabaceae	0	1	0
s6	*Polygonum aviculare* L.	Polygonaceae	1	1	1	s8	Vicia villosa subsp. varia (Host) Corb.	Fabaceae	0	1	0
s6	*Polygonum bistorta* L. subsp. Bistorta	Polygonaceae	1	0	0	s8	*Viola arvensis* Murray	Violaceae	1	1	1
s6	*Polygonum equisetiforme* Sm.	Polygonaceae	0	1	1	s8	*Viola canina* L. subsp. canina	Violaceae	1	0	1
s6	*Polygonum hydropiper* L.	Polygonaceae	1	1	0	s8	*Viola kitaibeliana* Schult. in Roem. et Schult.	Violaceae	1	1	1
s6	*Polygonum lapathifolium* L.	Polygonaceae	1	1	1	s8	Viola lactea Sm.	Violaceae	1	0	1
s6	*Polygonum maritimum* L.	Polygonaceae	0	0	1	s8	*Viola odorata* L.	Violaceae	0	1	0

S	Species	Family	MB	CP	FS	S	Species	Family	MB	CP	FS
s6	*Polygonum persicaria* L.	Polygonaceae	0	0	1	s8	*Viola palustris* L. subsp. palustri	Violaceae	1	1	1
s6	*Portulaca oleracea* L. subsp. granulatostellulata (Poelln.) Danin et H. G. Baker	Portulacaceae	0	1	0	s8	Viola riviniana Rchb.	Violaceae	1	1	1
s6	*Portulaca oleracea* L. subsp. oleracea	Portulacaceae	1	1	1	s8	Xolantha tuberaria (L.) Gallego, Muñoz Garm. et C. Navarro	Cistaceae	1	1	1
s6	Primula acaulis (L.) L. subsp. Acaulis	Primulaceae	0	1	0	s9	*Acanthus mollis* L.	Acanthaceae	0	0	1
s6	*Rumex acetosa* L. subsp. acetosa	Polygonaceae	1	1	0	s9	*Achillea ageratum* L.	Asteraceae	1	1	1
s6	*Rumex acetosella* L. subsp. angiocarpus (Murb.) Murb.	Polygonaceae	1	1	1	s9	*Achillea millefolium* L. subsp. millefolium	Asteraceae	1	1	1
s6	*Rumex bucephalophorus* L. subsp. gallicus (Steinh.) Rech. fil.	Polygonaceae	1	1	1	s9	*Achillea millefolium* L. subsp. ceretanica	Asteraceae	1	0	0
s6	*Rumex bucephalophorus* L. subsp. hispanicus (Steinh.) Rech. fil.	Polygonaceae	0	1	0	s9	Aetheorhiza bulbosa (L.) Cass. Subsp. bulbosa	Asteraceae	0	1	1
s6	*Rumex conglomeratus* Murray	Polygonaceae	1	1	1	s9	*Ageratina adenophora* (Spreng.) R. M. King et H. Rob.	Asteraceae	0	0	1
s6	*Rumex crispus* L.	Polygonaceae	1	1	1	s9	Ajuga iva (L.) Schreb. Var. iva	Lamiaceae	0	0	1
s6	Rumex induratus Boiss. et Reut.	Polygonaceae	1	1	1	s9	Ajuga pyramidalis L. subsp. meonantha (Hoffmanns. et Link) R. Fernandes	Lamiaceae	1	0	0
s6	Rumex intermedius DC.	Polygonaceae	0	0	1	s9	*Ajuga reptans* L.	Lamiaceae	1	0	0
s6	*Rumex obtusifolius* L.	Polygonaceae	1	1	1	s9	*Ammi majus* L.	Apiaceae	0	1	1
s6	*Rumex pulcher* L. subsp. pulcher	Polygonaceae	1	1	0	s9	*Ammi visnaga* (L.) Lam.	Apiaceae	0	0	1
s6	*Rumex pulcher* L. subsp. woodsii (De Not.) Arcang.	Polygonaceae	1	1	1	s9	Ammoides pusilla (Brot.) Breistr.	Apiaceae	0	0	1
s6	*Sagina apetala* Ard.	Caryophyllaceae	1	1	1	s9	*Anacyclus clavatus* (Desf.) Pers.	Asteraceae	0	0	1
s6	*Sagina procumbens* L.	Caryophyllaceae	1	1	0	s9	Anacyclus radiatus Loisel.	Asteraceae	0	0	1

S	Species	Family	MB	CP	FS	S	Species	Family	MB	CP	FS
s6	*Salicornia ramosissima* J. Woods	Amaranthaceae	0	0	1	s9	*Anagallis arvensis* L. subsp. arvensis	Primulaceae	1	0	1
s6	*Salsola kali* L.	Amaranthaceae	0	0	1	s9	*Anagallis monelli* L.	Primulaceae	0	0	1
s6	*Salsola soda* L.	Amaranthaceae	0	0	1	s9	*Anagallis tenella* (L.) L.	Primulaceae	1	0	0
s6	*Salsola vermiculata* L.	Amaranthaceae	0	0	1	s9	Anarrhinum bellidifolium (L.) Willd.	Plantaginaceae	1	1	1
s6	*Samolus valerandi* L.	Primulaceae	0	1	0	s9	*Anchusa azurea* Mill.	Boraginaceae	0	0	1
s6	*Saponaria officinalis* L.	Caryophyllaceae	1	1	1	s9	Anchusa calcarea Boiss.	Boraginaceae	0	0	1
s6	*Sarcocornia fruticosa* (L.) A. J. Scott	Amaranthaceae	0	0	1	s9	*Anchusa italica* Retz.	Boraginaceae	0	1	0
s6	*Sarcocornia perennis* (Mill.) A. J. Scott subsp. perennis	Amaranthaceae	0	0	1	s9	Anchusa undulata L. subsp. granatensis (Boiss.) Valdés	Boraginaceae	0	1	0
s6	*Scleranthus annuus* L.	Caryophyllaceae	1	1	0	s9	Anchusa undulata L. subsp. undulata	Boraginaceae	0	1	0
s6	Scleranthus polycarpos L.	Caryophyllaceae	1	0	0	s9	Andryala arenaria (DC.) Boiss. et Reut.	Asteraceae	0	0	1
s6	Silene acutifolia Link ex Rohrb.	Caryophyllaceae	1	1	0	s9	Andryala arenaria (DC.) Boiss. et Reut. subsp. parvipila Franco	Asteraceae	0	1	0
s6	Silene colorata Poir.	Caryophyllaceae	0	1	1	s9	Andryala integrifolia L.	Asteraceae	1	1	1
s6	Silene decipiens Barceló	Caryophyllaceae	0	0	1	s9	Andryala laxiflora DC.	Asteraceae	0	1	0
s6	*Silene dioica* (L.) Clairv.	Caryophyllaceae	1	0	0	s9	Andryala ragusina L.	Asteraceae	0	1	0
s6	*Silene gallica* L.	Caryophyllaceae	1	1	1	s9	*Angelica sylvestris* L.	Apiaceae	1	0	0
s6	*Silene inaperta* L. subsp. inaperta	Caryophyllaceae	0	1	1	s9	*Anthemis arvensis* L.	Asteraceae	1	1	0
s6	Silene laeta (Aiton) Godr.	Caryophyllaceae	0	0	1	s9	*Anthemis austriaca* Jacq.	Asteraceae	0	1	0
s6	*Silene latifolia* Poir.	Caryophyllaceae	1	1	1	s9	*Anthemis cotula* L.	Asteraceae	1	1	0
s6	Silene marizii Samp.	Caryophyllaceae	1	0	0	s9	Anthemis maritima L.	Asteraceae	0	0	1
s6	Silene micropetala Lag.	Caryophyllaceae	0	0	1	s9	*Anthriscus caucalis* M. Bieb.	Apiaceae	0	1	0
s6	Silene muscipula L.	Caryophyllaceae	0	1	0	s9	*Anthriscus sylvestris* (L.) Hoffm.	Apiaceae	0	1	0
s6	Silene niceensis All.	Caryophyllaceae	0	0	1	s9	Antirrhinum graniticum Rothm.	Plantaginaceae	0	1	0

S	Species	Family	MB	CP	FS	S	Species	Family	MB	CP	FS
s6	*Silene nocturna* L.	Caryophyllaceae	0	0	1	s9	Antirrhinum onubense (Fern.Casas) Fern.Casas	Plantaginaceae	0	0	1
s6	*Silene nutans* L. subsp. nutans	Caryophyllaceae	1	1	0	s9	*Antirrhinum orontium* L. var. orontium	Plantaginaceae	1	0	0
s6	Silene portensis L. subsp. portensis	Caryophyllaceae	0	1	1	s9	*Apium graveolens* L.	Apiaceae	0	0	1
s6	Silene psammitis Link ex Spreng. Subsp. Psammitis	Caryophyllaceae	1	0	0	s9	*Apium nodiflorum* (L.) Lag.	Apiaceae	1	1	1
s6	Silene scabriflora Brot. subsp. scabriflora	Caryophyllaceae	1	1	1	s9	*Arbutus unedo* L.	Ericaceae	1	0	1
s6	Silene scabriflora Brot. subsp. tuberculata (Ball) Talavera	Caryophyllaceae	0	0	1	s9	*Arctium minus* Bernh.	Asteraceae	1	1	0
s6	*Silene vulgaris* (Moench) Garcke subsp. vulgaris	Caryophyllaceae	1	1	1	s9	*Arctotheca calendula* (L.) Levyns	Asteraceae	0	0	1
s6	*Spergula arvensis* L.	Caryophyllaceae	1	1	1	s9	*Arnica montana* L. subsp. Atlantica	Asteraceae	1	0	0
s6	*Spergula morisonii* Boreau	Caryophyllaceae	1	0	0	s9	*Arnoseris minima* (L.) Schweigg. et Körte	Asteraceae	1	1	0
s6	*Spergula pentandra* L.	Caryophyllaceae	0	1	0	s9	Artemisia caerulescens L.	Asteraceae	0	0	1
s6	Spergularia capillacea (Kindb.) Willk.	Caryophyllaceae	1	0	0	s9	*Artemisia campestris* L. subsp. maritima Arcang.	Asteraceae	0	0	1
s6	Spergularia heldreichii Foucaud	Caryophyllaceae	0	0	1	s9	*Artemisia vulgaris* L.	Asteraceae	1	1	0
s6	*Spergularia marina* (L.) Besser	Caryophyllaceae	0	0	1	s9	Asperula hirsuta Desf.	Rubiaceae	0	0	1
s6	*Spergularia media* (L.) K. Presl	Caryophyllaceae	0	0	1	s9	*Aster lanceolatus* Willd.	Asteraceae	1	0	0
s6	*Spergularia purpurea* (Pers.) G. Don fil.	Caryophyllaceae	1	1	1	s9	*Aster squamatus* (Spreng.) Hieron.	Asteraceae	1	1	1
s6	*Spergularia rubra* (L.) J. Presl et K. Presl	Caryophyllaceae	1	0	1	s9	Asteriscus aquaticus (L.) Less.	Asteraceae	0	0	1
s6	Spergularia tangerina P. Monnier	Caryophyllaceae	0	0	1	s9	Asterolinon linum-stellatum (L.) Duby	Primulaceae	1	0	1
s6	*Stellaria alsine* Grimm	Caryophyllaceae	1	1	0	s9	Atractylis cancellata L. subsp. cancellata	Asteraceae	0	1	1

S	Species	Family	MB	CP	FS	S	Species	Family	MB	CP	FS
s6	*Stellaria graminea* L.	Caryophyllaceae	1	1	0	s9	Ballota hirsuta Benth.	Lamiaceae	0	1	0
s6	*Stellaria holostea* L.	Caryophyllaceae	1	0	0	s9	*Ballota nigra* L.	Lamiaceae	0	1	0
s6	*Stellaria media* (L.) Vill.	Caryophyllaceae	1	1	1	s9	*Bartsia trixago* L.	Orobanchaceae	0	1	1
s6	*Suaeda maritima* (L.) Dumort.	Amaranthaceae	0	0	1	s9	Bellis annua L.	Asteraceae	1	0	1
s6	Suaeda splendens (Pourr.) Gren. et Godr.	Amaranthaceae	0	0	1	s9	*Bellis perennis* L.	Asteraceae	1	1	1
s6	*Suaeda vera* Forssk. ex J. F. Gmel.	Amaranthaceae	0	0	1	s9	Bellis sylvestris Cirillo	Asteraceae	1	1	1
s6	*Tamarix africana* Poir.	Tamaricaceae	0	1	1	s9	*Bidens aurea* (Aiton) Scherff	Asteraceae	1	0	1
s6	*Tamarix canariensis* Willd.	Tamaricaceae	0	0	1	s9	*Bidens frondosa* L.	Asteraceae	1	1	1
s6	Tetragonia tetragonoides (Pall.) Kuntze	Aizoaceae	0	0	1	s9	*Bidens pilosa* L.	Asteraceae	0	0	1
s6	Thesium pyrenaicum Pourr.	Santalaceae	1	0	1	s9	*Bidens tripartita* L.	Asteraceae	1	0	0
s6	*Velezia rigida* Loefl. ex L.	Caryophyllaceae	0	1	1	s9	Blackstonia acuminata (W. D. J. Koch & Ziz) Domin subsp. acuminata	Gentianaceae	0	0	1
s7	Aceras anthropophorum (L.) W.T. Aiton	Orchidaceae	0	0	1	s9	Blackstonia acuminata (W.D.J. Koch & Ziz) Domin subsp. aestiva (K. Malý) Zeltner	Gentianaceae	0	1	1
s7	Acis autumnalis (L.) Sweet	Amaryllidaceae	1	0	1	s9	*Borago officinalis* L.	Boraginaceae	0	1	1
s7	Acis trichophylla G.Don	Amaryllidaceae	0	0	1	s9	*Buglossoides arvensis* (L.) I. M. Johnst.	Boraginaceae	0	0	1
s7	*Aegilops geniculata* Roth	Poaceae	0	1	1	s9	*Bunium bulbocastanum* L.	Apiaceae	0	0	1
s7	*Aegilops neglecta* Bertol.	Poaceae	0	1	1	s9	Bupleurum fruticosum L.	Apiaceae	0	0	1
s7	*Aegilops triuncialis* L.	Poaceae	0	1	1	s9	Bupleurum gerardi All.	Apiaceae	0	1	0
s7	*Agave americana* L. subsp. americana	Commelinaceae	0	1	1	s9	*Bupleurum lancifolium* Hornem.	Apiaceae	0	0	1
s7	*Agrostis capillaris* L.	Poaceae	1	0	0	s9	Bupleurum rigidum L. subsp. paniculatum (Brot.) H.Wolff	Apiaceae	0	0	1
s7	*Agrostis castellana* Boiss. et Reut.	Poaceae	1	1	1	s9	Bupleurum semicompositum L.	Apiaceae	0	0	1

DOI: http://dx.doi.org/10.5772/intechopen.1008900

S	Species	Family	MB	CP	FS	S	Species	Family	MB	CP	FS
s7	Agrostis curtisii Kerguélen	Poaceae	1	1	0	s9	Cachrys libanotis L.	Apiaceae	0	0	1
s7	Agrostis juressi Link	Poaceae	1	0	0	s9	Cachrys sicula L.	Apiaceae	0	0	1
s7	Agrostis pourretii Willd.	Poaceae	0	1	1	s9	*Calamintha sylvatica* Bromf. Subsp. ascendens (Jord.) P. W. Ball	Lamiaceae	0	1	0
s7	*Agrostis stolonifera* L.	Poaceae	1	1	1	s9	*Calendula arvensis* L.	Asteraceae	1	1	1
s7	*Agrostis tenerrima* Trin.	Poaceae	0	1	1	s9	*Calendula officinalis* L.	Asteraceae	1	0	0
s7	Agrostis truncatula Parl. subsp. commista Castrov. et Charpin	Poaceae	1	0	0	s9	Calendula suffruticosa Vahl subsp. algarbiensis (Boiss.) Nyman	Asteraceae	0	0	1
s7	Agrostis truncatula Parl. subsp. truncatula	Poaceae	1	1	0	s9	*Callitriche brutia* Petagna	Plantaginaceae	1	0	0
s7	Agrostis x fouilladei P. Fourn.	Poaceae	1	0	0	s9	*Callitriche hamulata* W. D. J. Koch	Plantaginaceae	1	0	0
s7	*Aira caryophyllea* L. subsp. caryophyllea	Poaceae	1	0	1	s9	*Callitriche stagnalis* Scop.	Plantaginaceae	1	1	0
s7	*Aira caryophyllea* L. subsp. uniaristata (Lag. et Rodr.) Maire	Poaceae	0	1	0	s9	Callitriche truncata Guss. subsp. occidentalis (Rouy) Schotsman	Plantaginaceae	0	0	1
s7	*Aira cupaniana* Guss.	Poaceae	0	1	1	s9	*Calluna vulgaris* (L.) Hull	Ericaceae	1	0	1
s7	*Aira praecox* L.	Poaceae	1	1	1	s9	*Calystegia sepium* (L.) R. Br. Subsp. Sepium	Convolvulaceae	0	1	1
s7	Airopsis tenella (Cav.) Asch. et Graebn.	Poaceae	0	1	1	s9	*Calystegia silvatica* (Kit.) Griseb. Subsp. Disjuncta	Convolvulaceae	1	0	0
s7	*Allium ampeloprasum* L.	Amaryllidaceae	1	1	1	s9	*Calystegia soldanella* (L.) R. Br.	Convolvulaceae	0	0	1
s7	Allium baeticum Boiss.	Alliaceae	0	1	0	s9	Campanula erinus L.	Campanulaceae	0	1	1
s7	Allium ericetorum Thore	Amaryllidaceae	1	0	0	s9	Campanula lusitanica L.	Campanulaceae	1	1	1
s7	Allium guttatum Steven subsp. sardoum (Moris) Stearn	Alliaceae	0	1	0	s9	Campanula rapunculus L.	Campanulaceae	1	1	1
s7	Allium massaessylum Batt. et Trab.	Alliaceae	0	1	1	s9	Carduncellus caeruleus (L.) K. Presl	Asteraceae	0	0	1
s7	*Allium neapolitanum* Cirillo	Alliaceae	0	1	1	s9	Carduus carpetanus Boiss. et Reut. Subsp. Carpetanus	Asteraceae	1	0	0

S	Species	Family	MB	CP	FS	S	Species	Family	MB	CP	FS
s7	Allium pallens L.	Alliaceae	0	1	0	s9	Carduus meonanthus Hoffmanns. & Link	Asteraceae	0	0	1
s7	*Allium paniculatum* L.	Amaryllidaceae	0	1	1	s9	Carduus platypus Lange subsp. platypus	Asteraceae	1	1	0
s7	Allium pruinatum Link ex Spreng.	Amaryllidaceae	0	0	1	s9	*Carduus pycnocephalus* L.	Asteraceae	1	1	1
s7	Allium roseum L.	Amaryllidaceae	0	1	1	s9	*Carduus tenuiflorus* Curtis	Asteraceae	1	1	1
s7	Allium schmitzii P. Cout.	Alliaceae	0	1	0	s9	Carlina corymbosa L.	Asteraceae	1	0	0
s7	*Allium schoenoprasum* L.	Amaryllidaceae	1	0	0	s9	Carlina hispanica Lam.	Asteraceae	1	1	1
s7	Allium scorzonerifolium DC. in Redouté	Amaryllidaceae	1	0	0	s9	Carlina racemosa L.	Asteraceae	0	1	1
s7	Allium sphaerocephalon L.	Amaryllidaceae	1	1	0	s9	*Carthamus lanatus* L.	Asteraceae	0	1	1
s7	Allium subvillosum Salzm. ex Schult. & Schult.f.	Amaryllidaceae	0	0	1	s9	*Carum verticillatum* (L.) W.D.J.Koch	Apiaceae	1	1	1
s7	*Allium triquetrum* L.	Alliaceae	0	1	0	s9	Centaurea alba L. subsp. alba	Asteraceae	0	1	0
s7	*Allium victorialis* L.	Amaryllidaceae	1	0	0	s9	Centaurea aristata Hoffm. et Link subsp. langeana	Asteraceae	1	0	0
s7	*Allium vineale* L.	Amaryllidaceae	1	1	0	s9	Centaurea aristata Hoffmanns. et Link subsp. exilis (Arènes) Dostál	Asteraceae	0	1	0
s7	*Alopecurus arundinaceus* Poir.	Poaceae	1	1	0	s9	*Centaurea aspera* L. subsp. aspera	Asteraceae	0	0	1
s7	*Ammophila arenaria* (L.) Link subsp. arundinacea H. Lindb.	Poaceae	0	0	1	s9	*Centaurea aspera* L. subsp. stenophylla (Dufour) Nyman	Asteraceae	0	0	1
s7	Anacamptis coriophora (L.) R.M.Bateman, Pridgeon & M.W.Chase	Orchidaceae	1	0	0	s9	*Centaurea benedicta* (L.) L.	Asteraceae	1	1	0
s7	Anacamptis coriophora (L.) R.M.Bateman, Pridgeon & M.W.Chase	Orchidaceae	0	0	1	s9	*Centaurea calcitrapa* L.	Asteraceae	0	1	0
s7	Anacamptis pyramidalis (L.) Rich.	Orchidaceae	0	0	1	s9	*Centaurea calcitrapa* L.	Asteraceae	0	0	1
s7	Anthericum liliago L.	Asparagaceae	1	0	0	s9	*Centaurea cyanus* L.	Asteraceae	0	1	0

S	Species	Family	MB	CP	FS	S	Species	Family	MB	CP	FS
s7	Anthoxanthum amarum Brot.	Poaceae	1	0	0	s9	Centaurea langei subsp. langei	Asteraceae	1	0	0
s7	*Anthoxanthum aristatum* Boiss. subsp. aristatum	Poaceae	1	1	1	s9	*Centaurea melitensis* L.	Asteraceae	0	1	0
s7	*Anthoxanthum odoratum* L.	Poaceae	1	1	0	s9	*Centaurea nigra* L. subsp. nigra	Asteraceae	1	0	0
s7	Antinoria agrostidea (DC.) Parl.	Poaceae	1	0	0	s9	*Centaurea nigra* L. subsp. rivularis	Asteraceae	1	0	0
s7	Antinoria agrostidea (DC.) Parl.	Poaceae	0	1	0	s9	*Centaurea paniculata* L. subsp. paniculata	Asteraceae	1	0	0
s7	Arrhenatherum album (Vahl) Clayton	Poaceae	0	1	1	s9	Centaurea pullata L.	Asteraceae	0	1	0
s7	*Arrhenatherum elatius* (L.) P. Beauv. ex J. et K. Presl subsp. baeticum Romero Zarco	Poaceae	1	0	1	s9	*Centaurium erythraea* Rafn subsp. grandiflorum (Biv.) Melderis	Gentianaceae	0	1	1
s7	*Arrhenatherum elatius* (L.) P. Beauv. ex J. et K. Presl subsp. bulbosum (Willd.) Schübl. et G. Martens	Poaceae	1	1	1	s9	Centaurium maritimum (L.) Fritsch	Gentianaceae	0	1	0
s7	*Arrhenatherum elatius* (L.) P. Beauv. ex J. et K. Presl subsp. elatius	Poaceae	1	0	0	s9	Centaurium maritimum (L.) Fritsch	Gentianaceae	0	1	1
s7	*Arundo donax* L.	Poaceae	1	1	1	s9	*Centaurium pulchellum* (Sw.) Druce subsp. tenuiflorum (Hoffmanns. et Link) Maire in Jahand.	Gentianaceae	0	1	1
s7	Asparagus acutifolius L.	Asparagaceae	0	1	1	s9	Centranthus calcitrapae (L.) Dufresne subsp. calcitrapae	Caprifoliaceae	1	1	1
s7	Asparagus albus L.	Asparagaceae	0	1	1	s9	Cephalaria leucantha (L.) Schrad. ex Roem. & Schult.	Caprifoliaceae	0	0	1
s7	Asparagus aphyllus L.	Asparagaceae	0	1	1	s9	Cerinthe gymnandra Gasp.	Boraginaceae	0	0	1
s7	Asphodelus aestivus Brot.	Xanthorrhoeaceae	0	1	0	s9	*Cerinthe major* L.	Boraginaceae	0	0	1
s7	*Asphodelus albus* Mill. subsp. albus	Xanthorrhoeaceae	1	1	0	s9	*Chaerophyllum temulum* L.	Apiaceae	1	1	0
s7	Asphodelus bento-rainhae P. Silva	Xanthorrhoeaceae	0	1	0	s9	Chamaeleon gummifer (L.) Cass.	Asteraceae	0	0	1
s7	Asphodelus lusitanicus P. Cout. Var. ovoideus (Merino) Z.Díaz & Valdés var. lusitanicus ver todas	Xanthorrhoeaceae	1	0	0	s9	*Chamaemelum fuscatum* (Brot.) Vasc.	Asteraceae	1	1	1

S	Species	Family	MB	CP	FS	S	Species	Family	MB	CP	FS
s7	Asphodelus macrocarpus Parl. Subsp. macrocarpus	Xanthorrhoeaceae	1	0	0	s9	*Chamaemelum mixtum* (L.) All.	Asteraceae	1	1	1
s7	Asphodelus ramosus L.	Xanthorrhoeaceae	0	1	1	s9	*Chamaemelum nobile* (L.) All.	Asteraceae	1	1	1
s7	Asphodelus serotinus Wolley-Dod	Xanthorrhoeaceae	0	0	1	s9	Cheirolophus sempervirens (L.) Pomel	Asteraceae	0	0	1
s7	*Asphodelus tenuifolius* Cav.	Xanthorrhoeaceae	0	0	1	s9	Cheirolophus uliginosus (Brot.) Dostál	Asteraceae	0	1	0
s7	*Avena barbata* Link in Schrad. Subsp. Barbata	Poaceae	1	1	1	s9	*Chondrilla juncea* L.	Asteraceae	1	0	1
s7	*Avena barbata* Link in Schrad. Subsp. lusitanica (Tab. Morais) Romero Zarco	Poaceae	0	1	0	s9	*Chrysanthemum segetum* L.	Asteraceae	1	1	1
s7	*Avena fatua* L.	Poaceae	0	1	0	s9	Cicendia filiformis (L.) Delarbre	Gentianaceae	0	1	1
s7	*Avena sativa* L. subsp. macrantha (Hack.) Rocha Afonso	Poaceae	0	1	0	s9	*Cichorium intybus* L.	Asteraceae	1	1	1
s7	*Avena sterilis* L. subsp. ludoviciana (Durieu) Nyman	Poaceae	0	1	0	s9	*Cirsium arvense* (L.) Scop.	Asteraceae	1	1	0
s7	*Avena sterilis* L. subsp. sterilis	Poaceae	0	1	1	s9	*Cirsium filipendulum* Lange	Asteraceae	1	1	0
s7	*Avena strigosa* Schreb.	Poaceae	1	1	0	s9	*Cirsium palustre* (L.) Scop.	Asteraceae	1	1	0
s7	Bellevalia dubia (Guss.) Rchb. Subsp. hackelii (Freyn) Feinbrun	Asparagaceae	0	0	1	s9	*Cirsium vulgare* (Savi) Ten.	Asteraceae	1	1	1
s7	*Bolboschoenus glaucus* (Lam.) S.G.Sm.	Cyperaceae	0	1	1	s9	Cistanche phelypaea (L.) Cout.	Orobanchaceae	0	0	1
s7	*Bolboschoenus maritimus* (L.) Palla in Koch	Cyperaceae	0	0	1	s9	Cleonia lusitanica (Loefl. ex L.) L.	Lamiaceae	0	1	1
s7	*Brachypodium distachyon* (L.) P. Beauv.	Poaceae	0	1	1	s9	*Clinopodium nepeta* (L.) Kuntze subsp. glandulosum (Req.) Govaerts	Lamiaceae	1	1	1
s7	*Brachypodium phoenicoides* (L.) Roem. et Schult.	Poaceae	0	1	1	s9	*Clinopodium vulgare* L.	Lamiaceae	1	1	0
s7	*Brachypodium pinnatum* (L.) P. Beauv.	Poaceae	1	1	0	s9	Coleostephus myconis (L.) Rchb. f.	Asteraceae	1	1	1
s7	*Brachypodium sylvaticum* (Huds.) P. Beauv.	Poaceae	1	1	0	s9	*Conium maculatum* L.	Apiaceae	0	1	1

S	Species	Family	MB	CP	FS	S	Species	Family	MB	CP	FS
s7	*Briza maxima* L.	Poaceae	1	1	1	s9	Conopodium majus (Gouan) Loret subsp. marizianum (Samp.) López Udias et Mateo	Apiaceae	1	1	0
s7	*Briza minor* L.	Poaceae	1	1	1	s9	Conopodium marianum Lange	Apiaceae	0	1	1
s7	*Bromus catharticus* Vahl	Poaceae	0	1	0	s9	Conopodium subcarneum (Boiss. et Reut.) Boiss. et Reut.	Apiaceae	0	1	0
s7	*Bromus diandrus* Roth	Poaceae	1	1	1	s9	*Convolvulus althaeoides* L.	Convolvulaceae	0	1	1
s7	*Bromus hordeaceus* L.	Poaceae	1	1	1	s9	*Convolvulus arvensis* L. var. arvensis	Convolvulaceae	1	1	1
s7	*Bromus lanceolatus* Roth	Poaceae	0	1	1	s9	Convolvulus meonanthus Hoffmanns. et Link	Convolvulaceae	0	0	1
s7	*Bromus madritensis* L.	Poaceae	0	1	1	s9	Convolvulus pentapetaloides L.	Convolvulaceae	0	0	1
s7	*Bromus rigidus* Roth	Poaceae	0	1	0	s9	Convolvulus siculus L.	Convolvulaceae	0	0	1
s7	*Bromus rubens* L.	Poaceae	0	1	0	s9	*Convolvulus tricolor* L.	Convolvulaceae	0	0	1
s7	*Bromus sterilis* L.	Poaceae	1	1	0	s9	*Conyza bonariensis* (L.) Cronquist	Asteraceae	0	1	1
s7	*Bromus tectorum* L.	Poaceae	0	1	0	s9	*Conyza canadensis* (L.) Cronquist	Asteraceae	1	1	1
s7	*Calamagrostis arundinacea* (L.) Roth	Poaceae	1	0	0	s9	Conyza sumatrensis (Retz.) E. Walker	Asteraceae	1	1	1
s7	Carex asturica Boiss. in Boiss. et Reut.	Cyperaceae	1	0	0	s9	Corema album (L.) D. Don	Ericaceae	0	0	1
s7	Carex binervis Sm.	Cyperaceae	1	1	0	s9	Cota segetalis (Ten.) Holub	Asteraceae	0	1	0
s7	Carex cuprina (I. Sándor ex Heuff.) Nendtv. Ex A. Kern.	Cyperaceae	1	0	1	s9	*Cotula australis* (Spreng.) Hook.	Asteraceae	0	1	1
s7	*Carex demissa* Hornem.	Cyperaceae	1	0	0	s9	*Cotula coronopifolia* L.	Asteraceae	0	0	1
s7	Carex depressa Link in Schrad. subsp. depressa	Cyperaceae	0	1	0	s9	*Crepis capillaris* (L.) Wallr. Var. capillaris	Asteraceae	1	1	1
s7	Carex distachya Desf.	Cyperaceae	0	1	0	s9	*Crepis foetida* L. subsp. foetida	Asteraceae	0	1	0
s7	*Carex divisa* Huds.	Cyperaceae	0	1	1	s9	*Crepis foetida* L. subsp. foetida	Asteraceae	0	0	1

S	Species	Family	MB	CP	FS	S	Species	Family	MB	CP	FS
s7	*Carex divulsa* Stokes in With. subsp. divulsa	Cyperaceae	0	1	1	s9	Crepis lampsanoides (Gouan) Tausch	Asteraceae	1	0	0
s7	*Carex divulsa* Stokes subsp. leersii	Cyperaceae	1	0	0	s9	*Crepis vesicaria* L. subsp. taraxacifolia	Asteraceae	1	1	1
s7	*Carex echinata* Murray	Cyperaceae	1	0	0	s9	Cressa cretica L.	Convolvulaceae	0	0	1
s7	Carex elata All. Subsp. reuterana	Cyperaceae	1	0	0	s9	Crithmum maritimum L.	Apiaceae	0	0	1
s7	*Carex extensa* Gooden.	Cyperaceae	0	0	1	s9	*Crucianella angustifolia* L.	Rubiaceae	1	1	1
s7	*Carex flacca* Schreb.	Cyperaceae	1	0	1	s9	Crucianella latifolia L.	Rubiaceae	0	0	1
s7	Carex halleriana Asso	Cyperaceae	0	0	1	s9	Crucianella maritima L.	Rubiaceae	0	0	1
s7	Carex hispida Willd. in Schkuhr	Cyperaceae	0	0	1	s9	Cruciata glabra (L.) Ehrend. Subsp. hirticaulis (Beck) Natali et Jeanm.	Rubiaceae	1	0	0
s7	Carex laevigata Sm.	Cyperaceae	1	1	0	s9	*Crupina vulgaris* Cass.	Asteraceae	1	1	1
s7	*Carex leporina* L.	Cyperaceae	1	0	0	s9	*Cuscuta approximata* Bab. Subsp. Approximata	Convolvulaceae	0	1	0
s7	*Carex muricata* L. subsp. lamprocarpa Celak.	Cyperaceae	0	1	0	s9	*Cuscuta campestris* Yunck.	Convolvulaceae	0	1	0
s7	*Carex muricata* L. subsp. pairae (F.W. Schultz) Celak.	Cyperaceae	1	0	0	s9	*Cuscuta epithymum* (L.) L.	Convolvulaceae	1	1	0
s7	Carex oedipostyla Duval-Jouve	Cyperaceae	0	0	1	s9	*Cuscuta planiflora* Ten.	Convolvulaceae	0	0	1
s7	*Carex otrubae* Podp.	Cyperaceae	0	1	0	s9	*Cyclospermum leptophyllum* (Pers.) Sprague	Apiaceae	0	0	1
s7	*Carex ovalis* Gooden.	Cyperaceae	1	1	0	s9	*Cymbalaria muralis* G. Gaertn.	Plantaginaceae	1	1	1
s7	*Carex panicea* L.	Cyperaceae	1	0	0	s9	Cynanchum acutum L.	Apocynaceae	0	0	1
s7	*Carex paniculata* L. subep. Lusitanica (Willd.) Maire	Cyperaceae	0	1	0	s9	Cynara algarbiensis Mariz	Asteraceae	0	0	1
s7	*Carex pendula* Huds.	Cyperaceae	0	1	1	s9	*Cynara cardunculus* L.	Asteraceae	0	0	1
s7	*Carex pilulifera* L. subsp. pilulifera	Cyperaceae	1	0	0	s9	Cynara humilis L.	Asteraceae	0	1	1
s7	*Catapodium rigidum* (L.) Dony subsp. rigidum	Poaceae	0	1	1	s9	Cynoglossum clandestinum Desf.	Boraginaceae	0	0	1

S	Species	Family	MB	CP	FS	S	Species	Family	MB	CP	FS
s7	Cephalanthera longifolia (L.) Fritsch	Orchidaceae	0	1	1	s9	Cynoglossum creticum Mill.	Boraginaceae	0	1	1
s7	Chaetopogon fasciculatus (Link) Hayek subsp. fasciculatus	Poaceae	0	1	1	s9	Daboecia cantabrica (Huds.) K. Koch	Ericaceae	1	0	0
s7	*Chamaerops humilis* L.	Arecaceae	0	0	1	s9	*Datura stramonium* L.	Solanaceae	1	1	1
s7	*Cladium mariscus* (L.) Pohl	Cyperaceae	0	0	1	s9	*Daucus carota* L. subsp. carota	Apiaceae	1	1	1
s7	Colchicum lusitanum Brot.	Colchicaceae	0	0	1	s9	*Daucus carota* L. subsp. halophilus (Brot.) A. Pujadas	Apiaceae	0	0	1
s7	*Cortaderia selloana* (Schult. et Schult. f.) Asch. et Graebn.	Poaceae	0	0	1	s9	*Daucus carota* L. subsp. maximus (Desf.) Ball	Apiaceae	0	1	1
s7	*Corynephorus canescens* (L.) P. Beauv.	Poaceae	1	0	1	s9	Daucus crinitus Desf.	Apiaceae	0	1	1
s7	*Corynephorus fasciculatus* Boiss. & Reut.	Poaceae	0	1	1	s9	Daucus muricatus (L.) L.	Apiaceae	0	0	1
s7	Corynephorus macrantherus Boiss. & Reut.	Poaceae	0	1	0	s9	Daucus setifolius Desf.	Apiaceae	0	1	0
s7	Crocus carpetanus Boiss. et Reut.	Iridaceae	1	1	0	s9	*Dichondra micrantha* Urb.	Convolvulaceae	0	0	1
s7	Crocus serotinus Salisb. Subsp. clusii (J.Gay) B.Mathew	Iridaceae	1	0	0	s9	*Digitalis purpurea* L. subsp. purpurea	Plantaginaceae	1	1	1
s7	Crocus serotinus Salisb. subsp. serotinus	Iridaceae	1	0	0	s9	Digitalis thapsi L.	Plantaginaceae	0	1	0
s7	*Crypsis aculeata* (L.) Aiton	Poaceae	0	0	1	s9	Dipsacus comosus Hoffmanns. et Link	Caprifoliaceae	0	0	1
s7	*Crypsis alopecuroides* (Piller & Mitterp.) Schrad.	Poaceae	0	0	1	s9	Distichoselinum tenuifolium (Lag.) F. García Mart. & Silvestre	Apiaceae	0	0	1
s7	*Crypsis schoenoides* (L.) Lam.	Poaceae	0	1	1	s9	*Dittrichia graveolens* (L.) Desf.	Asteraceae	0	1	0
s7	Cutandia maritima (L.) Barbey	Poaceae	0	0	1	s9	*Dittrichia viscosa* (L.) Greuter	Asteraceae	1	1	1
s7	*Cynodon dactylon* (L.) Pers.	Poaceae	1	1	1	s9	Echinops strigosus L.	Asteraceae	0	0	1
s7	*Cynosurus cristatus* L.	Poaceae	1	0	0	s9	*Echium lusitanicum* L. subsp. lusitanicum	Boraginaceae	1	1	1
s7	*Cynosurus echinatus* L.	Poaceae	1	1	1	s9	*Echium plantagineum* L.	Boraginaceae	1	1	1

S	Species	Family	MB	CP	FS	S	Species	Family	MB	CP	FS
s7	Cynosurus elegans Desf.	Poaceae	0	1	0	s9	Echium rosulatum Lag.	Boraginaceae	1	1	1
s7	Cyperus capitatus Vand.	Cyperaceae	0	0	1	s9	Echium tuberculatum Hoffm. et Link	Boraginaceae	0	1	0
s7	*Cyperus difformis* L.	Cyperaceae	0	0	1	s9	*Eclipta prostrata* (L.) L.	Asteraceae	0	0	1
s7	*Cyperus eragrostis* Lam.	Cyperaceae	0	1	1	s9	Elaeoselinum foetidum (L.) Boiss.	Apiaceae	0	1	0
s7	*Cyperus esculentus* L.	Cyperaceae	0	1	0	s9	Elaeoselinum foetidum (L.) Boiss.	Apiaceae	0	0	1
s7	*Cyperus fuscus* L.	Cyperaceae	0	1	0	s9	*Erica arborea* L.	Ericaceae	1	0	1
s7	*Cyperus involucratus* Rottb.	Cyperaceae	0	0	1	s9	Erica australis L.	Ericaceae	1	0	1
s7	Cyperus longus L.	Cyperaceae	1	1	1	s9	Erica ciliaris L.	Ericaceae	1	0	0
s7	Cyperus michelianus (L.) Link	Cyperaceae	0	1	1	s9	*Erica cinerea* L.	Ericaceae	1	0	0
s7	*Cyperus rotundus* L.	Cyperaceae	0	1	1	s9	*Erica lusitanica* Rudolphi	Ericaceae	0	0	1
s7	*Dactylis glomerata* L. subsp. hispanica (Roth) Nyman	Poaceae	1	1	1	s9	Erica scoparia L. subsp. scoparia	Ericaceae	1	0	1
s7	*Dactylis glomerata* L. subsp. lusitanica Stebbins et Zohary	Poaceae	1	1	0	s9	*Erica tetralix* L.	Ericaceae	1	0	0
s7	Dactylorhiza elata (Poir.) Soó	Orchidaceae	0	1	0	s9	Erica umbellata Loefl. ex L.	Ericaceae	1	0	1
s7	*Dactylorhiza maculata* (L.) Soó	Orchidaceae	1	0	0	s9	*Erigeron karvinskianus* DC.	Asteraceae	1	0	0
s7	*Danthonia decumbens* (L.) DC. in Lam. et DC.	Poaceae	1	1	0	s9	*Eryngium campestre* L.	Apiaceae	0	1	1
s7	*Deschampsia cespitosa* (L.) P. Beauv. Subsp. cespitosa	Poaceae	1	0	0	s9	Eryngium corniculatum Lam.	Apiaceae	0	0	1
s7	*Deschampsia flexuosa* (L.) Trin.	Poaceae	1	0	0	s9	Eryngium dilatatum Lam.	Apiaceae	0	0	1
s7	Deschampsia media (Gouan) Roem. et Shultes	Poaceae	1	0	0	s9	Eryngium duriaei J. Gay ex Boiss.	Apiaceae	1	0	0
s7	*Dichanthium ischaemum* (L.) Roberty	Poaceae	0	1	0	s9	Eryngium galioides Lam.	Apiaceae	0	0	1
s7	*Digitaria sanguinalis* (L.) Scop.	Poaceae	1	1	1	s9	*Eryngium maritimum* L.	Apiaceae	0	0	1

S	Species	Family	MB	CP	FS	S	Species	Family	MB	CP	FS
s7	Dipcadi serotinum (L.) Medik. subsp. serotinum	Hyacinthaceae	0	1	1	s9	Eryngium tenue Lam.	Apiaceae	0	1	0
s7	*Drimia maritima* (L.) Stearn	Hyacinthaceae	0	1	1	s9	*Eupatorium cannabinum* L. subsp. cannabinum	Asteraceae	1	0	0
s7	Echinaria capitata (L.) Desf.	Poaceae	0	0	1	s9	Evax pygmaea (L.) Brot. subsp. pygmaea	Asteraceae	1	0	0
s7	*Echinochloa crus-galli* (L.) P. Beauv.	Poaceae	1	1	1	s9	Exaculum pusillum (Lam.) Caruel in Parl.	Gentianaceae	0	1	1
s7	Eleocharis multicaulis (Sm.) Desv.	Cyperaceae	1	1	0	s9	Ferula communis L. subsp. catalaunica (Pau ex C. Vicioso) Sánchez Cuxart	Apiaceae	0	1	1
s7	*Eleocharis palustris* (L.) Roem. et Schult. subsp. palustris	Cyperaceae	0	1	1	s9	Filago carpetana (Lange) Chrtek & Holub	Asteraceae	0	0	1
s7	*Eleocharis palustris* (L.) Roem. et Schult. subsp. vulgaris Walters	Cyperaceae	0	0	1	s9	Filago lusitanica (Samp.) P. Silva	Asteraceae	0	0	1
s7	Eleogiton fluitans (L.) Link	Cyperaceae	0	1	0	s9	Filago lutescens Jord.	Asteraceae	0	0	1
s7	*Elymus farctus* subsp. boreo-atlanticus (Simonet & Guin.) Melderis	Poaceae	0	0	1	s9	Filago pygmaea L.	Asteraceae	0	0	1
s7	*Epipactis helleborine* (L.) Crantz	Orchidaceae	0	1	0	s9	*Filago pyramidata* L. subsp. pyramidata	Asteraceae	1	1	1
s7	Epipactis lusitanica Tyteca	Orchidaceae	0	0	1	s9	*Filago vulgaris* Lam.	Asteraceae	0	1	0
s7	Epipactis tremolsii Pau	Orchidaceae	0	1	1	s9	*Foeniculum vulgare* Mill.	Apiaceae	1	1	1
s7	*Eragrostis cilianensis* (All.) F. T. Hubb.	Poaceae	0	1	0	s9	Fraxinus angustifolia Vahl subsp. angustifolia	Oleaceae	1	1	1
s7	*Eragrostis minor* Host	Poaceae	0	1	1	s9	Galactites tomentosa Moench	Asteraceae	1	1	1
s7	*Eriophorum angustifolium* Honck.	Cyperaceae	1	0	0	s9	Galatella sedifolia (L.) Nyman	Asteraceae	1	0	0
s7	Festuca ampla Hack. subsp. ampla	Poaceae	1	1	1	s9	*Galeopsis tetrahit* L.	Lamiaceae	1	0	0
s7	*Festuca arundinacea* subsp. arundinacea	Poaceae	0	0	1	s9	*Galinsoga ciliata* (Raf.) S.F.Blake	Asteraceae	1	1	0

S	Species	Family	MB	CP	FS	S	Species	Family	MB	CP	FS
s7	Festuca durandoi Clauson in Billot subsp. livida (Hack.) Rivas Ponce et Cebolla	Poaceae	0	1	0	s9	*Galium aparine* L. subsp. aparine	Rubiaceae	1	1	1
s7	Festuca duriotagana Franco & Rocha Afonso	Poaceae	0	1	0	s9	Galium belizianum Ortega	Rubiaceae	1	0	1
s7	Festuca elegans Boiss.	Poaceae	1	0	0	s9	Galium broterianum Boiss. et Reut.	Rubiaceae	1	1	1
s7	Festuca indigesta Boiss. subsp. indigesta	Poaceae	1	0	0	s9	*Galium divaricatum* Lam.	Rubiaceae	1	0	1
s7	*Festuca nigrescens* Lam. subsp. microphylla (St-Yves) Markgr.-Dann.	Poaceae	1	0	0	s9	Galium lucidum All.	Rubiaceae	1	0	1
s7	Festuca rothmaleri (Litard.) Markgr.-Dann.	Poaceae	1	0	0	s9	*Galium mollugo* L. subsp. erectum Syme	Rubiaceae	0	1	0
s7	Festuca summilusitana Franco & Rocha Afonso	Poaceae	1	0	0	s9	*Galium mollugo* L. subsp. mollugo	Rubiaceae	1	0	1
s7	Fimbristylis bisumbellata (Forssk.) Bubani	Cyperaceae	0	1	0	s9	Galium murale (L.) All.	Rubiaceae	0	1	0
s7	Fuirena pubescens (Poir.) Kunth	Cyperaceae	0	0	1	s9	*Galium palustre* L. subsp. elongatum (K. Presl) Lange	Rubiaceae	0	1	0
s7	*Furcraea foetida* (L.) Haw.	Asparagaceae	0	0	1	s9	*Galium palustre* L. subsp. palustre	Rubiaceae	1	1	1
s7	*Gastridium ventricosum* (Gouan) Schinz et Thell.	Poaceae	1	1	1	s9	Galium papillosum Lapeyr. subsp. helodes (Hoffmanns. & Link) Ortega Oliv. & Devesa	Rubiaceae	1	1	1
s7	*Gaudinia fragilis* (L.) P. Beauv.	Poaceae	1	1	1	s9	*Galium parisiense* L. subsp. divaricatum (Pourr, ex Lam.) W. D. J. Koch	Rubiaceae	0	1	0
s7	Gennaria diphylla (Link) Parl.	Orchidaceae	0	0	1	s9	*Galium parisiense* L. subsp. parisiense	Rubiaceae	0	1	0
s7	*Gladiolus communis* L.	Iridaceae	1	1	1	s9	*Galium saxatile* L.	Rubiaceae	1	0	1
s7	*Gladiolus italicus* Mill.	Iridaceae	0	0	1	s9	*Galium verrucosum* Huds.	Rubiaceae	0	1	0
s7	*Glyceria declinata* Bréb.	Poaceae	1	1	0	s9	*Galium verum* L. subsp. verum	Rubiaceae	1	0	1
s7	*Hainardia cylindrica* (Willd.) Greuter	Poaceae	0	0	1	s9	Gentiana pneumonanthe L.	Gentianaceae	1	0	0

S	Species	Family	MB	CP	FS	S	Species	Family	MB	CP	FS
s7	Helictochloa marginata (Lowe) Romero Zarco	Poaceae	1	0	0	s9	Geropogon hybridus (L.) Sch.Bip.	Asteraceae	0	1	1
s7	Helictotrichon bromoides (Gouan) C.E. Hubb. Subsp. Bromoides	Poaceae	0	1	0	s9	Glandora prostrata (Loisel.) D.C. Thomas subsp. prostrata	Boraginaceae	1	1	1
s7	Helictotrichon marginatum (Lowe) Röser	Poaceae	0	1	0	s9	Glebionis coronaria (L.) Tzvelev	Asteraceae	0	0	1
s7	Holcus annuus C. A. Mey. Subsp. Annus	Poaceae	0	1	0	s9	Glossopappus macrotus (Durieu) Briq.	Asteraceae	0	0	1
s7	Holcus gayanus Boiss.	Poaceae	1	0	0	s9	Gnaphalium antillanum Urb.	Asteraceae	0	1	0
s7	*Holcus lanatus* L.	Poaceae	1	1	1	s9	*Gnaphalium uliginosum* L. subsp. uliginosum	Asteraceae	1	1	1
s7	*Holcus mollis* L.	Poaceae	1	1	0	s9	*Gomphocarpus fruticosus* (L.) W.T. Aiton	Apocynaceae	0	0	1
s7	Hordeum geniculatum All.	Poaceae	0	1	0	s9	*Gomphocarpus physocarpus* E.Mey.	Apocynaceae	0	0	1
s7	*Hordeum marinum* subsp. marinum Huds.	Poaceae	0	0	1	s9	Gratiola linifolia Vahl	Plantaginaceae	0	1	0
s7	*Hordeum murinum* L. subsp. leporinum (Link) Arcang.	Poaceae	0	1	1	s9	*Gratiola officinalis* L.	Plantaginaceae	0	1	0
s7	*Hordeum murinum* L. subsp. murinum	Poaceae	1	1	0	s9	*Hedera hibernica* (G. Kirchn.) Bean	Araliaceae	1	1	0
s7	*Hyacinthoides hispanica* (Mill.) Rothm.	Hyacinthaceae	1	1	1	s9	Hedera maderensis K.Koch ex A. Rutherf.	Araliaceae	0	1	0
s7	*Hyacinthoides non-scripta* (L.) Chouard ex Rothm.	Hyacinthaceae	1	0	0	s9	Hedypnois arenaria (Schousb.) DC.	Asteraceae	0	0	1
s7	Hyacinthoides paivae S. Ortiz & Rodr. Oubiña	Hyacinthaceae	1	0	0	s9	Hedypnois rhagadioloides (L.) F. W. Schmidt	Asteraceae	0	1	1
s7	*Hyparrhenia hirta* (L.) Stapf in Prain	Poaceae	0	1	1	s9	Helichrysum italicum (Roth) G. Don fil. in Loudon subsp. picardi (Boiss. & Reut.) Franco	Asteraceae	0	0	1
s7	*Imperata cylindrica* (L.) Raeusch.	Poaceae	0	1	1	s9	Helichrysum luteoalbum (L.) Rchb.	Asteraceae	1	0	0
s7	Iris albicans Lange	Iridaceae	0	0	1	s9	Helichrysum stoechas (L.) Moench	Asteraceae	1	1	1

S	Species	Family	MB	CP	FS	S	Species	Family	MB	CP	FS
s7	Iris boissieri Henriq.	Iridaceae	1	0	0	s9	*Heliotropium europaeum* L.	Boraginaceae	1	1	1
s7	*Iris foetidissima* L.	Iridaceae	0	0	1	s9	*Heliotropium supinum* L.	Boraginaceae	0	1	0
s7	*Iris germanica* L.	Iridaceae	1	1	1	s9	Helminthotheca comosa (Boiss.) Holub subsp. lusitanica (Schltdl.) P. Silva et Escud.	Asteraceae	0	1	1
s7	*Iris pseudacorus* L.	Iridaceae	1	1	1	s9	*Helminthotheca echioides* (L.) Holub	Asteraceae	0	0	1
s7	*Iris xiphium* L.	Iridaceae	0	1	1	s9	*Heracleum sphondylium* L. subsp. sphondylium	Apiaceae	1	1	0
s7	*Isolepis cernua* (Vahl) Roem. et Schult.	Cyperaceae	0	1	1	s9	Hieracium amplexicaule L.	Asteraceae	1	0	0
s7	*Isolepis fluitans* (C.B. Clarke) Muasya	Cyperaceae	1	0	0	s9	*Hieracium sabaudum* L. subsp. dumosum	Asteraceae	1	0	0
s7	*Isolepis pseudosetacea* (Daveau) Vasc.	Cyperaceae	0	0	1	s9	Hieracium umbellatum L.	Asteraceae	1	0	0
s7	*Isolepis setacea* (L.) R. Br.	Cyperaceae	1	1	0	s9	Hispidella hispanica Barnades ex Lam.	Asteraceae	1	0	0
s7	*Juncus acutiflorus* Hoffm. subsp. acutiflorus	Juncaceae	1	1	1	s9	*Hyoscyamus albus* L.	Solanaceae	0	0	1
s7	*Juncus acutus* L.	Juncaceae	0	0	1	s9	*Hypochaeris glabra* L.	Asteraceae	1	1	1
s7	*Juncus articulatus* L.	Juncaceae	1	1	1	s9	*Hypochaeris radicata* L. var. radicata	Asteraceae	1	1	1
s7	*Juncus bufonius* L.	Juncaceae	1	1	1	s9	*Ilex aquifolium* L.	Aquifoliaceae	1	0	0
s7	*Juncus bulbosus* L.	Juncaceae	1	1	0	s9	Inula conyza (Griess.) DC.	Asteraceae	0	1	0
s7	*Juncus capitatus* Weigel	Juncaceae	0	1	1	s9	*Ipomoea indica* (Burm.) Merr.	Convolvulaceae	0	0	1
s7	*Juncus conglomeratus* L.	Juncaceae	0	1	0	s9	Jacobaea minuta (Cav.) Pelser & Veldkamp	Asteraceae	0	1	1
s7	*Juncus effusus* L.	Juncaceae	1	1	1	s9	*Jacobaea vulgaris* Gaertn.	Asteraceae	1	0	0
s7	Juncus fontanesii Gay in Laharpe subsp. Fontanesii	Juncaceae	0	1	0	s9	*Jacobaea vulgaris* Gaertn.	Asteraceae	0	1	1
s7	Juncus heterophyllus Dufour	Juncaceae	1	1	0	s9	*Jasione montana* L.	Campanulaceae	1	1	1

DOI: http://dx.doi.org/10.5772/intechopen.1008900

S	Species	Family	MB	CP	FS	S	Species	Family	MB	CP	FS
s7	*Juncus hybridus* Brot.	Juncaceae	0	1	1	s9	Jasione sessiliflora Boiss. et Reut.	Campanulaceae	1	0	0
s7	*Juncus inflexus* L. var. inflexus	Juncaceae	0	1	1	s9	Jasminum fruticans L.	Oleaceae	0	1	1
s7	*Juncus maritimus* Lam.	Juncaceae	0	0	1	s9	Kickxia cirrhosa (L.) Fritsch	Plantaginaceae	0	1	1
s7	Juncus minutulus (Albert & Jahand.) Prain	Juncaceae	0	1	0	s9	Kickxia lanigera (Desf.) Hand.-Mazz.	Plantaginaceae	0	0	1
s7	Juncus pygmaeus Rich. in Thuill.	Juncaceae	0	1	1	s9	*Kickxia spuria* (L.) Dumort.	Plantaginaceae	0	0	1
s7	*Juncus squarrosus* L.	Juncaceae	1	1	0	s9	Klasea algarbiensis (Cantó) Cantó	Asteraceae	0	0	1
s7	*Juncus subnodulosus* Schrank	Juncaceae	0	1	1	s9	Klasea boetica (Boiss. ex DC.) Holub subsp. lusitanica (Cantó) Cantó & Rivas Mart.	Asteraceae	0	0	1
s7	Juncus subulatus Forssk.	Juncaceae	0	0	1	s9	Klasea flavescens (L.) Holub	Asteraceae	0	0	1
s7	Juncus tenageia L. f. subsp. tenageia	Juncaceae	1	1	0	s9	Klasea integrifolia (Vahl) Greuter subsp. monardii (Dufour) Cantó	Asteraceae	0	0	1
s7	Koeleria caudata (Link) Steud.	Poaceae	0	1	0	s9	Knautia nevadensis (M. Winkl. ex Szabó) Szabó var. nevadensis	Caprifoliaceae	1	0	0
s7	Koeleria crassipes Lange	Poaceae	0	1	0	s9	*Lactuca saligna* L.	Asteraceae	0	0	1
s7	*Lagurus ovatus* L.	Poaceae	0	0	1	s9	*Lactuca serriola* L.	Asteraceae	1	1	1
s7	*Lamarckia aurea* (L.) Moench	Poaceae	0	1	1	s9	Lactuca viminea (L.) J. et K. Presl subsp. chondrilliflora	Asteraceae	1	0	0
s7	*Leersia oryzoides* (L.) Sw.	Poaceae	0	1	0	s9	Lactuca viminea (L.) J.Presl & C.Presl subsp. viminea	Asteraceae	0	1	0
s7	Limodorum abortivum (L.) Sw.	Orchidaceae	0	1	1	s9	*Lactuca virosa* L.	Asteraceae	1	1	0
s7	*Lolium aristatum* (Willd.) Lang.	Poaceae	0	1	0	s9	*Lamium amplexicaule* L.	Lamiaceae	1	1	1
s7	*Lolium multiflorum* Lam.	Poaceae	1	0	0	s9	Lamium bifidum Cirillo	Lamiaceae	0	1	0
s7	*Lolium perenne* L.	Poaceae	1	1	0	s9	*Lamium maculatum* L.	Lamiaceae	1	1	0
s7	*Lolium rigidum* Gaudin	Poaceae	1	1	1	s9	*Lamium purpureum* L.	Lamiaceae	1	1	0

S	Species	Family	MB	CP	FS	S	Species	Family	MB	CP	FS
s7	*Lolium temulentum* L.	Poaceae	0	1	0	s9	*Lantana camara* L.	Verbenaceae	0	0	1
s7	*Luzula campestris* (L.) DC. in Lam. et DC.	Juncaceae	1	1	0	s9	Laphangium luteoalbum (L.) Tzvelev	Asteraceae	0	1	1
s7	*Luzula congesta* (Thuill.) Lej.	Juncaceae	0	1	0	s9	*Lapsana communis* L. subsp. communis	Asteraceae	1	1	0
s7	Luzula elegans Lowe	Juncaceae	0	1	0	s9	*Lavandula stoechas* L. subsp. luisieri (Rozeira) Rozeira in Agron. Lusit.	Lamiaceae	1	1	1
s7	*Luzula forsteri* (Sm.) DC. In Lam. & DC.	Juncaceae	1	1	0	s9	*Lavandula stoechas* L. subsp. stoechas	Lamiaceae	0	1	1
s7	Luzula lactea (Link) E.H.F. Meyer	Juncaceae	1	1	0	s9	Lavandula viridis L' Hér.	Lamiaceae	0	0	1
s7	*Luzula multiflora* (Retz.) Lej. Subsp multiflora	Juncaceae	1	1	0	s9	Legousia scabra (Lowe) Gamisans	Campanulaceae	0	1	0
s7	Luzula sylvatica (Hudson) Gaudin subsp. sylvatica	Juncaceae	1	1	0	s9	*Leontodon hispidus* L. subsp. hispidus	Asteraceae	1	0	0
s7	*Melica ciliata* L. subsp. magnolii (Gren. et Godr.) Husn.	Poaceae	0	1	1	s9	Leontodon longirostris (Vill.) Mérat	Asteraceae	1	0	1
s7	Melica minuta L. subsp. arrecta (Kunze) Breistr.	Poaceae	0	1	1	s9	*Leontodon saxatilis* Lam.	Asteraceae	1	0	0
s7	Merendera filifolia Cambess.	Colchicaceae	0	0	1	s9	Leontodon saxatilis subsp. rothii Maire	Asteraceae	0	1	0
s7	Merendera montana (L.) Lange in Willk. et Lange	Colchicaceae	1	0	0	s9	Leontodon taraxacoides Hoppe & Hornsch.	Asteraceae	0	1	0
s7	*Mibora minima* (L.) Desv.	Poaceae	1	1	0	s9	Leontodon tuberosus L.	Asteraceae	0	1	1
s7	Micropyrum patens (Brot.) Rothm. ex Pilg.	Poaceae	1	1	0	s9	Lepidophorum repandum (L.) DC.	Asteraceae	1	1	0
s7	Micropyrum tenellum (L.) Link	Poaceae	1	1	0	s9	Leucanthemopsis flaveola (Hoffmanns. et Link) Heywood subsp. Alpestris	Asteraceae	1	0	0
s7	*Molineriella laevis* (Brot.) Rouy	Poaceae	0	1	0	s9	Leucanthemopsis pulverulenta (Lag.) Heywood	Asteraceae	1	0	0
s7	Molineriella minuta (L.) Rouy subsp. australis (Paunero) Rivas Mart.	Poaceae	0	1	0	s9	Limbarda crithmoides (L.) Dumort.	Asteraceae	0	0	1

S	Species	Family	MB	CP	FS	S	Species	Family	MB	CP	FS
s7	*Molinia caerulea* (L.) Moench	Poaceae	1	0	0	s9	Linaria algarviana Chav.	Plantaginaceae	0	0	1
s7	Moraea sisyrinchium (L.) Ker Gawl.	Iridaceae	0	1	1	s9	Linaria amethystea (Lam.) Hoffmanns. et Link subsp. amethystea	Plantaginaceae	0	1	1
s7	*Muscari comosum* (L.) Mill.	Hyacinthaceae	1	1	1	s9	Linaria amethystea (Lam.) Hoffmanns. et Link subsp. multipunctacta (Brot.) Chater et D. A. Webb	Plantaginaceae	0	0	1
s7	*Muscari neglectum*	Hyacinthaceae	0	1	1	s9	Linaria elegans Cav.	Plantaginaceae	1	1	0
s7	Narcissus asturiensis (Jord.) Pugsley	Amaryllidaceae	1	0	0	s9	*Linaria incarnata* (Vent.) Spreng.	Plantaginaceae	0	1	0
s7	*Narcissus bulbocodium* L. subsp. bulbocodium	Amaryllidaceae	1	1	1	s9	Linaria munbyana Boiss. & Reut.	Plantaginaceae	0	0	1
s7	Narcissus calcicola Mendonça	Amaryllidaceae	0	0	1	s9	Linaria oblongifolia subsp. haenseleri (Boiss. & Reut.) Valdés	Plantaginaceae	0	1	1
s7	Narcissus gaditanus Boiss. & Reut.	Amaryllidaceae	0	0	1	s9	Linaria pedunculata (L.) Chaz.	Plantaginaceae	0	0	1
s7	*Narcissus jonquilla* L.	Amaryllidaceae	0	1	1	s9	Linaria polygalifolia Hoffmanns. & Link subsp. lamarckii (Rouy) D.A. Sutton	Plantaginaceae	0	0	1
s7	Narcissus obesus Salisb.	Amaryllidaceae	0	0	1	s9	Linaria saxatilis (L.) Chaz.	Plantaginaceae	1	0	0
s7	*Narcissus papyraceus* Ker Gawl.	Amaryllidaceae	0	0	1	s9	*Linaria spartea* (L.) Willd.	Plantaginaceae	1	1	1
s7	*Narcissus pseudonarcissus* L. subsp. major (Curtis) Baker	Amaryllidaceae	0	1	0	s9	Linaria triornithophora (L.) Willd.	Plantaginaceae	1	1	0
s7	*Narcissus pseudonarcissus* L. subsp. nobilis (Haw.) A. Fern.	Amaryllidaceae	1	0	0	s9	Linaria viscosa (L.) Dum. Cours.	Plantaginaceae	0	0	1
s7	*Narcissus pseudonarcissus* L. subsp. portensis (Pugsley) A. Fern.	Amaryllidaceae	1	0	0	s9	Lobelia urens L.	Campanulaceae	1	1	0
s7	Narcissus rupicola Dufour ex Schult. & Schult.f.	Amaryllidaceae	1	1	0	s9	*Logfia gallica* (L.) Coss. & Germ.	Asteraceae	1	1	1
s7	Narcissus serotinus L.	Amaryllidaceae	0	0	1	s9	*Logfia minima* (Sm.) Dumort.	Asteraceae	1	1	0

S	Species	Family	MB	CP	FS	S	Species	Family	MB	CP	FS
s7	*Narcissus tazetta* L.	Amaryllidaceae	0	0	1	s9	Lomelosia simplex (Desf.) Raf.	Caprifoliaceae	0	0	1
s7	Narcissus triandrus L. subsp. pallidus	Amaryllidaceae	1	1	0	s9	*Lomelosia stellata* (L.) Raf.	Caprifoliaceae	0	0	1
s7	Narcissus triandrus L. subsp. triandrus	Amaryllidaceae	1	0	0	s9	*Lonicera etrusca* Santi	Caprifoliaceae	0	1	1
s7	*Nardus stricta* L.	Poaceae	1	0	0	s9	Lonicera implexa Aiton	Caprifoliaceae	0	1	1
s7	Neotinea tridentata (Scop.) R.M.Bateman, Pridgeon & M.W.Chase subsp. conica (Willd.) R.M.Bateman, Pridgeon & M.W. Chase	Orchidaceae	0	1	1	s9	*Lonicera periclymenum* L. subsp. hispanica (Boiss. et Reut.) Nyman	Caprifoliaceae	1	1	0
s7	*Nothoscordum gracile* (Dryand. ex Aiton) Stearn	Amaryllidaceae	0	0	1	s9	*Lonicera periclymenum* L. subsp. peryclimenum	Caprifoliaceae	1	0	0
s7	Ophrys apifera Huds.	Orchidaceae	0	0	1	s9	Lycium europaeum	Solanaceae	0	0	1
s7	Ophrys bombyliflora Link	Orchidaceae	0	0	1	s9	*Lycopus europaeus* L.	Lamiaceae	1	1	1
s7	Ophrys fusca Link	Orchidaceae	0	0	1	s9	Lysimachia nemorum L.	Primulaceae	1	0	0
s7	Ophrys lutea Cav.	Orchidaceae	0	0	1	s9	*Lysimachia vulgaris* L.	Primulaceae	1	0	1
s7	Ophrys scolopax Cav.	Orchidaceae	0	0	1	s9	Magydaris panacifolia (Vahl) Lange in Willk. & Lange	Apiaceae	0	1	1
s7	Ophrys speculum Link in Schrad. Subsp. speculum	Orchidaceae	0	0	1	s9	*Mandragora autumnalis* Bertol.	Solanaceae	0	0	1
s7	Ophrys speculum Link subsp. lusitanica O. Danesch & E.Danesch	Orchidaceae	0	0	1	s9	*Mantisalca salmantica* (L.) Briq. et Cavill.	Asteraceae	0	1	0
s7	Ophrys tenthredinifera Willd.	Orchidaceae	0	1	0	s9	Margotia gummifera (Desf.) Lange	Apiaceae	0	1	1
s7	Ophrys tenthredinifera Willd.	Orchidaceae	0	0	1	s9	*Marrubium vulgare* L.	Lamiaceae	0	1	1
s7	Orchis italica Poir.	Orchidaceae	0	1	1	s9	*Matricaria discoidea* DC.	Asteraceae	1	0	1
s7	Orchis mascula K. Richt. Subsp. laxifloriformis Rivas Goday & B.Rodr.	Orchidaceae	0	1	0	s9	Melampyrum pratense L.	Orobanchaceae	1	0	0
s7	*Orchis morio* Loisel.	Orchidaceae	0	1	1	s9	*Melissa officinalis* L.	Lamiaceae	0	1	0

S	Species	Family	MB	CP	FS	S	Species	Family	MB	CP	FS
s7	Ornithogalum broteri M. Laínz	Hyacinthaceae	0	1	1	s9	Melittis melissophyllum L.	Lamiaceae	1	0	0
s7	Ornithogalum concinnum (Salisb.) P. Cout.	Hyacinthaceae	1	1	0	s9	*Mentha aquatica* L.	Lamiaceae	0	1	1
s7	Ornithogalum narbonense L.	Hyacinthaceae	0	1	1	s9	Mentha cervina L.	Lamiaceae	0	1	0
s7	Ornithogalum orthophyllum Ten. Subsp. baeticum (Boiss.) Zahar.	Hyacinthaceae	0	1	1	s9	*Mentha pulegium* L.	Lamiaceae	1	1	1
s7	*Ornithogalum pyrenaicum* L. subsp. pyrenaicum	Hyacinthaceae	1	1	1	s9	*Mentha suaveolens* Ehrh.	Lamiaceae	1	1	1
s7	*Pancratium maritimum* L.	Amaryllidaceae	0	0	1	s9	*Menyanthes trifoliata* L.	Menyanthaceae	1	0	0
s7	*Panicum repens* L.	Poaceae	0	0	1	s9	Micromeria graeca (L.) Benth. ex Rchb.	Lamiaceae	0	0	1
s7	Parapholis cylindrica (Willd.) Romero Zarco	Poaceae	0	1	0	s9	Micropus supinus L.	Asteraceae	0	0	1
s7	*Parapholis filiformis* (Roth) C.E.Hubb.	Poaceae	0	0	1	s9	Misopates calycinum (Lange) Rothm.	Plantaginaceae	0	0	1
s7	*Parapholis incurva* (L.) C.E.Hubb.	Poaceae	0	0	1	s9	*Misopates orontium* (L.) Raf.	Plantaginaceae	0	1	1
s7	*Paspalum dilatatum* Poir. in Lam.	Poaceae	0	1	1	s9	Myosotis debilis Pomel	Boraginaceae	0	1	0
s7	*Paspalum distichum* L.	Poaceae	0	1	1	s9	*Myosotis discolor* Pers. subsp. discolor	Boraginaceae	0	1	0
s7	*Paspalum vaginatum* Sw.	Poaceae	0	0	1	s9	*Myosotis discolor* Pers. subsp. dubia (Arrond.) Blaise	Boraginaceae	1	1	1
s7	*Pennisetum villosum* R.Br. ex Fresen.	Poaceae	0	0	1	s9	Myosotis personii Rouy	Boraginaceae	0	1	0
s7	Periballia involucrata (Cav.) Janka	Poaceae	0	1	0	s9	Myosotis secunda A. Murray	Boraginaceae	1	1	1
s7	*Phalaris coerulescens* subsp. coerulescens Desf.	Poaceae	0	1	1	s9	Myosotis stolonifera (DC.) Leresche et Levier subsp. stolonifera	Boraginaceae	1	0	1
s7	*Phalaris minor* Retz.	Poaceae	0	1	1	s9	Myosotis welwitschii Boiss. et Reut.	Boraginaceae	1	1	1
s7	*Phalaris paradoxa* L.	Poaceae	0	1	1	s9	Myrrhoides nodosa (L.) Cannon	Apiaceae	0	1	0
s7	*Phleum pratense* L.	Poaceae	0	1	0	s9	Nepeta tuberosa L.	Lamiaceae	0	0	1
s7	*Phragmites australis* (Cav.) Trin ex.Steud.	Poaceae	0	1	1	s9	*Nerium oleander* L.	Apocynaceae	0	0	1

S	Species	Family	MB	CP	FS	S	Species	Family	MB	CP	FS
s7	*Piptatherum miliaceum* subsp. miliaceum (L.) Coss.	Poaceae	0	0	1	s9	*Nicotiana glauca* Graham	Solanaceae	0	0	1
s7	*Poa annua* L.	Poaceae	1	1	1	s9	Nothobartsia asperrima (Link) Benedí & Herrero	Orobanchaceae	0	0	1
s7	*Poa bulbosa* L.	Poaceae	1	1	1	s9	Notobasis syriaca (L.) Cass.	Asteraceae	0	0	1
s7	*Poa infirma* Kunth	Poaceae	0	1	0	s9	*Odontites vernus* (Bellardi) Dumort.	Orobanchaceae	1	0	0
s7	*Poa pratensis* L.	Poaceae	0	1	0	s9	Oenanthe crocata L.	Apiaceae	1	1	1
s7	*Poa supina* Schrad.	Poaceae	1	0	0	s9	Oenanthe lachenalii C. C. Gmel.	Apiaceae	0	0	1
s7	*Poa trivialis* L. subsp. trivialis	Poaceae	1	1	0	s9	*Olea europaea* L.	Oleaceae	0	1	0
s7	Polygonatum odoratum (Mill.) Druce	Asparagaceae	1	1	0	s9	*Olea europaea* L. subsp. sylvestris (Mill.) Lehr	Oleaceae	1	1	1
s7	*Polypogon maritimus* Willd. subsp. maritimus	Poaceae	0	1	1	s9	*Omphalodes linifolia* (L.) Moench	Boraginaceae	0	1	0
s7	*Polypogon monspeliensis* (L.) Desf.	Poaceae	0	0	1	s9	Omphalodes nitida Hoffmanns. et Link	Boraginaceae	1	1	1
s7	*Polypogon viridis* (Gouan) Breistr.	Poaceae	0	1	0	s9	*Origanum vulgare* L. subsp. virens (Hoffmanns. & Link) Bonnier & Layens	Lamiaceae	1	1	1
s7	Pontederia crassipes Mart.	Ponteridaceae	0	0	1	s9	Orlaya grandiflora (L.) Hoffm.	Apiaceae	0	0	1
s7	Pseudarrhenatherum longifolium (Thore) Rouy	Poaceae	1	0	0	s9	Orobanche gracilis Sm.	Orobanchaceae	0	1	1
s7	Psilurus incurvus (Gouan) Schinz et Thell.	Poaceae	0	1	1	s9	*Orobanche hederae* Vaucher ex Duby	Orobanchaceae	0	0	1
s7	Puccinellia iberica (Wolley-Dod) Tzvelev	Poaceae	0	0	1	s9	Orobanche latisquama (F.W. Schultz) Batt. in Batt. et Trab.	Orobanchaceae	0	1	0
s7	*Pycreus flavescens* (L.) P. Beauv. ex Rchb.	Cyperaceae	0	1	1	s9	*Orobanche minor* Sm.	Orobanchaceae	0	1	1
s7	Romulea bulbocodium (L.) Sebast. et Mauri var. bulbocodium	Iridaceae	1	1	1	s9	*Orobanche ramosa* L. subsp. nana (Reut.) P. Cout.	Orobanchaceae	0	1	0
s7	Romulea columnae Sebast. et Mauri	Iridaceae	0	0	1	s9	*Orobanche ramosa* L. subsp. ramosa	Orobanchaceae	0	1	1

S	Species	Family	MB	CP	FS	S	Species	Family	MB	CP	FS
s7	Romulea ramiflora Ten. subsp. ramiflora	Iridaceae	0	1	1	s9	Orobanche rapum-genistae Thuill.	Orobanchaceae	1	1	1
s7	*Rostraria cristata* (L.) Tzvelev	Poaceae	0	0	1	s9	Orobanche schultzii Mutel	Orobanchaceae	0	0	1
s7	Ruscus aculeatus L.	Asparagaceae	1	1	1	s9	Otanthus maritimus (L.) Hoffmanns. & Link	Asteraceae	0	0	1
s7	*Schoenoplectus lacustris* (L.) Palla subsp. lacustris	Cyperaceae	0	1	1	s9	Pallenis spinosa (L.) Schulz Bip. subsp. spinosa	Asteraceae	1	1	1
s7	*Schoenus nigricans* L.	Cyperaceae	0	0	1	s9	*Parentucellia latifolia* (L.) Caruel	Orobanchaceae	0	1	0
s7	Scilla autumnalis L.	Hyacinthaceae	1	0	1	s9	*Parentucellia viscosa* (L.) Caruel	Plantaginaceae	1	1	1
s7	Scilla hyacinthoides L.	Hyacinthaceae	0	1	0	s9	*Pedicularis sylvatica* L. subsp. lusitanica (Hoffmann. & Link) Cout.	Orobanchaceae	1	0	0
s7	Scilla monophyllos Link in Schrad.	Hyacinthaceae	1	0	1	s9	*Pentaglottis sempervirens* (L.) Taush ex L. H. Bailey	Boraginaceae	1	0	1
s7	Scilla peruviana L.	Hyacinthaceae	0	0	1	s9	Petasites fragrans (Vill.) C. Presl	Asteraceae	0	1	0
s7	Scilla verna Huds. subsp. ramburei (Boiss.) K. Richt.	Hyacinthaceae	1	0	0	s9	Peucedanum gallicum Latourr.	Apiaceae	1	0	0
s7	Scilla verna Huds. subsp. verna	Hyacinthaceae	1	0	0	s9	Peucedanum lancifolium Lange	Apiaceae	1	0	0
s7	*Scirpoides holoschoenus* (L.) Sojak	Cyperaceae	1	0	0	s9	Phagnalon rupestre (L.) DC.	Asteraceae	0	0	1
s7	*Scirpoides holoschoenus* (L.) Sojak	Cyperaceae	0	1	1	s9	Phagnalon saxatile (L.) Cass.	Asteraceae	1	1	1
s7	Serapias cordigera L.	Orchidaceae	1	1	1	s9	Phalacrocarpum oppositifolium (Brot.) Willk.	Asteraceae	1	0	0
s7	Serapias lingua L.	Orchidaceae	1	1	1	s9	Phillyrea angustifolia L.	Oleaceae	1	1	1
s7	Serapias parviflora Parl.	Orchidaceae	0	1	1	s9	Phillyrea latifolia L.	Oleaceae	0	1	1
s7	Serapias strictiflora Welw. ex Veiga	Orchidaceae	0	1	1	s9	Phlomis lychnitis L.	Lamiaceae	0	1	1
s7	*Setaria adhaerens* (Forssk.) Chiov.	Poaceae	0	1	1	s9	Phlomis purpurea L.	Lamiaceae	0	0	1
s7	*Setaria faberi* Herrm.	Poaceae	0	1	0	s9	Physospermum cornubiense (L.) DC.	Apiaceae	1	1	0

S	Species	Family	MB	CP	FS	S	Species	Family	MB	CP	FS
s7	*Setaria italica* (L.) P. Beauv.	Poaceae	0	1	0	s9	*Picris echioides* L.	Asteraceae	0	1	0
s7	*Setaria pumila* (Poir.) Roem. et Schult.	Poaceae	0	1	1	s9	*Picris hieracioides* L.	Asteraceae	1	1	0
s7	*Setaria verticillata* (L.) P. Beauv.	Poaceae	0	1	0	s9	Pilosella pseudopilosella (Ten.) Soják	Asteraceae	1	0	0
s7	*Setaria viridis* (L.) P. Beauv.	Poaceae	0	1	0	s9	Pimpinella villosa Schousb.	Apiaceae	0	1	1
s7	Simethis mattiazzi (Vand.) Sacc.	Xanthorrhoeaceae	1	1	0	s9	Pinguicula lusitanica L.	Lentibulariaceae	1	0	1
s7	*Sorghum halepense* (L.) Pers.	Poaceae	0	1	0	s9	Plantago afra L.	Plantaginaceae	0	1	1
s7	*Sparganium emersum* Rehmann	Typhaceae	1	0	1	s9	Plantago albicans L.	Plantaginaceae	0	0	1
s7	*Sparganium erectum* L. subsp. oocarpum (Čelak.) Domin	Typhaceae	0	1	0	s9	Plantago algarbiensis Samp.	Plantaginaceae	0	0	1
s7	Sphenopus divaricatus (Gouan) Rchb.	Poaceae	0	0	1	s9	Plantago bellardii All.	Plantaginaceae	0	1	1
s7	Spiranthes spiralis (L.) Chevall.	Orchidaceae	0	0	1	s9	*Plantago coronopus* L. subsp. coronopus	Plantaginaceae	1	1	1
s7	Sporobolus maritimus (Curtis) P.M.Peterson & Saarela	Poaceae	0	0	1	s9	*Plantago coronopus* subsp. macrorhiza (Poir.) Arcang.	Plantaginaceae	0	0	1
s7	Sporobolus montevidensis (Arechav.) P.M. Peterson & Saarela	Poaceae	0	0	1	s9	Plantago holosteum Scop.	Plantaginaceae	1	0	0
s7	Sporobolus pumilus (Roth) P.M.Peterson & Saarela	Poaceae	0	0	1	s9	Plantago lagopus L.	Plantaginaceae	0	1	1
s7	Sporobolus pungens (Schreb.) Kunth	Poaceae	0	0	1	s9	*Plantago lanceolata* L.	Plantaginaceae	1	1	1
s7	*Stenotaphrum secundatum* (Walter) Kuntze	Poaceae	0	0	1	s9	*Plantago major* L.	Plantaginaceae	1	1	1
s7	*Stipa capensis* Thunb.	Poaceae	0	1	1	s9	Plantago serraria L.	Plantaginaceae	0	0	1
s7	*Stipa gigantea* Link	Poaceae	1	1	1	s9	Prangos trifida (Mill.) Herrnst. et Heyn	Apiaceae	0	0	1
s7	*Stipa tenacissima* L.	Poaceae	0	0	1	s9	Prasium majus L.	Lamiaceae	0	0	1
s7	*Taeniatherum caput-medusae* (L.) Nevski	Poaceae	0	1	1	s9	Primula acaulis (L.) L. subs. Acaulis	Primulaceae	1	0	0

S	Species	Family	MB	CP	FS	S	Species	Family	MB	CP	FS
s7	*Thinopyrum elongatum* (Host) D.R.Dewey	Poaceae	0	0	1	s9	Prunella grandiflora (L.) Scholler	Lamiaceae	1	0	0
s7	*Tradescantia fluminensis* Velloso	Commelinaceae	1	1	1	s9	*Prunella vulgaris* L.	Lamiaceae	1	1	0
s7	*Trichophorum cespitosum* (L.) Hartm.	Cyperaceae	1	0	0	s9	Pseudorlaya minuscula (Pau) M. Laínz	Apiaceae	0	0	1
s7	Trisetaria dufourei (Boiss.) Paunero	Poaceae	0	0	1	s9	Pseudorlaya pumila (L.) Grande	Apiaceae	0	0	1
s7	Trisetaria ovata (Cav.) Paunero	Poaceae	1	0	0	s9	Pterocephalidium diandrum (Lag.) G. López	Caprifoliaceae	0	0	1
s7	Trisetaria panicea (Lam.) Paunero	Poaceae	0	0	1	s9	*Pulicaria arabica* (L.) Cass. Subsp. hispanica (Boiss.) Murb.	Asteraceae	0	1	1
s7	*Typha angustifolia* L.	Typhaceae	0	1	0	s9	*Pulicaria dysenterica* (L.) Bernh.	Asteraceae	0	0	1
s7	*Typha domingensis* (Pers.) Steud.	Typhaceae	0	1	1	s9	Pulicaria odora (L.) Rchb.	Asteraceae	0	1	1
s7	*Typha latifolia* L.	Typhaceae	1	1	1	s9	Pulmonaria longifolia (Bastard) Boreau	Boraginaceae	1	0	1
s7	Vulpia alopecuros subsp. alopecuros (Schousb.) Dumort.	Poaceae	0	0	1	s9	Pycnocomon intermedium (Lag.) Greuter & Burdet	Caprifoliaceae	0	0	1
s7	*Vulpia bromoides* (L.) Gray	Poaceae	1	1	1	s9	Pycnocomon rutifolium (Vahl) Hoffmanns. & Link	Caprifoliaceae	0	0	1
s7	*Vulpia ciliata* Dumort.	Poaceae	0	1	0	s9	Reichardia gaditana (Willk.) Samp.	Asteraceae	0	0	1
s7	*Vulpia geniculata* (L.) Link	Poaceae	1	1	1	s9	Reichardia intermedia (Sch.Bip.) Samp.	Asteraceae	0	1	1
s7	Vulpia membranacea (L.) Dumort.	Poaceae	0	1	0	s9	*Reichardia picroides* (L.) Roth	Asteraceae	0	0	1
s7	*Vulpia muralis* (Kunth) Nees	Poaceae	1	1	0	s9	*Rhagadiolus edulis* Gaertn.	Asteraceae	0	1	0
s7	*Vulpia myuros* (L.) C. C. Gmel.	Poaceae	1	1	0	s9	*Rhagadiolus stellatus* (L.) Gaertn.	Asteraceae	0	1	1
s8	*Acacia dealbata* Link	Fabaceae	1	1	1	s9	Rhaponticum coniferum (L.) Greuter	Asteraceae	0	0	1
s8	*Acacia longifolia* (Andrews) Willd.	Fabaceae	0	0	1	s9	*Rhinanthus minor* L.	Orobanchaceae	1	0	0

S	Species	Family	MB	CP	FS	S	Species	Family	MB	CP	FS
s8	*Acacia melanoxylon* R. Br. in W. T. Aiton	Fabaceae	1	1	0	s9	Ridolfia segetum (L.) Moris	Apiaceae	0	0	1
s8	*Acacia pycnantha* Benth.	Fabaceae	0	1	1	s9	*Rosmarinus officinalis* L. var. officinalis	Lamiaceae	0	1	1
s8	*Acacia retinodes* Schltdl.	Fabaceae	0	1	0	s9	*Rubia peregrina* L. subsp. peregrina	Rubiaceae	1	1	1
s8	*Acacia saligna* (Labill.) H.L.Wendl.	Fabaceae	0	0	1	s9	Salvia sclareoides Brot.	Lamiaceae	0	1	1
s8	Acer monspessulanum L.	Sapindaceae	0	1	0	s9	*Salvia verbenaca* L.	Lamiaceae	0	1	1
s8	*Acer negundo* L.	Sapindaceae	0	1	0	s9	*Sambucus nigra* L.	Adoxaceae	1	0	0
s8	*Acer pseudoplatanus* L.	Sapindaceae	1	0	1	s9	*Samolus valerandi* L.	Primulaceae	0	0	1
s8	Adenocarpus complicatus (L.) J. Gay in Durieu	Fabaceae	1	1	0	s9	*Scabiosa atropurpurea* L.	Caprifoliaceae	0	0	1
s8	Adenocarpus lainzii (Castrov.) Castrov.	Fabaceae	1	1	0	s9	*Scabiosa columbaria* L. subsp. columbaria	Caprifoliaceae	1	0	0
s8	*Agrimonia eupatoria* L. sbsp. Eupatoria	Rosaceae	0	1	1	s9	Scabiosa galianoi Devesa, Ortega Oliv. & J.López	Caprifoliaceae	0	0	1
s8	*Ailanthus altissima* (Mill.) Swingle	Simaroubaceae	0	1	1	s9	Scandix australis L.	Apiaceae	0	0	1
s8	*Alcea rosea* L.	Malvaceae	0	1	0	s9	*Scandix pecten-veneris* L.	Apiaceae	0	1	1
s8	*Alliaria petiolata* (M. Bieb.) Cavara et Grande	Brassicaceae	1	1	1	s9	*Schenkia spicata* (L.) G. Mans.	Gentianaceae	0	0	1
s8	*Alnus glutinosa* (L.) Gaertn.	Betulaceae	1	1	1	s9	*Scolymus hispanicus* L.	Asteraceae	0	1	1
s8	Alyssum granatense Boiss. et Reut.	Brassicaceae	0	1	0	s9	*Scolymus maculatus* L.	Asteraceae	0	0	1
s8	*Alyssum simplex* Rudolphi	Brassicaceae	0	1	0	s9	Scorzonera angustifolia L. var. angustifolia	Asteraceae	0	1	1
s8	Anagyris foetida L.	Fabaceae	0	0	1	s9	Scorzonera humilis L.	Asteraceae	1	0	0
s8	*Anthyllis vulneraria* L. subsp. iberica (W. Becker) Jalas ex Cullen	Fabaceae	0	0	1	s9	*Scorzonera laciniata* Jacq.	Asteraceae	0	0	1
s8	*Aphanes arvensis* L.	Rosaceae	0	1	0	s9	*Scrophularia auriculata* L.	Plantaginaceae	1	1	1
s8	*Aphanes australis* Rydb.	Rosaceae	1	1	0	s9	*Scrophularia canina* L.	Plantaginaceae	1	1	0

S	Species	Family	MB	CP	FS	S	Species	Family	MB	CP	FS
s8	Aphanes cornucopioides Lag.	Rosaceae	0	1	0	s9	Scrophularia herminii Hoffmanns. et Link	Plantaginaceae	1	0	0
s8	*Arabidopsis thaliana* (L.) Heynh.	Brassicaceae	1	1	1	s9	Scrophularia lyrata Willd.	Scrophulariaceae	0	1	0
s8	*Argyrolobium zanonii* (Turra) P.W.Ball	Fabaceae	0	0	1	s9	Scrophularia sambucifolia L.	Scrophulariaceae	0	1	0
s8	*Astragalus boeticus* L.	Fabaceae	0	0	1	s9	Scrophularia scorodonia L.	Plantaginaceae	1	0	0
s8	Astragalus cymbaecarpos Brot.	Fabaceae	0	1	0	s9	*Scutellaria minor* Huds.	Lamiaceae	1	1	0
s8	*Astragalus echinatus* Murray	Fabaceae	0	0	1	s9	Senecio angulatus L. f.	Asteraceae	0	0	1
s8	*Astragalus epiglottis* L.	Fabaceae	0	0	1	s9	Senecio gallicus Vill. in Chaix	Asteraceae	1	0	1
s8	*Astragalus hamosus* L.	Fabaceae	0	1	1	s9	Senecio lividus L.	Asteraceae	1	1	1
s8	Astragalus pelecinus (L.) Barneby subsp. pelecinus	Fabaceae	0	1	1	s9	*Senecio sylvaticus* L.	Asteraceae	1	1	0
s8	*Barbarea vulgaris* R. Br.	Brassicaceae	0	1	0	s9	*Senecio vulgaris* L.	Asteraceae	1	1	1
s8	*Betula pubescens* Ehrh. var. glabrata	Betulaceae	1	0	1	s9	*Serratula tinctoria* L. subsp. Seoanei	Asteraceae	1	0	0
s8	Biscutella valentina (Loefl. ex L.) Heywood	Brassicaceae	0	1	0	s9	*Sherardia arvensis* L.	Rubiaceae	1	1	1
s8	*Bituminaria bituminosa* (L.) C.H.Stirt.	Fabaceae	0	1	1	s9	Sibthorpia europaea L.	Plantaginaceae	0	1	0
s8	Brassica barrelieri (L.) Janka	Brassicaceae	1	1	1	s9	Sideritis arborescens Salzm. ex Benth.	Lamiaceae	0	0	1
s8	*Calepina irregularis* (Asso) Thell.	Brassicaceae	0	1	0	s9	*Sideritis romana* L.	Lamiaceae	0	0	1
s8	Calicotome villosa (Poir.) Link	Fabaceae	0	1	0	s9	*Silybum marianum* (L.) Gaertn.	Asteraceae	1	1	1
s8	*Capsella bursa-pastoris* (L.) Medik.	Brassicaceae	1	1	1	s9	*Smyrnium olusatrum* L.	Apiaceae	0	1	1
s8	*Cardamine flexuosa* With.	Brassicaceae	1	1	1	s9	Smyrnium perfoliatum L.	Apiaceae	0	0	1
s8	*Cardamine hirsuta* L.	Brassicaceae	1	1	1	s9	*Solanum dulcamara* L.	Solanaceae	1	1	0
s8	*Cardamine pratensis* L. subsp. pratensis	Brassicaceae	1	0	1	s9	*Solanum linnaeanum* Hepper et P.-M. L. Jaeger	Solanaceae	0	0	1
s8	*Cardaria draba* (L.) Desv. Subsp. draba	Brassicaceae	0	1	0	s9	*Solanum nigrum* L.	Solanaceae	1	1	1

S	Species	Family	MB	CP	FS	S	Species	Family	MB	CP	FS
s8	*Castanea sativa* Mill.	Fagaceae	1	0	0	s9	*Solanum villosum* Mill.	Solanaceae	0	1	1
s8	*Celtis australis* L.	Cannabaceae	0	1	0	s9	Solenopsis laurentia (L.) C. Presl	Campanulaceae	0	1	1
s8	*Ceratonia siliqua* L.	Fabaceae	0	0	1	s9	Solidago virgaurea L.	Asteraceae	1	0	0
s8	*Chamaesyce canescens* (L.) Prokh.	Euphorbiaceae	0	1	1	s9	*Soliva stolonifera* (Brot.) R. Br. ex G. Don f.	Asteraceae	0	1	0
s8	*Chamaesyce maculata* (L.) Small	Euphorbiaceae	0	1	1	s9	*Sonchus asper* (L.) Hill subsp. asper	Asteraceae	1	1	1
s8	*Chamaesyce nutans* (Lag.) Small	Euphorbiaceae	0	0	1	s9	*Sonchus asper* (L.) Hill subsp. glaucescens (Jord.) Ball	Asteraceae	0	1	0
s8	*Chrozophora tinctoria* (L.) Rafin.	Euphorbiaceae	0	1	1	s9	Sonchus maritimus L.	Asteraceae	0	0	1
s8	Cistus albidus L.	Cistaceae	0	1	1	s9	*Sonchus oleraceus* L.	Asteraceae	1	1	1
s8	Cistus crispus L.	Cistaceae	0	1	1	s9	*Sonchus tenerrimus* L.	Asteraceae	0	1	1
s8	Cistus inflatus Pourr. ex J.-P. Denzoly	Cistaceae	1	1	0	s9	*Stachys arvensis* L.	Lamiaceae	1	1	1
s8	*Cistus ladanifer* L. subsp. ladanifer	Cistaceae	1	1	1	s9	*Stachys germanica* L.	Lamiaceae	0	0	1
s8	Cistus libanotis L.	Cistaceae	0	0	1	s9	*Stachys ocymastrum* (L.) Briq.	Lamiaceae	0	1	1
s8	*Cistus monspeliensis* L.	Cistaceae	0	0	1	s9	*Stachys officinalis* (L.) Trevis. subsp. officinalis	Lamiaceae	1	1	0
s8	Cistus populifolius L subsp. populifolius	Cistaceae	1	1	1	s9	Staehelina dubia L.	Asteraceae	0	0	1
s8	*Cistus psilosepalus* Sweet	Cistaceae	1	1	1	s9	Succisa pinnatifida Lange	Caprifoliaceae	1	0	0
s8	Cistus salvifolius L.	Cistaceae	1	1	1	s9	*Succisa pratensis* Moench	Caprifoliaceae	1	0	0
s8	Cleome violacea L.	Cleomaceae	0	1	1	s9	*Symphytum officinale* L.	Boraginaceae	0	1	0
s8	*Coincya monensis* (L.) Greuter et Burdet subsp. cheiranthos (Vill.) Aedo, Leadlay et Muñoz Garm.	Brassicaceae	1	1	1	s9	Tanacetum annuum L.	Asteraceae	0	0	1
s8	Coronilla glauca L.	Fabaceae	0	0	1	s9	*Tanacetum corymbosum* (L.) Sch. Bip.	Asteraceae	1	0	0
s8	*Coronilla juncea* L.	Fabaceae	0	0	1	s9	*Tanacetum parthenium* (L.) Sch. Bip.	Asteraceae	1	1	0

S	Species	Family	MB	CP	FS	S	Species	Family	MB	CP	FS
s8	*Coronilla repanda* (Poir.) Guss. Subsp. dura (Cav.) P. Cout.	Fabaceae	0	1	0	s9	*Tanacetum vulgare* L.	Asteraceae	1	0	0
s8	*Coronilla repanda* (Poir.) Guss. subsp. repanda	Fabaceae	0	0	1	s9	Taraxacum braun-blanquetii Soest	Asteraceae	1	0	0
s8	*Coronilla scorpioides* (L.) W.D.J.Koch	Fabaceae	0	0	1	s9	Taraxacum nordstedtii Dahlst.	Asteraceae	1	0	0
s8	*Corylus avellana* L.	Betulaceae	1	1	1	s9	Taraxacum panalpinum Soest	Asteraceae	1	0	0
s8	*Crambe hispanica* L. subsp. glabrata (DC.) P. Cout.	Brassicaceae	0	1	0	s9	Teucrium algarbiense (Cout.) Cout.	Lamiaceae	0	0	1
s8	*Crataegus monogyna* Jacq.	Rosaceae	1	1	1	s9	Teucrium capitatum L.	Lamiaceae	0	0	1
s8	Cytinus hypocistis (L.) L. subsp. macranthus Wettst.	Cytinaceae	1	0	1	s9	*Teucrium fruticans* L.	Lamiaceae	0	1	0
s8	Cytinus ruber Fourr. ex Fritsch	Cytinaceae	0	0	1	s9	Teucrium haenseleri Boiss.	Lamiaceae	0	0	1
s8	Cytisus arboreus (Desf.) DC. subsp. baeticus (Webb) Maire	Fabaceae	0	0	1	s9	Teucrium pseudochamaepitys L.	Lamiaceae	0	0	1
s8	Cytisus grandiflorus (Brot.) DC. subsp. grandiflorus	Fabaceae	0	0	1	s9	Teucrium scordium L.	Lamiaceae	0	1	1
s8	*Cytisus multiflorus* (L*Hér.) Sweet	Fabaceae	1	1	0	s9	*Teucrium scorodonia* L. subsp. scorodonia	Lamiaceae	1	1	1
s8	*Cytisus scoparius* (L.) Link subsp. scoparius	Fabaceae	1	1	0	s9	Thapsia garganica L.	Apiaceae	0	0	1
s8	*Cytisus striatus* (Hill) Rothm.	Fabaceae	1	1	0	s9	Thapsia minor Hoffmanns. et Link	Apiaceae	0	1	1
s8	Daphne gnidium L.	Thymelaeaceae	1	1	1	s9	Thapsia nitida Lacaita var. meridionalis A. Pujadas	Apiaceae	0	0	1
s8	Diplotaxis catholica (L.) DC.	Brassicaceae	0	1	0	s9	Thapsia transtagana L.	Apiaceae	0	1	1
s8	Diplotaxis virgata (Cav.) DC. Subsp. virgata	Brassicaceae	0	1	0	s9	Thapsia villosa L. var. dissecta Boiss.	Apiaceae	0	1	1
s8	*Dorycnium pentaphyllum* Scop.	Fabaceae	0	1	1	s9	Thapsia villosa L. var. villosa	Apiaceae	1	1	0
s8	*Dorycnium rectum* (L.) Ser.	Fabaceae	0	0	1	s9	Theligonum cynocrambe L.	Rubiaceae	0	0	1

S	Species	Family	MB	CP	FS	S	Species	Family	MB	CP	FS
s8	Dorycnopsis gerardi (L.) Boiss.	Fabaceae	0	1	0	s9	Thymbra capitata (L.) Cav.	Lamiaceae	0	0	1
s8	Draba muralis L.	Brassicaceae	1	0	1	s9	Thymus albicans Hoffmanns. & Link	Lamiaceae	0	0	1
s8	Echinospartum ibericum Rivas Mart., Sánchez Mata et Sancho	Fabaceae	1	0	0	s9	Thymus caespititius Brot.	Lamiaceae	1	0	0
s8	Elatine brochonii Clavaud	Elatinaceae	0	1	1	s9	Thymus carnosus Boiss.	Lamiaceae	0	0	1
s8	Elatine hexandra (Lapierre) DC.	Elatinaceae	0	1	0	s9	Thymus lotocephalus G.López & R. Morales	Lamiaceae	0	0	1
s8	Elatine macropoda Guss.	Elatinaceae	0	1	0	s9	Thymus mastichina L. subsp. mastichina	Lamiaceae	0	1	1
s8	*Epilobium brachycarpum* C.Presl.	Onagraceae	0	1	0	s9	*Thymus pulegioides* L.	Lamiaceae	1	0	0
s8	*Epilobium hirsutum* L.	Onagraceae	0	1	1	s9	Thymus villosus L. subsp. villosus	Lamiaceae	0	1	0
s8	*Epilobium obscurum* Schreb.	Onagraceae	1	1	0	s9	*Tolpis barbata* (L.) Gaertn.	Asteraceae	1	1	1
s8	*Epilobium parviflorum* Schreb.	Onagraceae	0	1	0	s9	*Tolpis umbellata* Bertol.	Asteraceae	1	1	1
s8	Epilobium tetragonum L. subsp. tetragonum	Onagraceae	0	1	0	s9	Tordylium maximum L.	Apiaceae	0	1	0
s8	Epilobium tetragonum L. subsp. tournefortii (Michalet) Rouy et E. G. Camus	Onagraceae	0	1	0	s9	*Torilis arvensis* (Huds.) Link subsp.				
neglecta (Spreng.) Thell. in Hegi	Apiaceae	0	1	1							
s8	*Erodium aethiopicum* (Lam.) Brumh. & Thell.	Geraniaceae	0	0	1	s9	*Torilis arvensis* (Huds.) Link subsp.				
purpurea (Ten.) Hayek	Apiaceae	1	1	1							
s8	*Erodium botrys* (Cav.) Bertol.	Geraniaceae	0	1	1	s9	*Torilis arvensis* (Huds.) Link subsp. recta Jury	Apiaceae	0	1	0

S	Species	Family	MB	CP	FS	S	Species	Family	MB	CP	FS
s8	Erodium chium (L.) Willd.	Geraniaceae	0	0	1	s9	Torilis elongata (Hoffmanns. et Link) Samp.	Apiaceae	0	1	0
s8	*Erodium cicutarium* (L.) L'Hér. Subs. Cicutarium	Geraniaceae	1	1	1	s9	*Torilis nodosa* (L.) Gaertn.	Apiaceae	0	0	1
s8	*Erodium laciniatum* (Cav.) Willd.	Geraniaceae	0	0	1	s9	*Tripolium pannonicum* (Jacq.) Dobrocz.	Asteraceae	0	0	1
s8	*Erodium malacoides* (L.) L'Hér.	Geraniaceae	0	1	1	s9	*Urospermum picroides* (L.) Scop. ex F. W. Schmidt	Asteraceae	0	1	1
s8	*Erodium moschatum* (L.) L*Hér.	Geraniaceae	1	1	1	s9	*Vaccinium myrtillus* L.	Ericaceae	1	0	0
s8	Erophaca baetica (L.) Boiss. subsp. baetica	Fabaceae	0	1	1	s9	Valeriana cornucopiae L.	Caprifoliaceae	0	0	1
s8	*Erophila verna* (L.) Chevall.	Brassicaceae	1	0	1	s9	Valeriana montana L.	Caprifoliaceae	1	0	0
s8	*Eucalyptus camaldulensis* Dehnh.	Myrtaceae	0	1	1	s9	Valeriana tuberosa L.	Caprifoliaceae	0	0	1
s8	*Eucalyptus globulus* Labill.	Myrtaceae	1	1	1	s9	Valerianella coronata (L.) DC.	Caprifoliaceae	0	1	0
s8	*Euphorbia amygdaloides* L. subsp. amygdaloides	Euphorbiaceae	1	1	0	s9	Valerianella discoidea (L.) Loisel.	Caprifoliaceae	0	0	1
s8	Euphorbia boetica Boiss.	Euphorbiaceae	0	0	1	s9	*Valerianella locusta* (L.) Laterrade	Caprifoliaceae	0	1	0
s8	*Euphorbia clementei* Boiss.	Euphorbiaceae	0	0	1	s9	*Verbascum pulverulentum* Vill.	Plantaginaceae	0	1	0
s8	*Euphorbia dulcis* L.	Euphorbiaceae	1	1	0	s9	*Verbascum sinuatum* L.	Plantaginaceae	0	1	1
s8	*Euphorbia exigua* L. subsp. exigua	Euphorbiaceae	0	1	1	s9	*Verbascum thapsus* L. subsp. thapsus	Plantaginaceae	1	0	0
s8	*Euphorbia falcata* L.	Euphorbiaceae	0	0	1	s9	*Verbascum virgatum* Stokes in With.	Plantaginaceae	0	1	0
s8	*Euphorbia helioscopia* L. subsp. helioscopia	Euphorbiaceae	0	1	1	s9	*Verbena bonariensis* L.	Verbenaceae	0	1	1
s8	*Euphorbia hirsuta* L.	Euphorbiaceae	0	0	1	s9	*Verbena officinalis* L.	Verbenaceae	0	1	0
s8	*Euphorbia matritensis* Boiss.	Euphorbiaceae	0	1	0	s9	*Verbena supina* L.	Verbenaceae	0	1	1
s8	*Euphorbia medicaginea* Boiss.	Euphorbiaceae	0	0	1	s9	Veronica anagallis-aquatica L. subsp. anagallis-aquatica	Plantaginaceae	1	1	1

S	Species	Family	MB	CP	FS	S	Species	Family	MB	CP	FS
s8	*Euphorbia oxyphylla* Boiss. in DC.	Euphorbiaceae	0	1	0	s9	Veronica anagalloides Guss. subsp. anagalloides	Plantaginaceae	1	1	0
s8	*Euphorbia paniculata* Desf. subsp. paniculata	Euphorbiaceae	0	1	1	s9	*Veronica arvensis* L.	Plantaginaceae	1	1	0
s8	*Euphorbia paralias* L.	Euphorbiaceae	0	0	1	s9	*Veronica beccabunga* L.	Plantaginaceae	1	0	0
s8	*Euphorbia peplis* L.	Euphorbiaceae	0	0	1	s9	*Veronica chamaedrys* L.	Plantaginaceae	1	0	0
s8	*Euphorbia peplus* L. var. peplus	Euphorbiaceae	0	1	1	s9	*Veronica cymbalaria* Bodard	Plantaginaceae	0	1	0
s8	*Euphorbia pterococca* Brot.	Euphorbiaceae	0	1	1	s9	*Veronica hederifolia* L.	Plantaginaceae	0	1	0
s8	*Euphorbia segetalis* L. var. pinea (L.) Lange in Willk. et Lange	Euphorbiaceae	0	0	1	s9	Veronica micrantha Hoffmanns. et Link	Plantaginaceae	1	0	0
s8	*Euphorbia serpens* Kunth	Euphorbiaceae	0	0	1	s9	*Veronica officinalis* L.	Plantaginaceae	1	0	0
s8	*Euphorbia serrata* L.	Euphorbiaceae	0	0	1	s9	*Veronica peregrina* L.	Plantaginaceae	0	1	0
s8	*Euphorbia sulcata* Lens ex Loisel.	Euphorbiaceae	0	0	1	s9	*Veronica persica* Poir.	Plantaginaceae	1	1	0
s8	*Euphorbia terracina* L.	Euphorbiaceae	0	0	1	s9	*Veronica scutellata* L.	Plantaginaceae	1	0	0
s8	*Euphorbia transtagana* Boiss.	Euphorbiaceae	0	0	1	s9	*Veronica serpyllifolia* L. subsp. serpyllifolia	Plantaginaceae	1	0	0
s8	*Ficus carica* L.	Moraceae	0	1	1	s9	*Viburnum tinus* L.	Caprifoliaceae	0	0	1
s8	*Flueggea tinctoria* (L.) G. L. Webster	Euphorbiaceae	0	1	0	s9	Vinca difformis Pourr.	Apocynaceae	1	0	1
s8	*Fragaria vesca* L. subsp. vesca	Rosaceae	1	0	0	s9	*Vinca major* L. subsp. major	Apocynaceae	0	1	0
s8	*Frangula alnus* Mill.	Rhamnaceae	1	1	0	s9	*Vincetoxicum hirundinaria* Medik.	Apocynaceae	1	0	0
s8	Fumana laevipes (L.) Spach	Cistaceae	0	0	1	s9	*Vincetoxicum nigrum* (L.) Moench	Asclepiadaceae	0	1	0
s8	Fumana thymifolia (L.) Spach	Cistaceae	0	0	1	s9	Wahlenbergia hederacea (L.) Rchb.	Campanulaceae	1	1	0
s8	Genista anglica L.	Fabaceae	1	1	0	s9	*Xanthium spinosum* L.	Asteraceae	0	1	1
s8	Genista cinerascens Lange	Fabaceae	1	0	0	s9	*Xanthium strumarium* L.	Asteraceae	0	1	1
s8	Genista falcata Brot.	Fabaceae	1	1	0						

Table A3.
Floristic catalog shorted by systematic group -S- (m, g, s2–9), species and family, per study area (MB, CP, and FS).

DOI: http://dx.doi.org/10.5772/intechopen.1008900

Var.	N	Range	Minimum	Maximum	Mean	Std. Deviation
FSAlt	1606	552,00	0,00	552,00	236,023	147,71,166
CPAlt	2777	898,00	130,00	1028,00	337,2182	122,6892
MBAlt	1755	1167,00	296,00	1463,00	886,4598	201,0507
FSBio	1606	19,51	96,23	115,74	106,4043	44,458
CPBio	2777	70,45	113,14	183,59	135,4471	12,2765
MBBio	1755	55,29	167,62	222,90	182,6374	98,110
FSP	1606	13,42	41,58	55,00	47,8697	31,791
CPP	2777	51,50	53,92	105,42	70,4628	92,624
MBP	1755	35,67	100,50	136,17	110,1942	64,085
FSTM	1606	2,80	20,01	22,81	21,4769	0,6446
CPTM	2777	6,38	15,72	22,10	20,8244	0,9181
MBTM	1755	6,31	12,14	18,45	15,2891	12,024
FSTmi	1606	3,27	9,85	13,12	11,4381	0,8571
CPTmi	2777	3,27	8,04	11,31	10,4523	0,4819
MBTmi	1755	4,57	4,33	8,90	65,838	0,8071

Table A4.
Statistical description for environmental variables (Var.): N, number of stations; Alt, elevation of station (for MB -MBAlt-, CP -CPAlt-, and FS -FSAlt-); Bio, average bioclimatic variables (MBBio, for MB; CPBio, for CP; FSBio for FS); P, average monthly precipitation (MBP, for MB; CPP, for CP; FSP, for FS); TM, average monthly maximum temperature (MBTM, for MB; CPTM, for CP; FSTM, for FS); Tmi, average monthly minimum temperature (MBTmi, for MB; CPTmi, for CP; FSTmi, for FS).

Author details

João Rocha[1]*, Margarida L.R. Liberato[2], Teresa Fonseca[3], Raul Morais[2], Nuno Silva[4] and António L. Crespí[5]

1 CoLAB ForestWISE® – Laboratório Colaborativo para a Gestão Integrada da Floresta e do Fogo, Vila Real, Portugal

2 Department of Engineering, School of Sciences and Technology, Centre for the Research and Technology of Agro-Environmental and Biological Sciences, University of Trás-os-Montes e Alto Douro, Vila Real, Portugal

3 Department of Forest Science and Landscape Architecture (CIFAP), School of Agricultural and Veterinary Sciences, Centre for the Research and Technology of Agro-Environmental and Biological Sciences, University of Trás-os-Montes e Alto Douro, Vila Real, Portugal

4 Engineering Department, School of Science and Technology, UTAD—University of Trás-os-Montes e Alto Douro, Quinta de Prados, Vila Real, Portugal

5 Biology and Environment Department, Herbarium, School of Life and Environmental Sciences, Centre for the Research and Technology of Agro-Environmental and Biological Sciences, University of Trás-os-Montes e Alto Douro, Vila Real, Portugal

*Address all correspondence to: acrespi@utad.pt

References

[1] Moreira F, Ascoli D, Safford H, Adams MA, Moreno JM, Pereira JMC, et al. Wildfire management in Mediterranean-type regions: Paradigm change needed. Environmental Research Letters. 2020;**15**:011001. DOI: 10.1088/1748-9326/ab541e

[2] Dupuy JL, Fargeon H, Martin-StPaul N, Pimont F, Ruffault J, Guijarro M, et al. Climate change impact on future wildfire danger and activity in southern Europe: A review. Annals of Forest Science. 2020;**77**:35. DOI: 10.1007/s13595-020-00933-5

[3] Aparício BA, Santos JA, Freitas TR, Sá ACL, Pereira JMC. Unravelling the effect of climate change on fire danger and fire behaviour in the transboundary biosphere Reserve of Meseta Ibérica (Portugal-Spain). Climatic Change. 2022;**173**:5. DOI: 10.1007/s10584-022-03399-8

[4] Amraoui M, Liberato MLR, Calado TJ, DaCamara CC, Coelho LP, Trigo RM, et al. Fire activity over Mediterranean Europe based on information from Meteosat-8. Forest Ecology and Management. 2013;**294**:62-75. DOI: 10.1016/j.foreco.2012.08.032

[5] Fernandes PM, Delogu GM, Leone V, Ascoli D. Wildfire policies contribution to foster extreme wildfires. In: Tedim F, Leone V, McGee TK, editors. Extreme Wildfire Events and Disasters. Elsevier; 2020. pp. 187-200. DOI: 10.1016/B978-0-12-815721-3.00010-2. ISBN 978012815721-3

[6] Gouveia CM, Bistinas I, Liberato MLR, Bastos A, Koutsias N, Trigo RM. The outstanding synergy between drought, heatwaves and fuel on the 2007 southern Greece exceptional fire season. Agricultural and Forest Meteorology. 2016;**218–219**:135-145. DOI: 10.1016/j.agrformet.2015.11.023

[7] Russo A, Gouveia CM, Páscoa P, DaCamara CC, Sousa PM, Trigo RM. Assessing the role of drought events on wildfires in the Iberian Peninsula. Agricultural and Forest Meteorology. 2017;**237**:50-59. DOI: 10.1016/j.agrformet.2017.01.021

[8] Turco M, von Hardenberg J, Agha Kouchak A, Llasat MC, Provenzale A, Trigo RM. On the key role of droughts in the dynamics of summer fires in Mediterranean Europe. Scientific Reports. 2017;**7**(1):81. DOI: 10.1038/s41598-017-00116-9

[9] Observatório Técnico Independente, Castro Rego F, Fernandes P, Sande Silva J, Azevedo J, Moura JM, et al. Análise de indicadores de desempenho do Sistema de Defesa da Floresta contra Incêndios na transição (2018-2020) para o Sistema de Gestão Integrada de Fogos Rurais. Lisboa: Observatório Técnico Independente, Assembleia da República; 2020. 42 p. Available from: https://www.parlamento.pt/Parlamento/Documents/oti/Estudotecnico-dez2020.pdf [Accessed: July 17, 2024]

[10] Sousa PM, Trigo RM, Pereira MG, Bedia J, Gutiérrez JM. Different approaches to model future burnt area in the Iberian Peninsula. Agricultural and Forest Meteorology. 2015;**202**:11-25. DOI: 10.1016/j.agrformet.2014.11.018

[11] Carvalho A, Flannigan MD, Logan KA, Gowman LM, Miranda AI, Borrego C. The impact of spatial resolution on area burned and fire occurrence projections in Portugal under climate change. Climatic Change. 2010;**98**:177-197. DOI: 10.1007/s10584-009-9667-2

[12] Collins RD, Neufville R, Claro J, Oliveira T, Pacheco AP. Forest fire management to avoid unintended consequences: A case study of Portugal using system dynamics. Journal of Environmental Management. 2013;**130**: 1-9. DOI: 10.1016/j.jenvman.2013.08.033

[13] Garnier E, Navas M-L, Grigulis K. Plant Functional Diversity: Organism Traits, Community Structure, and Ecosystem Properties. Oxford University Press; 2016. pp. 9-11. ISBN0198757360, 9780198757368

[14] Gillison AN, Carpenter G. A generic plant functional attribute set and grammar for dynamic vegetation description and analysis. Functional Ecology. 1997;**11**(6):775-783

[15] Barajas Barbosa MP, Craven D, Weigelt P, Denelle P, Otto R, Díaz S, et al. Assembly of functional diversity in an oceanic island flora. Nature. 2023;**619** (7970):545-550. DOI: 10.1038/s41586-023-06305-z

[16] McIntyre S, Lavorel S, Landsberg J, Forbes TDA. Disturbance response in vegetation–towards a global perspective on functional traits. Journal of Vegetation Science. 1999;**10**(5):621-630. DOI: 10.2307/3237077

[17] Pendry CA, Dick J, Pullan MR, Knees SG, Miller AG, Neale S, et al. In search of a functional flora-towards a greater integration of ecology and taxonomy. Plant Ecology. 2007;**192**:161-167. DOI: 10.1007/s11258-007-9304-y

[18] Wullschleger SD, Epstein HE, Box EO, Euskirchen ES, Goswami S, Iversen CM, et al. Plant functional types in earth system models: Past experiences and future directions for application of dynamic vegetation models in high-latitude ecosystems. Annals of Botany. 2014;**114**(1):1-16. DOI: 10.1093/aob/mcu077

[19] Hoyle GL, Steadman KJ, Good RB, McIntosh EJ, Galea LM, Nicotra AB. Seed germination strategies: An evolutionary trajectory independent of vegetative functional traits. Frontiers in Plant Science. 2015;**6**:731. DOI: 10.3389/fpls.2015.00731

[20] Díaz S, Kattge J, Cornelissen JH, Wright IJ, Lavorel S, Dray S, et al. The global spectrum of plant form and function. Nature. 2016;**529**(7585):167-171. DOI: 10.1038/nature16489

[21] Pluess M. Individual differences in environmental sensitivity. Child Development Perspectives. 2015;**9**(3): 138-143. DOI: 10.1111/cdep.12120

[22] Atkin OK, Bloomfield KJ, Reich PB, Tjoelker MG, Asner GP, Bonal D, et al. Global variability in leaf respiration in relation to climate, plant functional types and leaf traits. New Phytologist. 2015;**206**(2):614-636. DOI: 10.1111/nph.13253

[23] Funk JL, Larson JE, Ames GM, Butterfield BJ, Cavender-Bares J, Firn J, et al. Revisiting the holy grail: Using plant functional traits to understand ecological processes. Biological Reviews. 2017;**92**(2):1156-1173. DOI: 10.1111/brv.12275

[24] Barsoum N, Coote L, Eycott AE, Fuller L, Kiewitt A, Davies RG. Diversity, functional structure and functional redundancy of woodland plant communities: How do mixed tree species plantations compare with monocultures? Forest Ecology and Management. 2016;**382**:244-256. DOI: 10.1016/j.foreco.2016.10.005

[25] Aquilué N, Filotas É, Craven D, Fortin MJ, Brotons L, Messier C.

Evaluating forest resilience to global threats using functional response traits and network properties. Ecological Applications. 2020;**30**(5):e02095. DOI: 10.1002/eap.2095

[26] Bricca A, Carranza ML, Varricchione M, Cutini M, Stanisci A. Exploring plant functional diversity and redundancy of mediterranean high-mountain habitats in the Apennines. Diversity. 2021;**13**(10): 466. DOI: 10.3390/d13100466

[27] Garland G, Banerjee S, Edlinger A, Miranda Oliveira E, Herzog C, Wittwer R, et al. A closer look at the functions behind ecosystem multifunctionality: A review. Journal of Ecology. 2021;**109**(2): 600-613. DOI: 10.1111/1365-2745.13511

[28] Ricotta C, Carboni M, Acosta AT. Let the concept of indicator species be functional! Journal of Vegetation Science. 2015;**26**(5):839-847. DOI: 10.1111/jvs.12291

[29] Mouillot D, Graham NA, Villéger S, Mason NW, Bellwood DR. A functional approach reveals community responses to disturbances. Trends in Ecology & Evolution. 2013;**28**(3):167-177. DOI: 10.1016/j.tree.2012.10.004

[30] Oliver TH, Heard MS, Isaac NJ, Roy DB, Procter D, Eigenbrod F, et al. Biodiversity and resilience of ecosystem functions. Trends in Ecology & Evolution. 2015;**30**(11):673-684. DOI: 10.1016/j.tree.2015.08.009

[31] Gladstone-Gallagher RV, Pilditch CA, Stephenson F, Thrush SF. Linking traits across ecological scales determines functional resilience. Trends in Ecology & Evolution. 2019;**34**(12):1080-1091. DOI: 10.1016/j.tree.2019.07.010

[32] López DR, Brizuela MA, Willems P, Aguiar MR, Siffredi G, Bran D. Linking ecosystem resistance, resilience, and stability in steppes of North Patagonia. Ecological Indicators. 2013;**24**:1-11. DOI: 10.1016/j.ecolind.2012.05.014

[33] Le Provost G, Badenhausser I, Le Bagousse-Pinguet Y, Clough Y, Henckel L, Violle C, et al. Land-use history impacts functional diversity across multiple trophic groups. National Academy of Sciences of the United States of America. 2020;**117**(3):1573-1579. DOI: 10.1073/pnas.1910023117

[34] Newbold T, Bentley LF, Hill SL, Edgar MJ, Horton M, Su G, et al. Global effects of land use on biodiversity differ among functional groups. Functional Ecology. 2020;**34**(3):684-693. DOI: 10.1111/1365-2435.13500

[35] Forzieri G, Girardello M, Ceccherini G, Spinoni J, Feyen L, Hartmann H, et al. Emergent vulnerability to climate-driven disturbances in European forests. Nature Communications. 2021;**12**(1): 1081. DOI: 10.1038/s41467-021-21399-7

[36] Pilon NA, Cava MG, Hoffmann WA, Abreu RC, Fidelis A, Durigan G. The diversity of post-fire regeneration strategies in the cerrado ground layer. Journal of Ecology. 2021;**109**(1):154-166. DOI: 10.1111/1365-2745.13456

[37] Keith DA, Ferrer-Paris JR, Nicholson E, Bishop MJ, Polidoro BA, Ramirez-Llodra E, et al. A function-based typology for Earth's ecosystems. Nature. 2022;**610**(7932):513-518. DOI: 10.1038/s41586-022-05318-4

[38] Meira Castro AC, Nunes A, Sousa A, Lourenço L. Mapping the causes of forest fires in Portugal by clustering analysis. Geosciences. 2020;**10**(2):53. DOI: 10.3390/geosciences10020053

[39] Turco M, Jerez S, Augusto S, Tarín-Carrasco P, Ratola N, Jiménez-Guerrero P, et al. Climate drivers of the 2017

devastating fires in Portugal. Scientific Reports. 2019;**9**:13886. DOI: 10.1038/s41598-019-50281-2

[40] DR. Resolução do Conselho de Ministros n.° 49/2020 [Internet]. 2020. pp. 6-18. Available from: https://files.diariodarepublica.pt/1s/2020/06/12100/0000600018.pdf [Accessed: July 17, 2024]

[41] DR. Despacho n.° 7417/2019 [Internet]. 2019. pp. 86-88. Available from: https://diariodarepublica.pt/dr/detalhe/despacho/7417-2019-124098737 [Accessed: July 17, 2024]

[42] Messier C, Bauhus J, Doyon F, Maure F, Sousa-Silva R, Nolet P, et al. The functional complex network approach to foster forest resilience to global changes. Forest Ecosystems. 2019; **6**(1):1-16. DOI: 10.1186/s40663-019-0166-2

[43] DeRose RJ, Long JN. Resistance and resilience: A conceptual framework for silviculture. Forest Science. 2014;**60**(6): 1205-1212. DOI: 10.5849/forsci.13-507

[44] Béné C, Doyen L. From resistance to transformation: A generic metric of resilience through viability. Earth's Future. 2018;**6**(7):979-996. DOI: 10.1002/2017EF000660

[45] Harrison GW. Stability under environmental stress: Resistance, resilience, persistence, and variability. The American Naturalist. 1979;**113**(5): 659-669. DOI: 10.1086/283424

[46] Long JN, Windmuller-Campione M, DeRose RJ. Building resistance and resilience: Regeneration should not be left to chance. Forests. 2018;**9**(5):270. DOI: 10.3390/f9050270

[47] Mori AS, Furukawa T, Sasaki T. Response diversity determines the resilience of ecosystems to environmental change. Biological Reviews. 2013;**88**(2):349-364. DOI: 10.1111/brv.12004

[48] Enright NJ, Fontaine JB, Lamont BB, Miller BP, Westcott VC. Resistance and resilience to changing climate and fire regime depend on plant functional traits. Journal of Ecology. 2014;**102**(6):1572-1581. DOI: 10.1111/1365-2745.12306

[49] Ermitão T, Páscoa P, Trigo I, Alonso C, Gouveia C. Mapping the most susceptible regions to fire in Portugal. Fire. 2023;**6**(7):254. DOI: 10.3390/fire6070254

[50] Hersbach H, Bell B, Berrisford P, Hirahara S, Horányi A, Muñoz-Sabater J, et al. The ERA5 global reanalysis. Quarterly Journal of the Royal Meteorological Society. 2020;**146**:1-51. DOI: 10.1002/qj.3803

[51] Eyring V, Bony S, Meehl GA, Senior CA, Stevens B, Stouffer RJ. Taylor KE overview of the coupled model intercomparison project phase 6 (CMIP6) experimental design and organization. Geoscientific Model Development. 2016;**9**:1937-1958. DOI: 10.5194/gmd-9-1937-2016

[52] Kriegler E, O'Neill BC, Hallegatte S, Kram T, Lempert R, Moss R, et al. The need for and use of socio-economic scenarios for climate change analysis: A new approach based on shared socio-economic pathways. Global Environmental Change. 2012;**22**:807-822. DOI: 10.1016/j.gloenvcha.2012.05.005

[53] O'Neill BC, Kriegler E, Ebi KL, Kemp-Benedict E, Riahi K, Rothman DS, et al. The roads ahead: Narratives for shared socioeconomic pathways describing world futures in the 21st century. Global Environmental Change.

2017;**42**:169-180. DOI: 10.1016/j.gloenvcha.2015.01.004

[54] Riahi K, Schaeffer R, Arango J, Calvin K, Guivarch C, Hasegawa T, et al. 2022: Mitigation pathways compatible with long-term goals. In: Shukla PR, Skea J, Slade R, Al Khourdajie A, van Diemen R, McCollum D, et al., editors. IPCC, 2022: Climate Change 2022: Mitigation of Climate Change. Contribution of Working Group III to the Sixth Assessment Report of the Intergovernmental Panel on Climate Change. Cambridge, UK and New York, NY, USA: Cambridge University Press; 2022. DOI: 10.1017/9781009157926.005

[55] Fick SE, Hijmans RJ. World Clim 2: New 1km spatial resolution climate surfaces for global land areas. International Journal of Climatology. 2017;**37**(12):4302-4315. DOI: 10.1002/joc.5086

[56] Castro I, Rocha J, Martins M, Carnide V, Martín JP, Veiga P, et al. The redundancy effect under morphogenetic and environmental fluctuations. The case of the Dianthus pungens group. Plant Biosystems. 2022;**156**(1): 292-306. DOI: 10.1080/11263504.2020.1857864

[57] Windig JJ, Nylin S. How to compare fluctuating asymmetry of different traits. Journal of Evolutionary Biology. 1999;**13**(1):29-37. DOI: 10.1046/j.1420–9101.2000.00143.x

[58] Rasmuson M. Fluctuating asymmetry-indicator of what? Hereditas. 2002;**136**(3):177-183. DOI: 10.1034/j.1601-5223.2002.1360301.x

[59] Peel MC, Finlayson BL, McMahon TA. Updated world map of the Köppen–Geiger climate classification. Hydrology and Earth System Sciences. 2007;**11**: 1633-1644. DOI: 10.5194/hess-11-1633-2007

[60] Gallart F, Solé A, Puigdefábregas J, Lázaro R. Badland systems in the Mediterranean. In: Bull JL, Kirkby MJ, editors. Dryland Rivers: Hydrology and Geomorphology of Semi-Arid Channels. Chichester: Wiley; 2002. pp. 299-326

[61] Ramos AM, Trigo RM, Santo FE. Evolution of extreme temperatures over Portugal: Recent changes and future scenarios. Climate Research. 2011;**48**: 177-192. DOI: 10.3354/cr00934

[62] Andrade C, Fraga H, Santos JA. Climate change multi-model projections for temperature extremes in Portugal. Atmospheric Science Lettters. 2014;**15**: 149-156. DOI: 10.1002/asl2.485

[63] Soares PMM, Cardoso RM, Lima DCA, Miranda PMA. Future precipitation in Portugal: High-resolution projections using WRF model and EUROCORDEX multi-model ensembles. Climate Dynamics. 2017;**49**: 2503-2530. DOI: 10.1007/s00382-016-3455-2

[64] Cardoso RM, Soares PMM, Lima DCA, Miranda PMA. Mean and extreme temperatures in a warming climate: EURO CORDEX and WRF regional climate highresolution projections for Portugal. Climate Dynamics. 2019;**52**: 129-157. DOI: 10.1007/s00382-018-4124-4

[65] Soares PMM, Lima DCA. Water scarcity down to earth surface in a Mediterranean climate: The extreme future of soil moisture in Portugal. Journal of Hydrology. 2022;**615**:28731. DOI: 10.1016/j.jhydrol.2022.128731

[66] Liberato MLR, Montero I, Gouveia C, Russo A, Ramos AM, Trigo RM. Rankings of extreme and widespread dry

and wet events in the Iberian Peninsula between 1901 and 2016. Earth System Dynamics. 2021;**12**:197-210. DOI: 10.5194/esd-12-197-2021

[67] Ramos AM, Tomé R, Trigo RM, Liberato MLR, Pinto JG. Projected changes in atmospheric rivers affecting Europe in CMIP5 models. Geophysical Research Letters. 2016;**43**(17):9315-9323. DOI: 10.1002/2016gl070634

[68] Hevia V, Carmona CP, Azcárate FM, Torralba M, Alcorlo P, Ariño R, et al. Effects of land use on taxonomic and functional diversity: A cross-taxon analysis in a Mediterranean landscape. Oecologia. 2016;**181**:959-970. DOI: 10.1007/s00442-015-3512-2

[69] Coombes MA. Biogeomorphology: Diverse, integrative and useful. Earth Surface Processes and Landforms. 2016; **41**(15):2296-2300. DOI: 10.1002/esp.4055

[70] Vernham G, Bailey JJ, Chase JM, Hjort J, Field R, Schrodt F. Understanding trait diversity: The role of geodiversity. Trends in Ecology & Evolution. 2023;**38**(8):736-748. DOI: 10.1016/j.tree.2023.02.010

[71] López de Heredia U, Carrión JS, Jiménez P, Collada C, Gil L. Molecular and palaeoecological evidence for multiple glacial refugia for evergreen oaks on the Iberian Peninsula. Journal of Biogeography. 2007;**34**(9):1505-1517. DOI: 10.1111/j.1365-2699.2007.01715.x

[72] Rodríguez-Sánchez F, Hampe A, Jordano P, Arroyo J. Past tree range dynamics in the Iberian Peninsula inferred through phylogeography and palaeodistribution modelling: A review. Review of Palaeobotany and Palynology. 2010;**162**(3):507-521. DOI: 10.1016/j.revpalbo.2010.03.008

[73] Valencia E, Maestre FT, Le Bagousse-Pinguet Y, Quero JL, Tamme R, Börger L, et al. Functional diversity enhances the resistance of ecosystem multifunctionality to aridity in Mediterranean drylands. New Phytologist. 2015;**206**(2):660-671. DOI: 10.1111/nph.13268

[74] Gross N, Bagousse-Pinguet YL, Liancourt P, Berdugo M, Gotelli NJ, Maestre FT. Functional trait diversity maximizes ecosystem multifunctionality. Nature Ecology & Evolution. 2017;**1**(5):0132. DOI: 10.1038/s41559-017-0132

[75] Spasojevic MJ, Grace JB, Harrison S, Damschen EI. Functional diversity supports the physiological tolerance hypothesis for plant species richness along climatic gradients. Journal of Ecology. 2014;**102**(2):447-455. DOI: 10.1111/1365-2745.12204

[76] Stahl U, Reu B, Wirth C. Predicting species' range limits from functional traits for the tree flora of North America. National Academy of Sciences of the United States of America. 2014;**111**(38): 13739-13744. DOI: 10.1073/pnas.1300673111

[77] Andrew SC, Mokany K, Falster DS, Wenk E, Wright IJ, Merow C, et al. Functional diversity of the Australian flora: Strong links to species richness and climate. Journal of Vegetation Science. 2021;**32**(2):e13018. DOI: 10.1111/jvs.13018

[78] Kühn N, Tovar C, Carretero J, Vandvik V, Enquist BJ, Willis KJ. Globally important plant functional traits for coping with climate change. Frontiers of Biogeography. 2021;**13**(4): e53774. DOI: 10.21425/F5FBG53774

[79] Pender JE, Hipp AL, Hahn M, Starr JR. Trait evolution rates shape

continental patterns of species richness in North America's most diverse angiosperm genus (Carex, Cyperaceae). Journal of Systematics and Evolution. 2021;**59**(4):763-775. DOI: 10.1111/jse.12739

[80] Ordonez A, Svenning JC. Consistent role of quaternary climate change in shaping current plant functional diversity patterns across European plant orders. Scientific Reports. 2017;**7**(1): 42988. DOI: 10.1038/srep42988

[81] Olmedo-Cobo JA, Gómez-Zotano J, Serrano-Montes JL. *Pinus sylvestris* L. subsp. nevadensis (Christ) Heywood in southern Spain: An endangered endemic Mediterranean forest. Geographica Pannonica. 2017;**21**(3):151-165. DOI: 10.5937/GeoPan1703151O

[82] Rocha J, Ramírez R, Díaz M, Martins M, García-Cabral I, Amich F, et al. Morpho-environmental strategies in the genus *Ononis* L. (subsections Natrix and Viscosae) in Western Mediterranean. Plant Biosystems. 2018;**152**(1):14-30. DOI: 10.1080/11263504.2016.1244118

[83] Torres T, Valle M, Ortiz JE, Soler V, Araujo R, Rivas MR, et al. Evolución paleoambiental de la región suroccidental de Europa durante los últimos 800 ka. Journal of Iberian Geology. 2020;**46**:117-144. DOI: 10.1007/s41513-020-00123-2

[84] Urruty N, Tailliez-Lefebvre D, Huyghe C. Stability, robustness, vulnerability and resilience of agricultural systems. A review. Agronomy for Sustainable Development. 2016;**36**:1-15. DOI: 10.1007/s13593-015-0347-5

[85] Sun N, Liu N, Zhao X, Zhao J, Wang H, Wu D. Evaluation of spatiotemporal resilience and resistance of global vegetation responses to climate change. Remote Sensing. 2022;**14**(17):4332. DOI: 10.3390/rs14174332

[86] Morelli F, Benedetti Y, Su T, Zhou B, Moravec D, Šímová P, et al. Taxonomic diversity, functional diversity and evolutionary uniqueness in bird communities of Beijing's urban parks: Effects of land use and vegetation structure. Urban Forestry & Urban Greening. 2017;**23**:84-92. DOI: 10.1016/j.ufug.2017.03.009

[87] Li W, He S, Cheng X, Zhang M. Functional diversity outperforms taxonomic diversity in revealing short-term trampling effects. Scientific Reports. 2021;**11**(1):18889. DOI: 10.1038/s41598-021-98372-3

[88] Accatino F, Sabatier R, De Michele C, Ward D, Wiegand K, Meyer KM. Robustness and management adaptability in tropical rangelands: A viability-based assessment under the non-equilibrium paradigm. Animal. 2014;**8**(8):1272-1281. DOI: 10.1017/S1751731114000913

[89] Hillebrand H, Kunze C. Meta-analysis on pulse disturbances reveals differences in functional and compositional recovery across ecosystems. Ecology Letters. 2020;**23**(3): 575-585. DOI: 10.1111/ele.13457

[90] Deák B, Kovács B, Rádai Z, Apostolova I, Kelemen A, Kiss R, et al. Linking environmental heterogeneity and plant diversity: The ecological role of small natural features in homogeneous landscapes. Science of the Total Environment. 2021;**763**:144199. DOI: 10.1016/j.scitotenv.2020.144199

[91] Guo T, Weise H, Fiedler S, Lohmann D, Tietjen B. The role of landscape heterogeneity in regulating plant functional diversity under different precipitation and grazing regimes in

semi-arid savannas. Ecological Modelling. 2018;**379**:1-9. DOI: 10.1016/j.ecolmodel.2018.04.009

[92] Lloret F, Zedler PH. The effect of forest fire on vegetation. In: Cerda A, editor. Fire Effects on Soils and Restoration Strategies. CRC Press; 2009. pp. 273-312. DOI: 10.1201/9781439843338-c9. ISBN 9780429063596

Chapter 6

Quantitative Methods for the Monitoring and Evaluation of Vegetation Survey Data

Damien Hicks

Abstract

This chapter reviews a range of analytical tools available to the vegetation scientist, together with the nature of the raw data which inform choice and optimisation of survey protocols. In the context of big data and influential mass media, the interpretation and communication of such analyses are often pivotal. Vegetation data are collected for objectives varying from land management to ecological research, and conceptualized either as the central subject of interest or as an independent habitat variable. Several powerful quantitative methods are to hand, which are widely applicable to other subjects sharing the standard dataframe format. The development of open source coding languages, together with recent advances in computational power, have revolutionized the analyst's toolkit.

Keywords: quantitative biology, vegetation, monitoring, evaluation, survey data, statistical sampling, ecological communities

1. Introduction

This chapter begins from a consideration of definitions, the nature of the data and its collection (Sections 1–4) before concentrating on specific methodologies and tools for quantitative evaluation (Sections 5–10).

How do we draw robust inferences from vegetation survey data? Evidence-led policy and management of vegetation is in high demand, while the subject matter tend to be (interestingly) continuous, non-normal, vertically structured, spatially clustered and/or temporally autocorrelated.

The focus here is on vegetation data, yet the same considerations and methods could be applied to the quantitative monitoring and evaluation of a vast range of other subjects. The concept of monitoring followed in this chapter is the systematic measurement of variables and processes over time [1], and of evaluation is the calculation of quality, amount or value [2].

A broad definition of vegetation is followed, as a community of plant species (or taxa, given identification challenges of Section 3) which are spatially located in three dimensions across a given area of land or water. The scale of study can vary from a few square centimeters (e.g. bryophytes on a wall, filamentous algae on a littoral stone) to strictly demarcated stands of urban plantings (e.g. floral meadows, residential gardens), to the vegetation types of entire continents.

Vegetation is variously studied as the central subject of interest or as an independent 'habitat' or 'environmental factor' variable. Its data are sampled for a wide range of activities including ecological research; geographical mapping; Environmental Impact Assessment; farm yield survey; forestry; water quality assessment; taxonomic collections; land planning and land-take; restoration; ground-truthing of remote sensing; habitat classification; and protected area design.

All graphs and tables in this work have been generated from artificial data. Remote sensing (e.g. LiDAR, satellite imagery) and environmental sensor network (ESN) data are valuable external subjects to this chapter. With apologies there will inevitably a bias in this work to methods used in Britain and Ireland, and to the author's preferred coding language of R (with packages cited by its standard {...}), and future collaborations to expand on this are most welcomed.

2. The nature of the data

The individual species, and their biological dynamics and interactions in the field, are usually of prime interest: the focus of this chapter is on the data that we collect and attempt in various ways to make interpretations about the former. In theory we should expect to be able to detect signals of a range of independent variables (e.g. topography, soil type, rainfall, etc.) which constrain the assemblage of species present at a given site.

This is made more complex by the likely interactions of those independent variables, as well as ecological processes in train at the site (e.g. succession, disturbance, competition) and characteristic to that vegetation. In common with other natural systems, vegetation exhibits marked heterogeneity which varies in intensity and scale.

2.1 Units in unit space

Raw vegetation data are usually stored in the format of a sparse matrix (i.e. a dataframe, with a large number of empty cells) of species in sample space (e.g. **Table 1**), and vice-versa. Species tend to occur in this matrix at starkly differing frequencies, often with the majority being infrequent and one or few being dominant. The empty cells are taken to be 'true zeroes', in other words that the search effort has been sufficient even for the most rare and/or cryptic species.

As mentioned in Section 1, this structure of data is similar or identical to several other fields of study, for example international trade flow (commodities in location space, and vice versa), medical trials (illnesses in risk factor space, and vice versa), and genetic interactions (phenotypes in sequence space, and vice versa). All of these subjects typically present statistically noisy data, requiring the analyst to embrace considerable variability when collecting the same data variables from the same individual unit. That variability can be an inconvenience, or in itself of interest. Thus even within an apparently homogenous stand of vegetation, replicates of quadrat samples often return a variety of species and abundances.

2.2 Presence/absence and abundance

From a study of a range of plant community data, Bastow Wilson [3] concluded that 'the primary assembly rule control on communities is on the presence of species,

		Samples (e.g. quadrats)							
		1	2	3	4	5	6	7	8
Units (e.g. taxa)	Species a	4			1	2		1	
	Species b	7	27	12	18	3	11	15	8
	Species c			2		1			1
	Species d	9			54				
	Species e					92		56	
	Species f			76					13
	Species g		3		24	1			
	Species h	53		8					
	Species i							2	
	Species j	1			8				
	Species k			1					
	Species l		6			1			
	Species m						9		
	Species n	1				1			2
	Species o			1					
	Species p		4			1		5	
	Species q				3				2

Table 1.
Example matrix of vegetation data, with orientation of species-in-sample-space.

not their abundance'. This appears to suggest that showing up (or not) conveys higher information value than being dominant (or rare). Or perhaps more pessimistically that the measurements which we collect as proxies for abundance are not robust. Regardless, this would be a convenient finding for surveyors as presence/absence data can be substantially less time-consuming to obtain.

The measurement of abundance is a philosopher's goldmine: (i) Percentage Cover seems particularly appropriate to sedentary light-seeking organisms, yet suffers from wide observer-estimation error (different humans, and same humans on different days, perceive clusters and patterns differently) and quadrat totals greater than 100% for (the majority) vertically-structured vegetation. (ii) Density (count per quadrat) is intuitively attractive but often far from Gaussian and, more fundamentally, becomes questionable when the units being counted at the site diverge in morphology (e.g. a woodland containing large trees as well as small herbaceous plants) or mode of reproduction (e.g. a grassland containing seed-grown Dandelions as well as grass tillers originating from a single plant). (iii) Frequency (summed presence/absence of all quadrats sampled) is fairly unequivocal (assuming the surveyor is thorough) and quickly recorded in the field, but highly sensitive to the density and spatial distribution of the plants present [4]. (iv) Occupancy (sub-quadrat frequency) partially addresses some of the shortfalls of density and frequency. (v) Biomass solves some of the above problems of subjectivity but is destructive of the subject material.

Chen et al. [5] echoed from their evaluation of grassland vegetation, that 'because species composition is intrinsically qualitative, quantitatively measuring species

Figure 1.
*The Ouvrard megaquadrat, deployed here for the detection of scale of pattern of California poppy (*Eschscholzia californica*) by sub-counts in herbaceous urban vegetation.*

composition can be difficult'. In conclusion abundance is a surprisingly challenging measure to pin down, and hard to remove the subjective human element from. An optimal strategy may be to collect data of more than one measure of abundance at each sampling point, though sufficient time in the field can be rare.

2.3 Data sampling

Optimization of sampling window (or quadrat) size is also likely to be species-specific and often therefore a compromise in the field. The data collected from differing scales of sampling unfortunately does not also scale neatly, being considerably affected by a range of variables including plant size, dispersal mechanism, density and soil interactions. Following Grieg-Smith [4], a pilot strategy of data collection by sub-counts of different sizes (e.g. **Figure 1**) may be informative, revealing a peak in sampled density at the block size which corresponds to the spatial abundance pattern of the species.

3. Taxonomic identification

As a routine part of survey work, it is hard to overstate the value of (rapidly dried and pressed) voucher plant specimens. Even with that benefit, the presence or absence of key plant identification characters (e.g. flowers or fruits) at time of survey can be decisive, and impacts of complex taxonomy (e.g. hybridization, polyploidy, introgression from ornamental plants) on the survey data is also to be expected.

Figure 2.
Multiple hybrid ornamental plants of the genus Primula *in urban vegetation, UK.*

For instance the genus *Taraxacum*, in which sexual reproduction in Britain and Ireland is rare, if indeed occurs at all [6]. For such apomictic taxa, the plants that the surveyor encounters in the field are a population of genetically invariable seed-clones, or microspecies. To compound further the difficulty of identification, these clones produce widely differing phenotypes according to different age, local environment, and season of the year. Such challenges in the field are common with plants of other apomictic taxa such as *Hieracium, Alchemilla* and *Sorbus*.

Natural hybridization between species is the norm in some genera (e.g. *Epilobium, Salix*, *Verbascum*, *Euphrasia, Rumex*), and further complicated in those that are anthropogenically selected in cultivation (e.g. *Rosa*, *Cotoneaster, Narcissus, Primula*). The recent UK authoritative publication on the subject [7] inventoried 909 hybrids reliably recorded in the wild, with notes on an additional 156, for a country with less than 1700 native plant species. The authors list for instance the popular garden plant *Primula* x*polyantha* (see **Figure 2**) as one of the commonest of all wild British hybrids.

The importance of sufficient geographic sampling to encompass the variability across the genetic population was emphasized by Ouvrard et al. [8] while targeting 63 plant species in 15 families, through collecting datapoints from 14 countries across their distributional range in Europe.

For collection of vegetation survey data, the most practical way to deal with identification would seem to be to harmonize taxonomic lists between different surveyors and timepoints, so maximizing comparability. In this way the analyst is required to be decisive on what is taxonomically possible, and potentially to reject some information. For instance records of taxa which are differentially identified, e.g. to species at one site and to genus at another, are all subsumed to genus level.

4. Statistical considerations

Ideally the sample of data enables the analyst to make robust generalizations to the true population (i.e. the sampling universe), together with some measure of the goodness of those generalizations [9].

4.1 Monitoring and evaluation

For monitoring, the ask is generally to be able to detect any change (e.g. across sites, time, pre- and post-intervention, phenology), while for evaluation it is usually to measure useful attributes (e.g. physical structure, community type, floral/fruit density, pest load) of vegetation. Both activities require that the data are collected and analyzed in a way that tracks the reality on the ground. There is perhaps a tendency in less scientific projects to settle for subjectively conjuring up metrics, rather than use variables whose response to the question in hand has been established.

4.2 Sampling distributions

Communities of plants are strongly dependent on a range of drivers such as seasonality, succession, and response to environmental gradients. The resultant variability is grappled with by replications of measurement, to produce a statistical distribution of observations.

Plant abundance data are often zero-inflated, a feature which is challenging to iron out to normality by transformations. Particularly for observations of count data, an assumption that the sample mean is automatically representative of the true population can be misleading, and it is advisable therefore to inspect the shape and spread of its statistical distribution. This could be done on-the-fly or by obtaining pilot study data. A cumulative approach to such a sensitivity analysis was developed in [10] by trialing the impact of transect sampling strategy on estimates of floral resources. Sampling of 1 m^2 quadrats in this way pointed to a graphical 'elbow' of decrease in the variance in estimates of nectar and pollen resources with increasing numbers of randomly sampled quadrats.

4.3 Experimental error

As with the inherent variability, statistical error is to be embraced: our ability to detect any change in a system requires that the change is greater than the estimation error associated with the variable being measured.

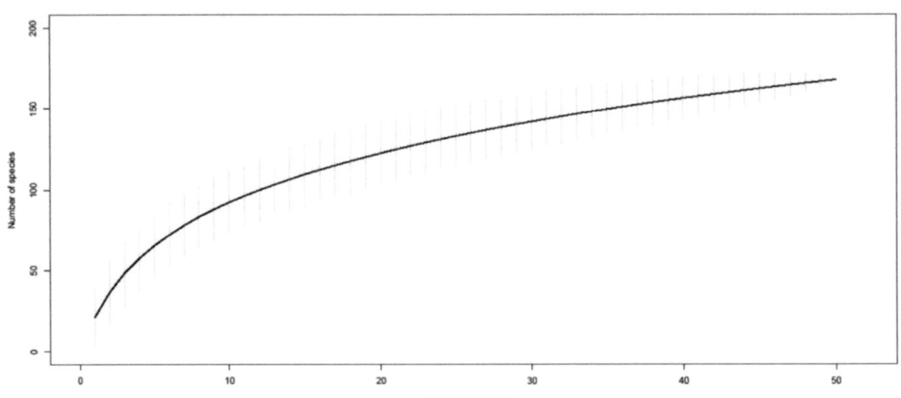

Figure 3.
Cumulative species richness curve of 50 samples of synthetic vegetation data, with their standard deviations represented by blue vertical bars. Clearly in this case the asymptote is a way off to the right, but with a sharply increasing cost-per-return of sampling effort.

If the aim is to infer cause and effect, strictly we should expect to have to impose experimental conditions including controls and randomization. For some questions however, the ability to state that the effect presented from the data is consistent or not with a range of competing scenarios can be sufficiently powerful.

The data of multiple sites or quadrats can be pooled and their cumulative species richness plotted to assess the characteristic curve to horizontal asymptote (e.g. **Figure 3**). The curves are calculated from random permutations of the data (subsampling without replacement) with standard deviations (blue bars) from the moment-based estimation of Ugland et al. [11]. Such a plot may provide an objective basis to follow the stand-delimitation rules of Andrade et al. [12], to be (i) large enough to contain all species of that plant community, (ii) uniform within the chosen area, and (iii) as homogeneous as possible in plant coverage.

5. Multivariate tools

Vegetation community data are often interpreted through a multivariate analysis, on the grounds that pattern or order is sought in more than one dependent variable (in this case, species or site) and that the joint behavior of these variables is also important. Oddly in this sense, analysis of vegetation surveys can be comparable to that of movie-provider platforms and internet marketing —clickbait— where an optimal fit of multiple axes is sought in order to maximize explainability (for the latter, conversion rate of internet consumers) from the data.

Handling the simultaneous effects of the wide range of explanatory factors (Section 2) on many individual actors (species, in our case) is a challenge, in particular the statistical errors associated with sampling, non-standard distributions, and the ogre of multiple comparisons. For example a study on a Mediterranean island [13] concluded that vegetation succession is a complex multi-factor process in which the final (or stable) stage of the process is not predefined, but depends on the factors affecting succession.

The goal of this approach is to analyze the data in an optimal way based on similarity or 'distance', much like the mileage charts between cities in twentieth century road maps (e.g. [14]). The survey data are converted to a dissimilarity matrix (generated by one of a range of different criteria, e.g. 'Euclidean', 'Manhattan') so to provide a measure of how different any plot is to another based on its species and presence or abundances. That matrix is the starting point for all of the methods discussed in this section.

5.1 Cluster analysis

The {mlr3} package [15] provides functions for a range of methods including regression, classification and cluster analysis. Performance of the clustering method chosen can be gauged by ground-truthing to actual class membership (where this is known), or alternatively by quantification of the intrinsic similarity of the data that have been placed within each cluster.

In K-means clustering [16] for instance, the analyst specifies in advance their desired number of clusters (the 'k'). The algorithm iteratively seeks for centroids of clusters (i.e. their mean positions) and assigns each point to its nearest centroid until a convergence condition is met.

Alternatively, instead of creating centroids and measuring distances, a density-based method can be used such as DBSCAN (Density-Based Spatial Clustering of

Applications with Noise, [17]). An important difference to K-means is that the analyst is not required to choose the number of clusters. Instead this approach takes into account the number of points in the surrounding area of a core sample (a sample that is within an area of high density): distance between points inside and outside of the core sample is calculated, and that process is iteratively spread throughout the entire dataset. As a result, density-based methods can cluster in shapes that are perhaps more realistic to the field data.

A third type of approach, hierarchical clustering, generates several levels of the data analogous to the tree-building methods used in phylogenetic analysis. Each data point at the base is treated as an individual cluster, leading upwards to a single cluster at the top containing all data points. The informative part in the middle allows us to examine for a varying number of clusters and their associated different subsets of the data.

The National Vegetation Classification (NVC: [18]) is the standard definition of clusters of vegetation types in Britain, built from around 35,000 samples across over 80% of the 10 × 10 km grid squares in England, Scotland and Wales. The R Shiny web application RMAVIS [19] enables the automated assignment of vegetation plot data to NVC units.

5.2 Ordination

Another strategy is to try to arrange the species or samples of the matrix along a gradient, termed ordination. The goal of this is to reduce the dimensions of the dataset to a smaller one, while still preserving as much of the original information value (i.e. relationship to explanatory factors) as possible. In recent years ordination has also been adopted by population geneticists, for example to infer from equivalent stretches of DNA sequences how individuals and populations of a species are structured (e.g. [20]).

Ordination is detailed at length elsewhere in this series [21], and in summary features two main classes of methods—those which aim for restoration of species composition (e.g. Nonmetric Multidimensional Scaling, NMDS), and those which seek analysis of gradients (e.g. Detrended Correspondence Analysis, DCA). By way of example we will briefly look at the former.

The {vegan} package [22] contains functions for ordination (and other multivariate analyses) as well as indices for diversity analysis of ecological communities. Its metaMDS function iteratively searches for optimal 'axes' in the dissimilarity matrix described in Section 5, above. This is similar in approach to searching for 'best fit' lines on a bivariate graph, which in each case condense the information of the matrix to a single column. Several iteration runs of the function are recommended, each using different optimization criteria. NMDS can handle non-linear species responses, and rectifies the 'arch' effect of some other ordination techniques (e.g. Principal Coordinates Analysis, PCoA) by maximizing the rank order correlation. A criticism however is that the information on species identities is obscured when the distance matrix is created.

A standard way to visualize an ordination is to plot the values of the first two axes against each other (e.g. **Figure 4**). This can be interpreted as 'plots in species space' (or vice versa, if the aim is to study species rather than plots). The direction of the axes, and their numeric scale, is arbitrary and so should not affect the interpretation.

If a reasonable ordination fit is achieved, we may expect the ranking of the datapoints on its axes, and placement of the dots in its plots, to have ecological meaning.

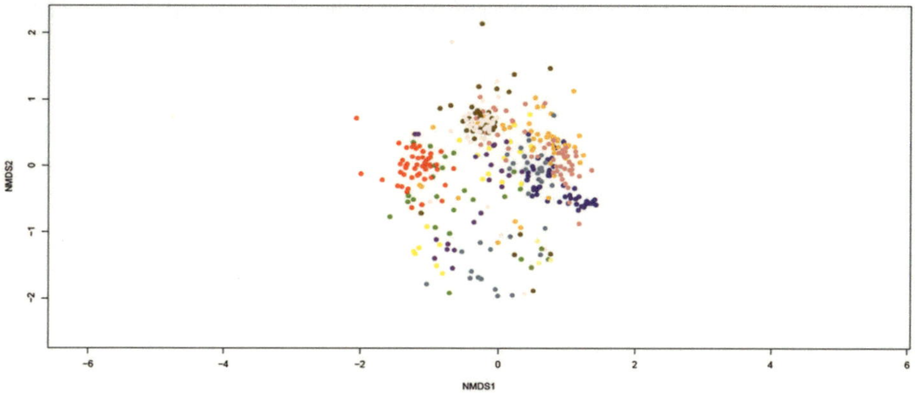

Figure 4.
Non-metric multidimensional scaling ordination of synthetic vegetation plot data: first two axes. Quadrats from respective sites are represented by the same color.

In **Figure 5**, site environmental variables have been plotted on the same NMDS diagram. The arrow points to the direction of most rapid change in the environmental variable, termed the direction of the gradient. The length of the arrow is proportional to the correlation between ordination and environmental variable, termed the strength of the gradient.

5.3 A model-based approach

In contrast to the distance-based methods above, the {mvabund} package enables a model-based approach to the analysis of multivariate abundance data [23]. A central aim of this is to account better for the mean-variance relationship alluded to in Section 2, resulting from the sparse and zero-inflated nature of vegetation data. A single Generalized Linear Model (GLM) is fit to each response variable, then resampling is used to test for community-level or species-level responses to the predictors. As ever when using GLM, the model's residuals should be plotted to check for random

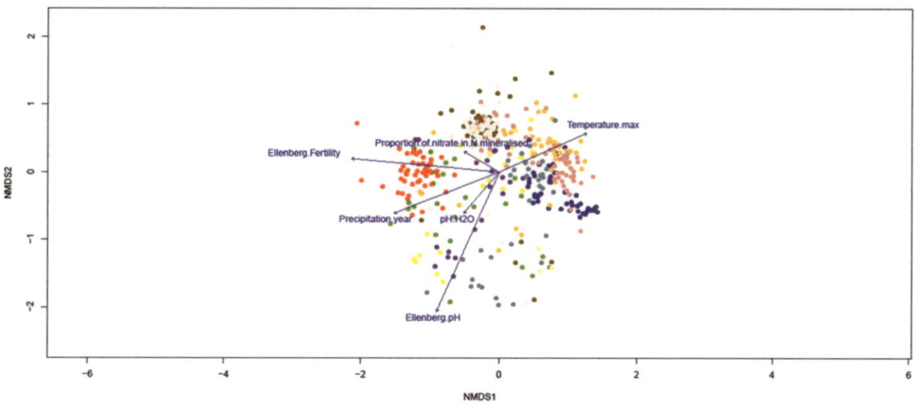

Figure 5.
Superimposition of environmental variables (shown in light blue type) over **Figure 4***, the non-metric multidimensional scaling ordination of synthetic vegetation plot data.*

scatter, and a statistical distribution chosen (the 'family' function in {mvabund}) which best matches the data.

An analysis of species compositional variation in urban flora [10] used the manyglm function of {mvabund}. A negative binomial error distribution was specified, and assumptions of mean-variance and log-linearity checked by plotting directly and by plotting residuals versus fitted values. In this way it was possible to point to key species accounting for the majority of differences (i.e. the model's deviance explained) between sites surveyed. Interestingly in this case that was a small suite of herbaceous taxa including weeds.

6. Robust indicators for monitoring

6.1 Type diversity

As number of plant species rarely increases linearly with number of individuals present, it would be misleading to attempt to correct for different sample sizes simply by dividing the number of species by the number of individuals. The challenge is to quantify how many different types (in this case species) exist in a dataset whilst reliably adjusting for both sampling effort and starkly non-normal distributions of abundance.

Information science such as on words and letters (e.g. [24]) and number of political parties (e.g. [25]) has made available a range of variance measures of type abundance distribution. These methods can also be applied to surveys of biological species—data which commonly share those statistical properties of non-normality and incomplete sampling. The intention is to count types while also taking account of their relative abundance, so to produce a more faithful representation of diversity that is not confounded by total number of types (i.e. richness), essentially to be independent of sampling intensity. These indices tend to behave differently, accordingly with differing interpretations of their tracking of underlying phenomena, and are often correlated [26]. For example Shannon H', a measure of entropy, is calculated as:

$$H' = -\Sigma\left[p_i \, x \ln(p_i)\right] \quad (1)$$

where p_i in our case would be the proportion of quadrats belonging to each species, of the total number of quadrats sampled. As unevenness is penalized by Shannon H', this index is intuitively useful for comparing site species lists, where the total number of species may be similar yet one site is dominated by few species, and the other is not. Typical values of H' for ecological data range between 1.5 and 3.5. H' can be translated to Effective Number of Species (ENS) by calculating:

$$ENS = e^{H'} \quad (2)$$

which is interpreted as the amount of equally-common species which could be supported at that site, given the survey data recorded. ENS is at its maximum per site (the actual number of species recorded) when all species were recorded as equally common.

Pielou's J [27] is instead a measure of evenness, which may be useful alongside simple richness, and is calculated simply from Shannon H' as:

$$J = \frac{H'}{H_{\max}} \tag{3}$$

where $H_{\max}$ is the maximum possible value of H', i.e. if every species was equally likely.

6.2 Resource provision

The species and abundance data collected at survey can be used to estimate ecological resources being provided by unit area of the vegetation, for example winter bird seed, frugivore food, structural nesting resource, biomass for detritivores, and pollinator food (e.g. **Figure 6**). Aside from a reliable sampling protocol to return representative data (discussed above), what is required is a plausible means of translating estimated counts to summed resource.

For example in research of nectar and pollen provision by urban vegetation [10], a preliminary study obtained species-specific values of nectar sugar and pollen mass per flower. The former was estimated by using microcapillary tubes to draw up known volumes of nectar from individual flowers, then obtaining a sucrose concentration of that nectar with a refractometer [28], the same equipment used by car mechanics to measure the concentration of antifreeze in radiator coolant. Pollen mass was estimated by opening mature anthers of individual flowers into solution under the microscope, estimating each pollen grain's mass from measurements of its dimensions, and finally estimating the total mass per flower using the grids of a Lund cell.

6.3 Other indices

Fortunately for analysts of vegetation data in Britain and Ireland there is a comprehensive atlas of plant attributes [29]. Valuably this also contains climatic values for plant species (calculated as the mean temperature and rainfall of the 10-km squares

Figure 6.
*Pollen beetle (*Brassicogethes *sp., centre of image) and Potato bug (*Closterotomus norwegicus*, middle right of image) feeding on the floral resources of Corn marigold (*Glebionis segetum*).*

where they occur in Britain, Ireland and the Channel Islands, averaged over the squares enumerated for the counts).

Conveniently when summed to the scale of plot vegetation, the expectation is that these indices will be reasonably robust to inter-surveyor variability, given that any disagreement of a few species (which would be from a similar ecological niche, so have a similar index value) is unlikely to have a large effect on the overall mean.

For instance, Ellenberg et al. [30] defined a series of scales (including Light (L), Fertility (N), pH (R) and Wetness (F)) and assigned values to each species based on their characteristic environmental response to that variable. Conversely Grime's indices [31] classify plant species in terms of the three established strategies, which were originally used to plot as a coordinate in a triangle whose points are Competitors (C), Stress-tolerators (S) and Ruderal (R). Note therefore that the values of Grime's index are not independent: for any site an average increase in one will lead to an average decrease in one or other strategy.

A range of similarity indices are available (e.g. Jaccard, Sørensen, Czekanowski coefficient) which aim to gauge how much the (species-, in our case) composition of two samples is alike or unalike. These indices are calculated from the species that are alternatively common to both samples or feature only in one of them.

The Raunkiær system [32] listed life forms for most species, based on the location of the bud of the plant during adverse conditions, for example phanerophytes (having resting bud more than 50 cm above the soil surface), chamaephytes (bud a maximum of 25 cm above the soil surface) and cryptophytes (buds subterranean or under water resting buds). In a study of dry grassland vegetation these values were used with a range of functional indices by Schnoor et al. [33], to conclude that such a trait-based approach informs on both the vegetation response and the abiotic conditions created.

Studying vegetation with indices derived from the species therefore appears to provide something more and different to the analysis of species composition alone.

7. Autocorrelation in time and space

7.1 Temporal change

Long-term vegetation monitoring can be a challenge to win funding for [34], which seems counterintuitive in that it often yields the most powerful results (e.g. [35, 36]). For example, spatial and temporal scale of plant-pollinator networks have been demonstrated to be pivotal to their interpretation [37], yet accounting for these variables in a statistical model brings its own challenges.

Quantification of the national vegetation stock and change in Britain has been provided by Countryside Survey data [38], which is invaluable to the tracking of change in plant species diversity and composition.

A common workaround in the absence of a single long-term dataset is to extend the time series backwards by using different datasets, and endeavor to control for their methodological differences. For example in a study of remnant native riparian vegetation, Jones et al. [39] approached this challenge as one of data calibration using 'double sampling'. A large sized, low-resolution dataset was collected in conjunction with a more detailed one of a subset of the same sampling characteristics. The latter was modeled (with a hierarchical Bayesian framework, see Section 10, below), and the calibration achieved by applying that model's parameters to convert the historical, low-resolution dataset.

With luck, historical references may exist and be informative. In a study of montane vegetation, Trant et al. [40] used old photographs together with modern imagery to conclude that tree-line at higher altitudes and further north had a greater probability of advancing. Critically, regional climate factors in their Generalized Linear Mixed Models (GLMM) did not significantly explain the results.

7.2 Spatial patterns

Kent [41] determined the incorporation of spatial analysis and geostatistics into vegetation science to be a key development for the future. As elsewhere, increasing CPU power and the ubiquitous collection of big (spatial, in this case) data have stimulated research on pattern and process. The range of associated statistical tests have strengths and weaknesses, and the {spatstat} family of packages [42] is an excellent resource for such analysis of spatial point patterns in vegetation. From the outset, its quadrat count test (used by calling the function quadrat.test) can be used as a simple statistical check on whether points exhibit Complete Spatial Randomness (CSR, i.e. a homogeneous Poisson process) or in other words are uniformly random across the area surveyed.

The spatial configuration of plants in a stand is of importance both as a covariate to explain responses (e.g. herbivory, pollination, nesting habitat) and for examination of the result of vegetation processes (e.g. competition, soil, succession), yet such point process data are challenging for traditional statistical tests. Spatial measures are often analyzed using simulation procedures [43], such as the examination of farm survey data for spatial autocorrelation [44] by Monte Carlo permutation of Moran's I statistic.

{spatstat} can be used for simulation of patterns of plant spatial distribution, often with the intent of determining their contagiousness or repulsion. Spatial and temporal patterns are a consideration in themselves due to the implications of autocorrelation, i.e. residual non-random structure in the data. Perry et al. [45] noted the importance of appreciating how spatial dependence may affect non-spatial inferential statistical techniques, which typically assume that model errors are independent [46].

Spatial interpolation of vegetation data, including kriging and variogram modeling, can be done simply with the package {gstat} [47]. This was used for example in a 20-year study [48] of wetland vegetation to plot interpolated maps of quadrat data. Overall functional changes became apparent, and the spatial mapping indicated that areas of the site were exhibiting varying characteristics in space and time.

More powerful spatial analyses can be done with an ensemble of models, such as available in the package {biomod2} [49]. This allows the analyst to simultaneously run up to 10 single models on a presence/absence dataset, examine for model performance and plot the subsequent ensemble projections.

8. Network structure

Another way to examine vegetation data and generate indices is to borrow from methodologies developed in the field of graph theory. Emergent properties of complex sets of binary interactions can be tracked by the mathematical structure of nodes and edges (i.e. connections between the nodes) which comprise their network (e.g. **Figure 7**), with similar patterns found in a wide range of systems including metabolic pathways [51], food webs [52], film actors [53], world trade [54], social networks [55] and the spread of contagious diseases [56].

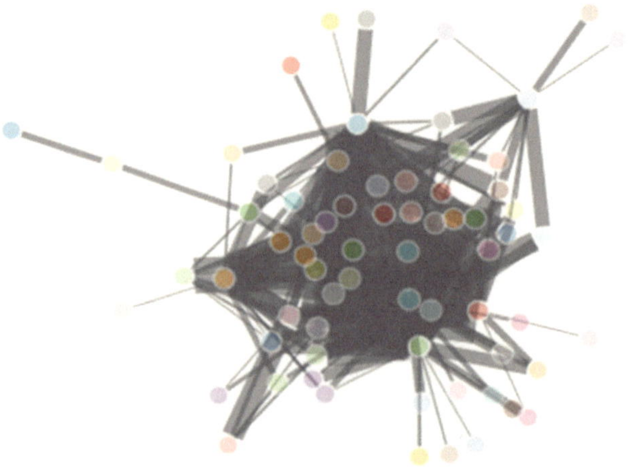

Figure 7.
Interactive visualization plotted using {networkD3} [50] of its network structure of nodes and edges. With this package the analyst can manipulate their data in three dimensions to view weights (i.e. strength of edges) and degrees (i.e. number of edges joining each node).

Characteristics of self-construction, departure from randomness, and a capacity for robustness against random failure have been observed across these different networks. Indeed, the original 'PageRank' search engine used network structure to randomly pick nodes (i.e. websites) and traverse to other nodes with a given probability, enabling it to assess websites which exist on the periphery.

A bipartite network is a common type of network in which the nodes fall into two groups, which interact with the other group but not themselves. Dozens of indices of network topology (e.g. nestedness, linkage density, modularity) of vegetation data can be calculated with {bipartite} [57] and the networks visualized in graphical form (see **Figure 8**). In our experience these indices tend to display a useful robustness to sample size, and are therefore appropriate candidates to monitor a form of change across time.

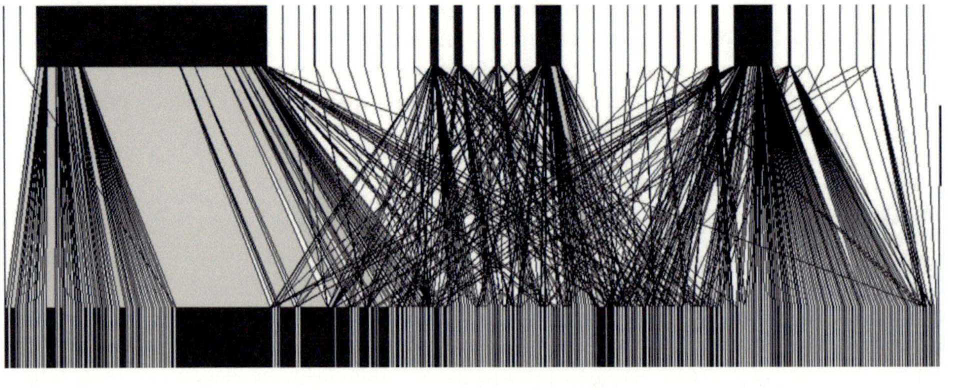

Figure 8.
Projection of a bipartite graph of synthetic vegetation data, with species comprising the lower level and sites comprising the upper level. Even before generating numerical indices of the network, the analyst can perceive general patterns of structure in the data such as specialist, generalist and dominant nodes.

9. Machine learning

In contrast to the theoretical and asymptotic principles of twentieth century statistics, machine learning methods thrive on the wealth of available data of the twenty-first century. Its supervised methods split data into a 'training' set and a 'testing' set then do trial and error (lots of times) with a feedback mechanism from each error to the next trial. The use of a testing set is to provide an unbiased estimate of a model's performance, which is important in providing a realistic approximation for how well the model will perform in the wild. The aim is that this model generalizes well in a practical, theory-decoupled way to new, unseen data, and the predictive power of this approach is often unassailable.

A good example is the Random Forest method [58], which is primarily a classification tool but can also easily deal with regression tasks. The algorithm builds multiple decision trees then merges them to obtain a more accurate and stable prediction. The 'forest' that is built is an ensemble of decision trees, whose parameters can be tuned to optimize predictivity. The general and intuitive logic of this approach is that a combination of learning models improves the overall performance of the model. This deceptively simple tool often performs exceptionally on predictivity, and some light can be shed on the 'black box' for example in the {randomForest} package [59], by comparing importance measures for each variable in the dataset (e.g. **Figure 9**).

Equivalent to traditional cluster analysis, unsupervised learning is used to discover underlying structure in the raw data. For example with a dataset of 22,000 labeled images of native weed-rich grassland [60], a Convolutional Neural Network (CNN) was trained in Python's PyTorch deep learning framework [61] based on the library developed by Facebook AI Research. A model was built which identifies and counts floral units 35 times more rapidly than an expert human surveyor. Errors abound of

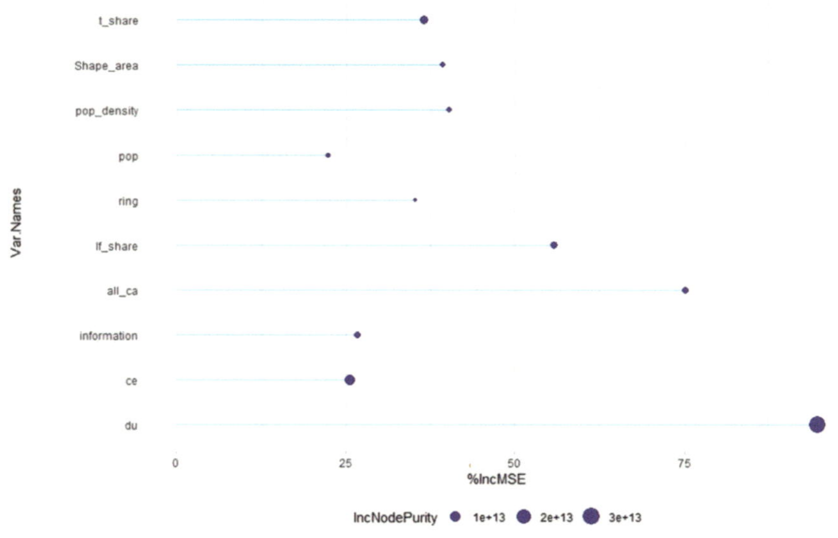

Figure 9.
Assessment of the importance of the candidate variables in a Random Forest model of synthetic vegetation data. This is represented via the horizontal axis of permutation differences in prediction error that a variable contributes, divided by its variability (%IncMSE); and via the blue ball size of change in homogeneity of the groups created by the trees using the Gini coefficient ('IncNodePurity'), i.e. how well the permuted trees split the data.

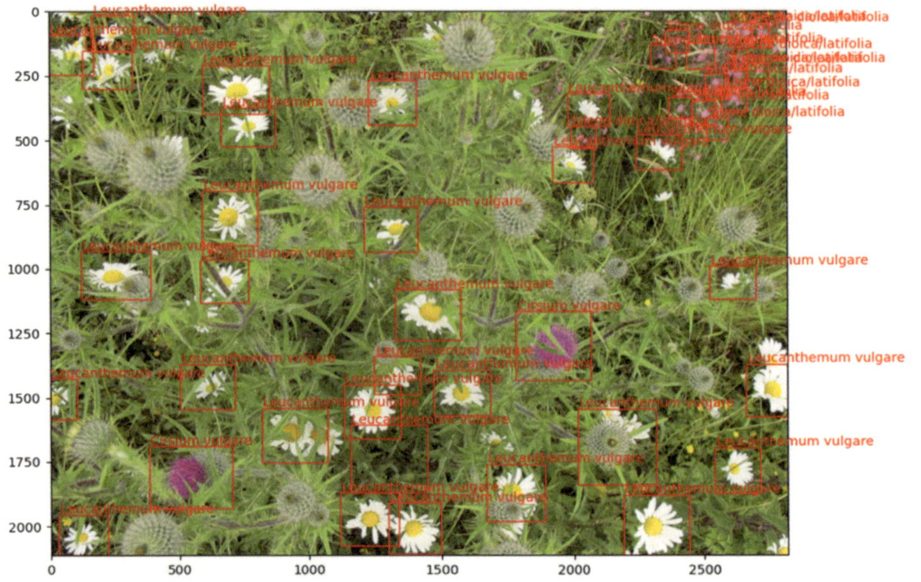

Figure 10.
Object detection of floral resources on a model-unseen image of rough grassland using a Convolutional Neural Network Model [60], allowing rapid conversion to resource provision per square metre of this vegetation.

course, but they are within that of human surveyors, and in contrast to humans the estimate of the model has known subjectivity (see **Figure 10**).

10. Bayesian hierarchical modeling

The view of Kent [41] on vegetation science was that 'one of the most significant developments for the future almost certainly lies in Bayesian statistics'. The Bayesian approach derives from the theorem posthumously published by Reverend Thomas Bayes in 1763 [62]. This was overshadowed by the popularity of the frequentist methods developed in the first half of the twentieth century, and indeed was prohibitively intractable to implement until the development of new computational power in the late 1980s.

In the alternative, frequentist, framework a parameter of interest is assumed to be unknown but fixed. That is, it is assumed that in the population there is only one true population parameter, for example, one true mean or one true regression coefficient. Weaknesses in this approach may appear when only limited or variable data are available, and according to Kruschke [63], Bayesian models bring an advantage of being 'designed to be appropriate to the data structure, without having to make approximation assumptions typical in Null Hypothesis Significance Testing'.

In the Bayesian view of subjective probability [62], all unknown parameters are treated as uncertain and therefore are described by a probability distribution. Every parameter is unknown, and everything unknown receives a distribution. In this framework there is the opportunity to incorporate domain expertise, as prior information, into the inferences.

While in frequentist statistics the data alone are used to model unknown parameters and quantities, Bayesian statistics also incorporate any external information, as priors. Generally, priors are specified as an assumed observational model, informed by those variables which can be directly observed and those which cannot.

10.1 Bayesian priors

A general strategy could be to select priors which are non-informative, effectively deciding that there is no existing information about the variables of interest, and assigning equal weight to all values within the prior distribution. But a Bayesian analyst might argue that usually something is known in advance, even if it is the order of magnitude, and that a correct model therefore should incorporate these bounds.

10.2 Stochastic simulation

For a system that is too complex to study directly (as often the case for vegetation data, not being in the form of a standard probability distribution), it is possible to create a simulation and study the properties of that instead. One general and powerful method is Markov chain Monte Carlo (MCMC), which is capable of sampling from almost any shape of distribution.

MCMC is a class of techniques for solving integrals, in this case involving the specification of a transition model with the aim to draw samples from an intractable posterior distribution. A 'memoriless' random walker, the Markov chain, is set in motion which has as its equilibrium distribution (i.e. the stable phase following convergence) the product of the given priors and likelihoods. The walker is programmed to make local steps which are accepted or rejected, to move in the parameter space towards areas of higher probability. In this way the posterior distribution is estimated by sampling, while all that is known about the distribution is how to calculate its density for different samples.

10.3 Coding

Increases in computing power together with probabilistic programming languages such as WinBUGS [64], JAGS [65] and Stan [66] facilitate the user to construct and run Bayesian models, largely without having to grapple with the complexities of the stochastic simulations.

Deploying Stan for its estimation and sampling, the package {brms} [67] provides Bayesian equivalents of many popular R modeling packages such as {lme4} [68]. Stan implements Hamiltonian Monte Carlo [69] together with the No-U-Turn (NUTS) Sampler [70] to converge faster than other MCMC algorithms, particularly for the high-dimensional scenarios [71] common in multilevel modeling.

The model formula, data, and family (with additional optional arguments, including number of iterations and chains) are defined by the user in R, then brms generates the model code in Stan language and prepares the data for use. The Stan code and data are passed to functions of the {rstan} package (the R interface of Stan — [72]) then translated and compiled in C++. The returned fitted model object is processed in brms for manipulation in R.

Bayesian inferences of unknown parameters are outputted as posterior distributions which can be plotted and summarized by measures of location (e.g. **Figure 11**).

By default four MCMC chains are started independently from each other, each one searching for samples of the posterior distribution in a multidimensional space corresponding to each of the model parameters, priors and likelihood specified by the

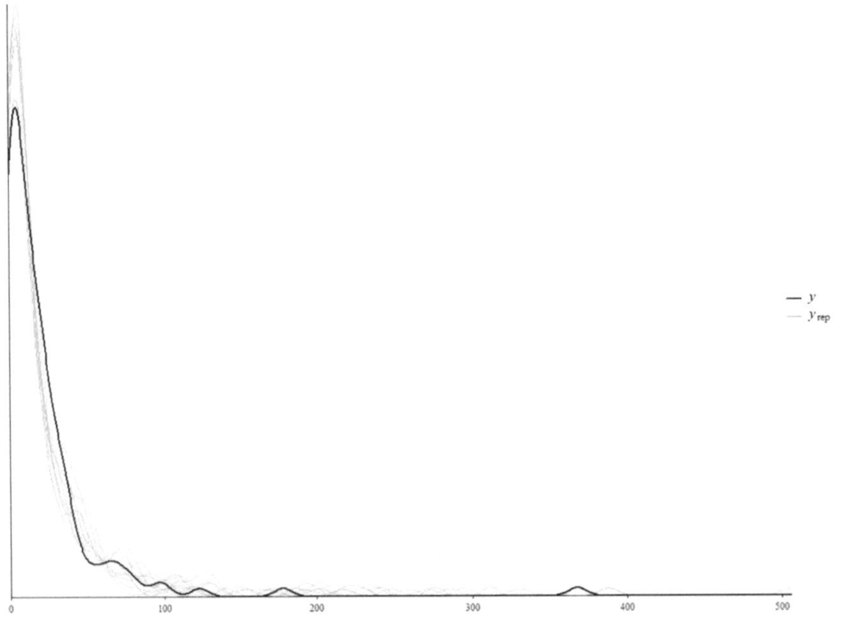

Figure 11.
Posterior distribution plot of synthetic vegetation abundance data in Stan, illustrating how the model faithfully tracks field survey abundance data (y, in black) by 20 model replicates (y_{rep} in gray). This is achieved despite the apparently intractable distribution of the black line of raw data.

user. Beginning at random locations, eventually (assuming successful convergence) the chains reach the region of the posterior distribution and sample the posterior thereafter.

The mixing of chains are viewed as trace plots, and initial burn-in set of samples removed to eliminate them from the starting path to the posterior distributions. As the MCMC samples are autocorrelated within a chain, estimations of effective sample size (ESS) are made as the number of independent samples with the same estimation power as the number of autocorrelated samples. Low ESS indicates sampling problems.

Performing vegetation analysis in Stan allows the specification of a more complex model than would be possible in standard modeling software, at once accounting for the non-independence of variables such as survey time and site, use of Bayesian priors for the predictor variables, and trials of a range of data distributions.

11. Conclusion

This chapter founded on the nature of vegetation data and considerations arising from its capture by survey (Section 2) and biological and statistical factors intrinsic to the subject material (Sections 3 and 4). Quantitative methods proposed to address those features progressed from analysis of the joint behavior of variables (Section 5), indices of overall attributes such as type diversity and resource provision (Section 6), temporal and spatial patterns (Section 7), emergent properties of network structure (Section 8), to machine learning (Section 9) and Bayesian modeling (Section 10).

It is a real pleasure to review a range of powerful methods and open source coding tools available to the analyst of vegetation data, and all signs suggest that innovative techniques and further developments are just around the corner.

Acknowledgements

Many thanks to Nina Miocevic and Ana Cristina Gonçalves for guidance and support.

Author details

Damien Hicks
Kings Pool, York, United Kingdom

*Address all correspondence to: damien.hicks1@defra.gov.uk

References

[1] Spellerberg IF. Monitoring Ecological Change. 2nd ed. Cambridge: Cambridge University Press; 2005. DOI: 10.1017/CBO9780511614699

[2] Cambridge Dictionary. Cambridge University Press & Assessment. 2024. Available from: https://dictionary.cambridge.org/dictionary/english/evaluation

[3] Bastow Wilson J. Species presence/absence sometimes represents a plant community as well as species abundances do, or better. Journal of Vegetation Science. 2012;**23**:1013-1023. DOI: 10.1111/j.1654-1103.2012.01430.x

[4] Grieg-Smith P. Quantitative Plant Ecology. Oxford: Blackwell Scientific Publications; 1983. ISBN 0-632-01084-3

[5] Chen J, Shiyomi M, Huang D, Yu H. Quantitative evaluation of species composition dissimilarity within a community and among communities. Ecological Research. 2021;**36**:152-160. DOI: 10.1111/1440-1703.12192

[6] Richards AJ. Field handbook to British and Irish Dandelions. BSBI Handbook 23. Durham: Botanical Society of Britain and Ireland; 2021

[7] Stace CA, Preston CD, Pearman DA. Hybrid Flora of the British Isles. Bristol: Botanical Society of Britain & Ireland; 2015

[8] Ouvrard P, Hicks DM, Mouland M, Nicholls JA, Baldock KCR, Goddard MA, et al. Molecular taxonomic analysis of the plant associations of adult pollen beetles (Nitidulidae: Meligethinae), and the population structure of *Brassicogethes aeneus*. Genome. 2016;**59**(12):1101-1116. DOI: 10.1139/gen-2016-0020

[9] McCall CH. Sampling and Statistics Handbook for Research. Ames, IA: The Iowa State University Press; 1982

[10] Hicks DM, Ouvrard P, Baldock KCR, Baude M, Goddard MA, Kunin WE, et al. Food for pollinators: Quantifying the nectar and pollen resources of urban flower meadows. PLoS One. 2016;**11**(6):e0158117. DOI: 10.1371/journal.pone.0158117

[11] Ugland KI, Gray JS, Ellingsen KE. The species-accumulation curve and estimation of species richness. Journal of Animal Ecology. 2003;**72**:888-897. DOI: 10.1046/j.1365-2656.2003.00748.x

[12] Andrade B, lob Boldrini I, Cadenazzi M, De Patta Pillar V, Overbeck GE. Grassland vegetation sampling - a practical guide for sampling and data analysis. Acta Botânica Brasílica. 2019;**33**(4):2-7. DOI: 10.1590/0102-33062019abb0160

[13] Tzanopoulos J, Mitchley J, Pantis JD. Vegetation dynamics in abandoned crop fields on a Mediterranean island: Development of succession model and estimation of disturbance thresholds. Agriculture, Ecosystems & Environment. 2007;**120**(2-4):370-376. DOI: 10.1016/j.agee.2006.10.011

[14] Rand McNally: Transcontinental mileage chart of the United States [Internet]. 1948. Available from: https://archive.org/details/dr_rand-mcnally--transcontinental-mileage-chart-of-the-united-states-14003053

[15] Lang M, Binder M, Richter J, Schratz P, Pfisterer F, Coors S, et al.

mlr3: A modern object-oriented machine learning framework in R. Journal of Open Source Software. 2019. Available from: https://joss.theoj.org/papers/10.21105/joss.01903

[16] Steinhaus H. Sur la division des corps matériels en parties. Bulletin L'Académie Polonaise des Science. 1957;**4**(12):801-804 MR 0090073. Zbl 0079.16403

[17] Ester M, Kriegel H, Sander J, Xu X. A density-based algorithm for discovering clusters in large spatial databases with noise. In: Proceedings of the Second International Conference on Knowledge Discovery and Data Mining. Portland, Oregon: AAAI Press; 1996. pp. 226-231. CiteSeerX 10.1.1.121.9220. ISBN 1-57735-004-9

[18] Rodwell JS. National Vegetation Classification User's Handbook. Peterborough: Joint Nature Conservation Committee; 2006. ISBN: 13 978 1 86107 574 1

[19] Marshall Z, Smart SM, Harrower C, Marrs R. RMAVIS. Zenodo; 2024. Available from: https://zenodo.org/doi/10.5281/zenodo.10818640

[20] Jombart T, Pontier D, Dufour AB. Genetic markers in the playground of multivariate analysis. Heredity. 2009;**102**:330-341. DOI: 10.1038/hdy.2008.130

[21] Zare Chahouki MA. Classification and Ordination Methods as a Tool for analyzing of Plant Communities. Multivariate Analysis in Management, Engineering and the Sciences. London, UK: IntechOpen; 2013. DOI: 10.5772/54101

[22] Oksanen FJ, Blanchet G, Kindt R, Legendre P, Minchin PR, O'Hara RB, et al. Vegan: Community Ecology Package. R package Version 2.4-3. 2019. Available from: https://CRAN.R-project.org/package=vegan

[23] Wang Y, Naumann U, Wright ST, Warton DI. Mvabund– An R package for model-based analysis of multivariate abundance data. Methods in Ecology and Evolution. 2012;**3**:471-474. DOI: 10.1111/j.2041-210X.2012.00190.x

[24] Shannon CE. A mathematical theory of communication. Bell System Technical Journal. 1948;**27**(3):379-423. DOI: 10.1002/j.1538-7305.1948.tb01338.x

[25] Laakso M, Taagepera R. "Effective" number of parties: A measure with application to West Europe. Comparative Political Studies. 1979;**12**(1):3-27. DOI: 10.1177/001041407901200101

[26] Hill MO. Diversity and evenness: A unifying notation and its consequences. Ecology. 1973;**54**(2):427-432. DOI: 10.2307/1934352

[27] Pielou EC. The measurement of diversity in different types of biological collections. Journal of Theoretical Biology. 1966;**13**:131-144. DOI: 10.1016/0022-5193(66)90013-0

[28] Corbet SA. Nectar sugar content: estimating standing crop and secretion rate in the field. Apidologie. 2003;**34**:1-10. DOI: 10.1051/apido:2002049

[29] Hill MO, Preston CD, Roy DB. PLANTATT: Attributes of British and Irish Plants. Cambridgeshire: Centre for Ecology and Hydrology; 2004

[30] Ellenberg H, Weber HE, Düll R, Wirth V, Werner W, Paulissen D. Zeigerwerte von Pflanzen in Mitteleuropa. Scripta Geobotanica. 1991;**18**:1-248

[31] Grime JP. Plant Strategies and Vegetation Processes. Chichester: John Wiley; 1979

[32] Raunkiær C. Types biologiques pour la géographie botanique. Oversigt over Det Kongelige Danske Videnskabernes Selskabs Forhandlinger. 1905;**5**:347-438

[33] Schnoor T, Bruun HH, Olsson PA. Soil disturbance as a grassland restoration measure—Effects on plant species composition and plant functional traits. PLoS One. 2015;**10**(4):e0123698. DOI: 10.1371/journal.pone.0123698

[34] Breeze TD, Bailey AP, Balcombe KG, Brereton T, Comont R, Edwards M, et al. Pollinator monitoring more than pays for itself. Journal of Applied Ecology. 2021;**58**:44-57

[35] Song DS. Spatial and temporal variation in wild pollination service in the Mongolian Steppe. Publicly Accessible Penn Dissertations. 2015. Available from: https://repository.upenn.edu/edissertations/1142

[36] Herrera CM. Complex long-term dynamics of pollinator abundance in undisturbed Mediterranean montane habitats over two decades. Ecological Monographs. 2019;**89**(1):1-28. Available from: https://www.jstor.org/stable/26641228

[37] Prendergast KS, Ollerton J. Spatial and temporal scale of analysis alter conclusions about the effects of urbanisation on plant–pollinator networks. Arthropod-Plant Interactions. 2022;**16**:553-565. DOI: 10.1007/s11829-022-09925-w

[38] Wood CM, Smart SM, Bunce RGH, Norton LR, Maskell LC, Howard DC, et al. Long-term vegetation monitoring in Great Britain – The countryside survey 1978-2007 and beyond. Earth System Science Data. 2017;**9**:445-459. DOI: 10.5194/essd-9-445-2017

[39] Jones CS, Duncan DH, Morris WK, Robinson D, Vesk PA. Using data calibration to reconcile outputs from different survey methods in long-term or large-scale studies. Environmental Monitoring and Assessment. 2022;**194**(3):185. DOI: 10.1007/s10661-021-09727-2

[40] Trant A, Higgs E, Starzomski BM. A century of high elevation ecosystem change in the Canadian Rocky Mountains. Scientific Reports. 2020;**10**:9698. DOI: 10.1038/s41598-020-66277-2

[41] Kent M. Vegetation Description and Data Analysis : A Practical Approach. Hoboken: John Wiley & Sons, Incorporated; 2011. ProQuest Ebook Central. Available from: http://ebookcentral.proquest.com/lib/ed/detail.action?docID=7104526

[42] Baddeley A, Rubak E, Turner R. Spatial Point Patterns: Methodology and Applications with R. London: Chapman and Hall/CRC Press; 2016. ISBN 9781482210200. Available from: https://www.routledge.com/Spatial-Point-Patterns-Methodology-and-Applications-with-R/Baddeley-Rubak-Turner/p/book/9781482210200/

[43] Diggle PJ. Statistical Analysis of Spatial Point Patterns. London: Academic Press; 2003

[44] Betts CG, Hicks D, Reader M, Wilson P. Nitrogen balance is a predictor of farm business performance in the English Farm Business Survey. Frontiers in Sustainable Food Systems. 2023;**7**:1106196. DOI: 10.3389/fsufs.2023.1106196

[45] Perry G, Rolls ET, Stringer SM. Spatial vs temporal continuity in view invariant visual object recognition learning. Vision Research.

2006;**46**(23):3994-4006. DOI: 10.1016/j.visres.2006.07.025

[46] Carroll SS, Pearson DL. Detecting and modeling spatial and temporal dependence in conservation biology. Conservation Biology. 2000;**14**(6):1893-1897

[47] Pebesma EJ. Multivariable geostatistics in S: The gstat package. Computers & Geosciences. 2004;**30**:683-691

[48] Hicks D, Abraham F, Bardsley L, Cousins M, Webster E, Whitman J. Spatial and temporal vegetation analysis of Amberley Wild Brooks over two decades. British & Irish Botany. 2019;**1**(4):9-13. DOI: 10.33928/bib.2019.01.309

[49] Thuiller W, Georges D, Gueguen M, Engler R, Breiner F, Lafourcade B, et al. biomod2: Ensemble Platform for Species Distribution Modeling. R Package Version 4.2-6. 2024. Available from: https://biomodhub.github.io/biomod2/

[50] Allaire JJ, Ellis P, Gandrud C, Kuo K, Lewis BW, Owen J, et al. networkD3: D3 JavaScript Network Graphs from R. 2019. Available from: https://cran.r-project.org/web/packages/networkD3/index.html

[51] Ravasz E, Somera AL, Mongru DA, Oltvai ZN, Barabási AL. Hierarchical organization of modularity in metabolic networks. Science. 2002;**297**(5586):1551-1555. DOI: 10.1126/science.1073374

[52] Dunne JA, Williams RJ, Martinez ND. Food-web structure and network theory: The role of connectance and size. Proceedings of the National Academy of Science USA. 2002;**99**:12917-12922

[53] Collins J, Chow C. It's a small world. Nature. 1998;**393**:409-410. DOI: 10.1038/30835

[54] Serrano MA, Boguná M. Topology of the world trade web. Physical Review E. 2003;**68**(1):015101. DOI: 10.1103/PhysRevE.68.015101

[55] Jackson MO, Rogers BW, Zenou Y. The economic consequences of social-network structure. Journal of Economic Literature. 2017;**55**(1):49-95. DOI: 10.1257/jel.20150694

[56] Canright GS, Engoe-Monsen K. Spreading on networks: A topographic view. Complexus. 2006;**3**:131-146

[57] Dormann CF, Gruber B, Fruend J. Introducing the bipartite package: Analysing ecological networks. R News. 2008;**8**(2):8-11

[58] Breiman L. Random forests. Machine Learning. 2001;**45**(1):5-32. DOI: 10.1023/A:1010933404324

[59] Liaw A, Wiener M. Classification and regression by randomForest. R News. 2002;**2**(3):18-22. Available from: https://CRAN.R-project.org/doc/Rnews/

[60] Hicks D, Baude M, Kratz C, Ouvrard P, Stone G. Deep learning object detection to estimate the nectar sugar mass of flowering vegetation. Ecological Solutions and Evidence. 2021;**2**:e12099. DOI: 10.1002/2688-8319.12099

[61] Paszke A, Gross S, Chintala S, Chanan G. PyTorch. Facebook AI Research (FAIR); 2016. Available from: https://pytorch.org/

[62] Bayes T. An essay towards solving a problem in the doctrine of chances. Philosophical Transactions of the Royal Society of London. 1763;**53**:370

[63] Kruschke J. Doing Bayesian Data Analysis: A Tutorial with R, JAGS, and Stan. 2nd ed. Boston: Academic Press; 2015

[64] Lunn DJ, Thomas A, Best N, Spiegelhalter D. WinBUGS-A Bayesian modelling framework: Concepts, structure, and extensibility. Statistics and Computing. 2000;**10**(4):325-337

[65] Plummer M. JAGS Version 4.2.0. User Manual; 2016. Available from: https://people.stat.sc.edu/hansont/stat740/jags_user_manual.pdf

[66] Carpenter B, Gelman A, Hoffman MD, Lee D, Goodrich B, Betancourt MJ, et al. Stan: A probabilistic programming language. Journal of Statistical Software. 2017;**76**(1):2-27

[67] Bürkner P. brms: Bayesian Regression Models Using "Stan". 2019. Available from: https://CRAN.R-project.org/package=brms

[68] Bates DM, Mächler M, Bolker B, Walker S. Fitting linear mixed-effects models using lme4. Journal of Statistical Software. 2015;**67**(1):1-48. DOI: 10.18637/jss.v067.i01

[69] Neal RM. MCMC using Hamiltonian dynamics. In: Brooks S, Gelman A, Jones G, Meng X-L, editors. Handbook of Markov Chain Monte Carlo. New York: Taylor & Francis; 2011. DOI: 10.1201/b10905-10

[70] Hoffman MD, Gelman A. The no-U-turn sampler: Adaptively setting path lengths in Hamiltonian Monte Carlo. Journal of Machine Learning Research. 2014;**15**(1):1593-1623. Available from: http://dl.acm.org/citation.cfm?id=2627435.2638586

[71] Betancourt M. A conceptual introduction to Hamiltonian Monte Carlo. arXiv. 2017. DOI: 10.48550/arXiv.1701.02434

[72] Stan Development Team. Stan Modeling Language Users Guide and Reference Manual 2.32. 2023. Available from: https://mc-stan.org

Chapter 7

Pseudotsuga menziesii ((Mirb.) Franco) in Portugal: A Review of Species Characteristics, Silviculture, and Management

Teresa Fidalgo Fonseca, Maria Emília Silva, José Luís Lousada, Jaime Sales Luis and Ana Cristina Gonçalves

Abstract

Pseudotsuga menziesii ((Mirb.) Franco), known as Douglas fir, is a prominent conifer species native to North America. It is known for its adaptability, rapid growth, and high-quality wood. It was introduced to Europe in the nineteenth century and is currently the most abundant non-native tree species cultivated in Central European forests. In Portugal, the species was initially cultivated as an ornamental plant and used for experimental purposes. However, its notable qualities, particularly its high volume production, have garnered significant interest. This has led to the implementation of reforestation programs, which expanded the forested areas by the late twentieth century. This study aims to review the state-of-the-art on the species, with a focus on forest stands outside its natural distribution area, specifically in Portugal. The authors present information on the species' traits, growth dynamics, silviculture and management, available models to support its management, and wood properties. The main specific areas covered are: (*i*) the botanical and ecological characteristics of the species, including its distribution, genetic diversity, and responses to biotic and abiotic stresses; (*ii*) silvicultural practices and management strategies adapted to the species in Portugal, along with an overview of models and decision support tools; and (*iii*) characteristics of wood, covering its biometry, physical and mechanical properties.

Keywords: Douglas fir, species traits, distribution, silviculture, models, wood technology

1. Introduction

Pseudotsuga menziesii ((Mirb.) Franco), also known as Douglas fir, is a conifer of great economic and ecological importance in Western British Columbia and the United States of America, representing one of the largest natural fiber resources in

North America [1, 2]. The species has been recognized for its silvicultural and economic value in its natural distribution area and when cultivated outside its habitat [3]. In Europe, it has become an economically important species due to its rapid growth rate and high-quality wood [4–6], and it is currently the most abundant exotic tree cultivated in Central European forests [1].

In Portugal, the species covers a small area of about 4200 ha [7], in comparison to the 713,300 ha occupied by the country's most abundant conifer, the maritime pine (*Pinus pinaster* Ait.) [8]. Despite its small area the potential for afforestation with *Pseudotsuga menziesii* in Portugal has been estimated at 235,000 ha (Forestry Project Co-ordinating Group—World Bank 1986, cit. in [7]). Based on assessments of the suitability of the territory for the species, with the first reports dating back to the 1970s [9, 10], it is currently one of the species prioritized for planting in subregions of North and Central Portugal, as indicated by the legal documents of the Portuguese Government approving the Regional Forest Management Programmes, PROF (e.g., PROF de Trás-os-Montes e Alto Douro, Order no. 57/04/2019 [11]).

The purpose of this review is to advance the comprehension of *Pseudotsuga menziesii* beyond its native range, thereby facilitating the incorporation of the species into sustainable forest management practices in Portugal. This review aims to examine the *Pseudotsuga menziesii*'s botanical and ecological characteristics, including distribution and genetic diversity (Section 2); present current silvicultural practices and management strategies (Section 3); and provide information on wood, including properties and applications (Section 4). The review encompassed an investigation into relevant sources pertaining to the utilization of the species in Portugal. In instances where national information was scarce, the scope was expanded to encompass the European context or the species' area of origin.

2. Species traits and distribution

Pseudotsuga menziesii is a conifer belonging to the *Pinaceae* family and *Pseudotsuga* genus. It is an evergreen species that reaches large dimensions, up to 100 m in height and 4 m in diameter [12, 13]. It has a cylindrical trunk, and a conic crown, which becomes flat or irregular with aging [1, 12]. The species is monoecious and unisexual, starting fruiting at about 12–15 years of age [1]. It is naturally distributed in North America, the United States of America, Canada, and Mexico (**Figure 1**). In its natural distribution area, it develops from sea level up to 3200 m of elevation. The species is fast growing and tolerant to shade [1]. Two subspecies are found in its natural distribution area *Pseudotsuga menziesii* subsp. *menziesii* and *Pseudotsuga menziesii* subsp. *glauca* [12]. It is sensitive to wind and frost, especially the spring ones, and prefers light, deep, not compacted soils, with pH between 4.5 and 6, and does not tolerate active calcareous soils [14, 15].

Outside its natural range (**Figure 2**), *Pseudotsuga menziesii* was introduced to several countries in Europe, Oceania, South America, and Africa [15, 16]. A more detailed distribution of *Pseudotsuga menziesii* outside its natural range can be found in Van Loo [17]. Europe has the largest area share, due to its good adaptation to a variety of edaphic and climatic conditions, good quality wood, fast growth, and low damages caused by pests and diseases [1, 15].

Pseudotsuga menziesii was introduced to Europe [15, 18] in the nineteenth century. In Portugal, it was introduced in the nineteenth century as an ornamental plant near

Figure 1.
Natural distribution of Pseudotsuga menziesii. *(source: https://github.com/wpetry/USTreeAtlas).*

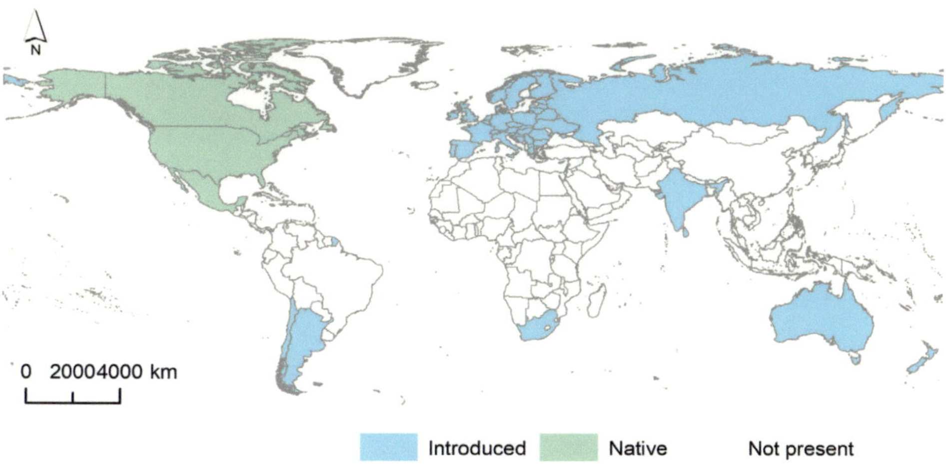

Figure 2.
Distribution of Pseudotsuga menziesii *outside its natural range.*

Lisbon (Sintra) followed by the plantation of a few individuals in Central Portugal (Buçaco and Pedras Salgadas). It was from the beginning of the twentieth century onward that the first plantations were established by the Portuguese Forest Service in Central (Serra da Estrela) and North (Serra do Gerês) Portugal ([19] and references therein). Until the mid-1970s of the twentieth century, the area of plantation of *Pseudotsuga menziesii* in Portugal was small (300 ha). After that, a program of afforestation resulted in circa 7000 ha of plantation of *Pseudotsuga menziesii* in the 1990s of the twentieth century [20]. Some seed sources of the older plantations are not known, whereas other sources are known. The latter are likely related to three provenances, from California (2) and Oregon, Washington, or British Columbia (1). The Portuguese provenances have a considerable level of genetic diversity, similar to that of their native range [19]. In Europe, several provenance trials were made in the middle of the last century onward, where survival and growth were evaluated. The seed source from the west of Cascade (USA) of elevations inferior to 600 m was the best suited. Other seed sources are the European provenances, where seed orchards are pivotal to seed production [18].

In Europe, precipitation has been described as the most limiting factor for the development of *Pseudotsuga menziesii* and both the annual and the growing season precipitation are prevalent. Annual precipitation should be higher than 600–750 mm and growing season precipitation above 350–400 mm. In Portugal, the areas better suited to the country's edaphic and climatic conditions for *Pseudotsuga menziesii* subsp. *menziesii* (**Figure 3**) are those under Atlantic climatic influence, with elevation from 400 to 1000 m [7] up to a maximum of 1350 m [14, 19], mean annual temperature of 8–9°C, mean temperature from April to September of 14–15°C [21], and mean annual precipitation larger than 700 mm [19]. Areas with suitable sites for the species growth are located in the north and center at 700–1000 m elevation and a

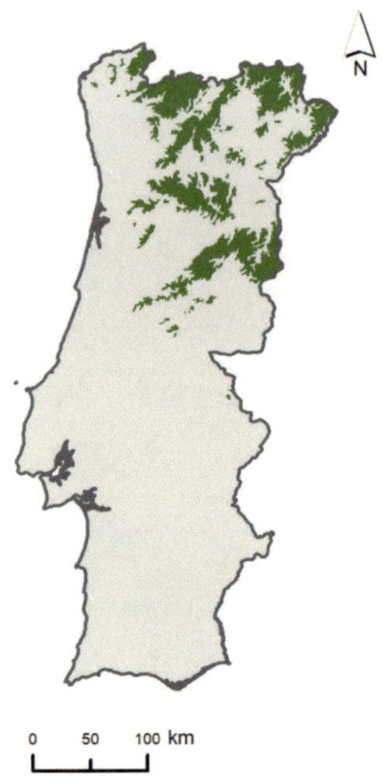

Figure 3.
Suitability map for Pseudotsuga menziesii *in Portugal [14].*

net precipitation above 400 mm [7]. Currently, it is one of the species prioritized for planting in inner central and northern Portugal (subregions of Trás-os-Montes e Alto Douro, Entre Douro e Minho, Litoral Centro, and Interior Centro).

In some countries (Germany, Switzerland, Czech Republic, and Romania), *Pseudotsuga menziesii* seem to have less drought sensitivity than other conifer and broadleaved species (e.g., *Picea abies*, *Abies alba*, *Pinus sylvestris*, *Larix* spp., *Quercus* spp., and *Fagus sylvatica*). However, under extreme drought, growth reduction and mortality have been observed in European stands [22]. Moreover, the drought response of *Pseudotsuga menziesii* depends on stand structure. Resistance is enhanced by low density and resilience is hindered, whereas their admixtures with other conifers (e.g., *Picea abies* and *Abies alba*) have a neutral effect [23–25]. *Pseudotsuga menziesii* has higher tolerance to drought when compared to *Larix decidua* [26] or *Picea abies* [26, 27]. It has higher resistance [27–29] due to the well-developed root system and higher resilience due to its position in the upper canopy [15, 23]. It is sensitive to snow damages. Its crowns are dense and favor the retention of snow, which makes them prone to stem breakage [22]. Its heavy crowns also make them prone to wind throw [30].

In its native range, fire is a common disturbance. Moderate-intensity fires favor the survival and regeneration of *Pseudotsuga menziesii* with thick bark [13]. Browsing damages, either in their natural or outside their natural range, can originate damages to individuals up to 1–2 m height. The protection of the young trees by fencing or with individual tree shelters is an alternative to reduce the damage of the game [22].

3. Silvicultural characteristics and management

3.1 Tree species characteristics

Pseudotsuga menziesii develops a long tap root up to 1.5–3.2 m in soils without limitations [15]. It develops about 50% of the total root length in the first 3–5 years and 90% in the following 6–8 years [31]. This species develops a wide root system with secondary roots that develop from the tap root obliquely into the deeper layers of the soil. Thus, when grown in deep well-drained soils anchorage is enhanced resulting in trees less prone to wind throw, while in shallow and/or poorly drained soils more superficial root systems do not provide good anchorage, making trees prone to wind throw [22].

In its natural range, *Pseudotsuga menziesii* is shade tolerant when young and intermediate or shade intolerant in mature stages of development [22]. Studies in Europe reported that it has intermediate shade tolerance [32]. When young (during recruitment), trees are more shade tolerant than *Abies alba* and *Picea abies* [33] and at maturity, trees are more tolerant to shade than other species (e.g., *Quercus petraea*, *Quercus robur*, and *Pinus sylvestris*) [15]. For the establishment and growth of seedlings, a direct light intensity between 15% and 40% is needed [34, 35]. The shade tolerance of *Pseudotsuga menziesii* enables its establishment under the canopy, though higher success is probable under canopies with lower crown cover [36].

Fruiting starts between 30 and 40 years of age, and the full production is reached at 50–60 years onward [22]. Masting occurs with shorter or longer periodicity. Cycles of 2–3 years, 4–6 years, 7 years as well as irregular cycles have been reported [22, 37]. It has also been observed that there is a trend toward higher fruiting in years following dry summers [22]. Its seeds are wind dispersed to distances between 20 and 100 m, with the largest distances of dispersal reaching 150 m [22].

3.2 Silviculture and management

3.2.1 Stand establishment

Pseudotsuga menziesii stands have been established, either from natural or from artificial regeneration. Natural regeneration has been observed in several countries (United Kingdom, Romania, Belgium, France, Czech Republic, Bulgaria, Slovakia, Italy, and Portugal) [22, 38]. The natural regeneration of the Portuguese stands is, in general, abundant [38]. The success of germination and seedling establishment is promoted by low competition between *Pseudotsuga menziesii* seedlings and other vegetation [34] and by suitable light environments [15, 39]. The favorable light conditions can be enhanced by silvicultural practices such as thinnings; by a low crown cover that occurs in shade-intolerant forest species stands (e.g., *Pinus sylvestris* and *Larix decidua*); or by the creation of gaps, either by silvicultural practices or by natural disturbances (i.e., wind) [39]. Artificial regeneration is mostly done by plantation of bare root or containerized seedlings. Most plantations of bare root use seedlings 2–3 years old with 30–60 cm height and 5-mm root collar diameter. The seedlings produced in containers are 1–2 years old [22]. Initial densities in *Pseudotsuga menziesii* plantations range from 600 to 4000 plantsha^{-1}. In its native range, the most frequent initial densities are 750–1500 plantsha^{-1}, whereas outside its native range is 1000–2000 plantsha^{-1}. In plantations with high densities, cleaning should be done when height is 2–5 m to a density of 2000 plantsha^{-1} and a thinning when height is 5–10 m to a density of about 1000–1500 plantsha^{-1} [22]. In Central, Northern, and Eastern Europe, plantations are done in the spring [22]. The plantations have survival rates equal to or larger than 70–80% [40]. The failure of its plantation is related to the seedlings' sensitivity to root dissection and dealing with the seedlings inadequately [22]. In Portugal, it is recommended to plant during the vegetative dormancy period, usually in the fall. Since it is sensitive to late frosts, it is advisable to use late-budding varieties in regions where the risk of frost exists, or to delay planting until late enough in the spring to avoid frost damage [21].

The effect of density on mortality and growth of a *Pseudotsuga menziesii* plantation was carried out for spacing between 1 and 6 m. The best-suited spacing was 3 m as it presented low mortality, trees maintained vigor and stability, and reached suitable growth [41]. In the early stages of development, it is quite sensitive to competition, when height growth is slow. A revision study identified that 90% of its plantations had mortality rates larger than 50%, due to competition, especially in mixed stands with *Fagus sylvatica* that outcompeted the *Pseudotsuga menziesii*. Thus, pre- and commercial thinnings that reduce competition both above and below ground are pivotal to their survival and growth [42]. However, in some situations, shelter provided by the spontaneous vegetation benefits its survival and growth by reducing direct sunlight high intensities [43]. In plantations with densities equal to or lower than 2000 plantsha^{-1}, cleaning is not necessary [22]. Moreover, when the initial stand density is low, trees have better stability, and social differentiation is enhanced [42].

The species suitability to the site is reflected in the regeneration. A study evaluating the effects of a temperature and precipitation gradient on germination reported that at optimal temperature germination proportion had smaller variance and variance increased toward the lower and upper thresholds of temperature. Also, time to seed germination was shorter at optimal temperature, though with some variability, which was related to the dormancy (physiological) of the seeds. Moreover, the variability in seed germination cannot be explained with temperature alone, but rather in combination with precipitation [44].

3.2.2 Forest systems and silvicultural practices

Stands of *Pseudotsuga menziesii* range from pure even aged to mixed uneven aged. The even-aged stands management depends on the target wood products and thus on the silvicultural practices (thinning and pruning) and rotation length. The prescription of early thinning is recommended to promote stand and tree stability and to reduce the risk of wind throw [22]. Ideally, it should be done before the stand reaches 20 years of age [45]. In Europe, several thinning methods, such as thinning from below, from above, line thinning, and selective thinning, have been used. Moderate intensity (20–35% of the number of trees) is recommended for the first thinning. The cycle of thinning has a periodicity of 5–10 years and the number of thinning interventions depends on the rotation length, being reported 3 to 5 interventions [22]. In some models of silviculture, the second thinning is recommended when the mean stand diameter reaches 15 cm, prior to which the selection of the future trees is done. The selection of the future trees is done as a function of tree vigor, quality of the stem, and species [40]. The goal is to favor future trees by removing the competition from their neighbors and promoting growth [46]. The number of future trees increases from extensive to intensive silviculture and decreases with the increase of the target diameter. The final number of future trees varies between 150 and 300 treesha^{-1} [40].

Pseudotsuga menziesii branches have a slow self-pruning, thus originating wood with knots. Pruning is a pivotal silvicultural practice to improve the quality of the logs. The number of branches is dependent on its provenance and its diameter on stand density. As pruning is an expensive practice, it is recommended to prune the trees that will be cut at the end of the production cycle. In Europe, the recommendations are to prune 100–400 treesha^{-1}, to a height of 5–15 m in two interventions. The latter is influenced by target diameter, that is, for a target diameter of 60 cm the pruning height should be 5 m, for 80 cm 10 m, and less frequently for 100 cm 15 m [22]. The selection of the trees to prune should be done at the early stages of development, but after social differentiation has occurred. If future trees are selected, those are the candidates to be pruned. Moreover, pruning should be done before the tree diameter reaches one-third of the target diameter, which corresponds to a diameter at breast height of 10–15 cm [47]. Pruning should be carried out in spring to enable a faster and better healing of the wounds, and ideally, the branches should have a diameter smaller than or equal to 3 cm [22]. The proportion of live crown pruned should be 25–30% of the total crown to minimize the losses in growth, improve stem taper, and promote rapid wound healing [30, 48].

The more frequently used rotation length in Europe is 40–80 years, though longer ones have been reported. The silvicultural method more frequent in European countries is clear felling and the regeneration method is plantation [22].

In Europe, mixed stands with *Pseudotsuga menziesii* and native conifer and broadleaved species have been established. The goals are to increase biodiversity and thus the resilience and resistance to disturbances as well as to manage stand by close to nature silviculture [40]. In Europe, the composition of mixed stands includes conifers, such as *Picea abies*, *Picea sitchensis*, *Abies alba*, *Pinus nigra*, *Larix decidua*, *Larix × eurolepsis*, *Larix kaempferi*, *Tsuga heterophyla* and broadleaved species, such as *Fagus sylvatica*, *Alnus cordata*, and *Castanea sativa* [22, 49]. The proportion of the *Pseudotsuga menziesii* in the mixture varies between 20% and 75% and the spatial arrangement can be made by row or by group [22, 49].

In a study in Switzerland on mixed stands with *Abies alba*, *Fagus sylvatica*, and *Picea abies*, Frei et al. [34] observed the seedlings of *Pseudotsuga menziesii* in the majority of the stands in 45% of the plots, while saplings were present in 30% of the stands and 8% of the plots. Regeneration of Douglas fir was not dominant in any stand, and the limiting factors for regeneration were the understory vegetation, the thick litter layer, seed source distance, shade, and calcareous soils. Inversely, the top soil disturbances with the removal of the litter layer and sites poor in nutrients or dry favored its regeneration [34]. The understory vegetation outcompetes *Pseudotsuga menziesii* seedlings and saplings for growing space, namely light, water, and nutrients, and can also promote higher predator density (e.g., mice). The litter layer can be a mechanical barrier to the access of the roots to water ([34] and references therein).

Mixed and pure plantations of *Pseudotsuga menziesii* and *Castanea sativa* with mixing proportions of 75% and 50% of *Pseudotsuga menziesii* with different spatial arrangements in the plantation lines were studied in Portugal. The mixed stands at 15 years old had higher productivity than the pure stands [49]; 28 years after plantation, there was a vertical stratification in mixed stands with *Pseudotsuga menziesii* in the upper layer and *Castanea sativa* in the lower layer. Moreover, the diameter at breast height of *Pseudotsuga menziesii* was larger than that of *Castanea sativa* [50]. The mixed stands' higher productivity when compared to the pure stands is indicative of facilitation interactions between the two species [50].

The admixture of *Pseudotsuga menziesii*, *Picea abies*, and *Abies alba* has positive effects on the growth of *Pseudotsuga menziesii* under years with temperature and precipitation within the normal range and negative effects under years where drought occurs. However, *Pseudotsuga menziesii* showed positive complementarity with *Picea abies* and *Abies alba* and drought indices were higher for monospecific neighbors than for mixtures. The effect of drought on *Pseudotsuga menziesii* growth on mixtures was small and may be explained by its position in the upper canopy layer. Its radial growth was similar in monospecific and mixed stands [25].

The ecological traits of *Pseudotsuga menziesii* are considered suitable to uneven aged silviculture. The silvicultural systems include group shelterwood [45], single tree selection [33], group selection [51], and irregular shelterwood [35, 45, 52].

In its native range, there are mixed uneven aged stands of *Pseudotsuga menziesii* and *Tsuga heterophylla*, which can also include other conifer and broadleaved species. The variability in composition of the stands is related to edaphic and climatic conditions as well as to infrequent wildfires. The young stands are dense with *Pseudotsuga menziesii* in the upper canopy layers due to their high growth rates. At this stage, competition is intense, resulting in the mortality of dominated individuals. Also, due to the high shading in the understory, diversity of shrubs and herbaceous plants is low. As the stand ages, it enters the mature stage, frequently after thinning practices. In the mature stage, mortality is the result of disturbances and tends to occur in patches creating gaps in the canopy and, thus, growing space for the regeneration and recruitment of species more tolerant to shade (e.g., *Tsuga heterophylla*). The old growth stage is characterized by a complex structure, where tree spatial arrangement is heterogeneous with the horizontal distribution of trees in gaps and dense patches and a continuous vertical distribution. Mortality at this stage can be derived both from competition and from disturbances. The management of these stands under ecological silviculture is based on variable density thinning, variable retention harvest, and natural regeneration. Artificial regeneration is considered a complement to natural regeneration, particularly for species for which seed bearers are scarce or to introduce seeds that are considered better adapted [53].

In complex stand structures, i.e., uneven aged pure or mixed stands, a pivotal factor is the maintenance of the balance between the number of trees in each cohort, with the increase in the number of trees from the oldest to the youngest cohort.

The most important aspect is in implementing silvicultural practices that enable recruitment [51]. In these stands, selective thinning of light to moderate intensity is suggested by Schütz et al. [33], as trees are frequently slender and thus with low stability [33]. Thinning interventions should start at about 25–30 years old [33] after the differentiation of the future trees [54]. A basal area of about 30-35 m^2 ha^{-1} and 30–40% full sunlight are indicated to promote suitable recruitment while enabling growth and enhance tree and stand stability [33, 51, 55–57].

Two advantages have been reported for the uneven aged structure stands. On one hand, in the selection system, the timber of small size is less than 30% of that of the even-aged structures. Providing more timber of large size enables higher revenues while productivity is similar. On the other hand, uneven aged structures rely on natural regeneration, which is costless [57].

Even-aged stands have been converted and transformed into mixed uneven aged stands. The process encompasses a set of thinning and cutting interventions. The silviculture practices are done to maintain a continuous crown cover at small scale with a grid pattern. The system implementation relies on management continuity and flexibility to approach the target uneven aged structure gradually as a function of the trees and species spatial arrangement and growth patterns [56]. In stands where the transformation from even aged to uneven aged structure has not yet reached a balanced structure, there might occur an unbalanced number of trees in some cohorts [51, 54]. In this case, active management with thinning interventions will enable to reach the target uneven aged structure gradually [54]. Furthermore, the thinning regime should be planned for the short-, medium-, and long term in a dynamic way with changes, whenever necessary. This allows dealing with disturbances (e.g., storms) that might affect the development of regeneration, recruitment, and main stand and thus the target uneven aged structure [54, 57]. Studies on uneven aged structures reported for a balanced structure with a q value (ratio between the number of individuals of one diameter class and the number of individuals of the next class of diameter class, dimensionless) of 1.3 [56] and between 1.4 and 1.6 [55].

3.2.3 Silvicultural model for the species in Portugal

In general, the models of silviculture for pure even-aged stands of *Pseudotsuga menziesii* in Portugal typically consider its planting with initial densities above 800 treesha^{-1} and below 1300 treesha^{-1} [21], with cleaning after 10–15 years, followed by 3 thinnings. Pruning is recommended to be carried out on the trees selected as future trees. Harvest age is around 55–60 years. **Table 1** shows two silvicultural models proposed for use in Portugal. Model 1, presented by Louro et al. [58], is currently recommended for Central Portugal (PROF Centro Interior and PROF Centro Litoral). Model 2 is recommended for the Northern regions (PROF Trás-os Montes and PROF Alto Douro e Minho). Due to their similar characteristics, the description is given for Model 1. The model of silviculture for pure even-aged stands of *Pseudotsuga menziesii* in Portugal by Louro et al. [58] considers its plantation, an initial density of 1100–1300 treesha^{-1}; cleaning at 10–15 years of age to reduce density to 800–1000 treesha^{-1}, being coincident with the first selection of the future trees (200–300 treesha^{-1}) and their pruning. It considers 3 thinning interventions, the first two selective and the latter from below of moderate intensity. The first thinning should start at 20–30 years of age and the remaining with a periodicity of 10 years. The second pruning should be done at the time of the first thinning in 150–300 treesha^{-1}. The considered rotation length is 60 years and the final density of 150–200 treesha^{-1}.

Year	Model 1	Model 2
0	Stand establishment: artificial regeneration (plantation) 1100–1300 $treesha^{-1}$.	Stand establishment: artificial regeneration (plantation) 1100–1300 $treesha^{-1}$.
2–10	2–3 release operations.	Control of spontaneous vegetation.
10–15	Release operation. Reduction of stand density to 800–1000 $treesha^{-1}$. First selection of the future trees (200–300 $treesha^{-1}$) and their pruning.	Density reduction with selective criteria, removing dead trees and those of poorer quality (poor shape). Reduction of stand density to 800–1000 $treesha^{-1}$. First pruning of 200–300 $treesha^{-1}$.
15–30		Second pruning in trees pre-selected as the future trees (200–300 $treesha^{-1}$).
20–30	First thinning (selective), with removal of 35–45% $treesha^{-1}$ and second pruning (150–220 $treesha^{-1}$)	First thinning. High to mixed thinning. Removal of 30–40% $treesha^{-1}$.
30–40	Second thinning. Removal of 30–40% $treesha^{-1}$.	Second thinning. High to mixed thinning. Removal of 30–40% $treesha^{-1}$.
40–50	Third thinning. Thinning from below of moderate intensity. Removal of 30–40% $treesha^{-1}$.	Third thinning. Thinning from below. Removal of 30–40% $treesha^{-1}$.
55–60		Harvest with a final density of 150–250 $treesha^{-1}$.
60	Harvest with a final density of 150–220 $treesha^{-1}$.	

Table 1.
Silvicultural models proposed for Pseudotsuga menziesii in Portugal.

Similar to studies conducted for other conifer species in Portugal, such as *Pinus pinaster* [59], it would be interesting to recommend silvicultural prescriptions to regulate stand density (thinning intensity prescription) based on the allometric relationship between tree size and density. For *Pseudotsuga menziesii*, approaches based on the identification of the maximum density line have been proposed by Drew [60], and more recently, by Heiderman and Kimsey [61]. Interestingly, in their study of forests in the United States of America, Heiderman and Kimsey [61] conclude that the self-thinning line for the species has a slope of −1.607, which is similar to the constant value proposed by Reineke [62] of −1.605. However, this line showed deviations when considering site-specific environmental factors, so the use of this approach for the species in Portugal will require validation of the identified line or the development of a specific one for national stands.

3.2.4 Diversity

Generally, diversity increases from pure even-aged stands to mixed uneven aged stands [63–66]. This can be related to the light levels in the understory, which are higher in mixed stands than in pure ones [64, 66]. Silvicultural practices that create canopy gaps enable higher levels of light in the lower stand layers, which promotes the increase of diversity, both in the number of species and individuals [63, 66–68]. For example, in mixed stands litter of *Pseudotsuga menziesii* and *Fagus sylvatica*, litter decomposition originated higher availability of nitrogen and other nutrients, which promoted higher diversity in the understory [64, 69]. However, understory species composition was different in pure and mixed stands of *Pseudotsuga menziesii* and *Fagus sylvatica*, either by the change in species or by the variation of each specie density [64]. In mixed stands of *Pseudotsuga menziesii* and *Fagus sylvatica*, the increase in *Pseudotsuga menziesii* proportion originated a change toward more nitrogen demanding understory species, due to the decrease of acidic species [64].

Overall, for pure *Pseudotsuga menziesii* (and also for *Picea abies*) stands diversity (alpha, beta, and gamma) was higher for a wide set of species (from fungi to mammals) than for pure ones or *Fagus sylvatica* stands, whereas mixed stands presented diversity levels between the conifer and the broadleaved species. Moreover, alpha and gamma diversity was higher for *Pseudotsuga menziesii* stands than for *Picea abies ones* [70].

Diversity is impacted by aridity. As aridity increases diversity decreases, as well as *Pseudotsuga menziesii* productivity, above ground biomass, and below ground biomass, probably due to the decrease in water availability. Inversely, herbaceous species tend to increase with aridity, likely because of the lower crown cover of the *Pseudotsuga menziesii* stands and higher light levels in their understory [71].

3.3 Growth and yield

In Europe, *Pseudotsuga menziesii* growth rates are higher than those of native European forest species, such as *Picea abies*, *Pinus sylvestris*, *Fagus sylvatica*, or *Quercus robur* ([15] and references therein), varying between 7.6 and 26.7 m^3 ha^{-1} y^{-1} [15, 40]. In even-aged stands, *Pseudotsuga menziesii* shows an increasing mean annual volume increment until 50–80 years, followed by a gradual decrease afterward [33]. It is considered one of the most productive forest tree species in Portugal achieving growth rates of 15–18 m^3 ha^{-1} y^{-1} [38]. Diniz [9] provides stand tables, indicating that the average volume growth for the species in Portugal ranges from 18.7 to 27.9 m^3 ha^{-1} y^{-1}, depending on site quality. The productivity of mixed stands when compared to pure stands has been reported to be higher [15, 50, 72], similar [15, 73, 74], and lower [24]. In Portugal, initial development of *Pseudotsuga menziesii* is slow, up to circa 5 years of age, after which growth rates increase [75].

The diameter growth correlates positively with water availability and negatively with competition, whereas composition has no effect [23]. In a study of mixed stands of *Pseudotsuga menziesii* and *Tsuga heterophylla* (50:50) and pure stands of each species with three different densities, the two lower densities in pure stands the productivity was higher than in mixed stands, whereas at the highest density, mixed stands attained higher productivities than the pure stands. The higher productivity in the mixed stands was explained by the development of a stratified canopy and a more efficient use of the growing space [76]. Another study [77] reported that for mixed stands of *Quercus petraea* and *Pseudotsuga menziesii*, density was the most important growth factor and that the dominant tree's growth was higher and sensitive to climatic stresses (e.g., drought) when compared to the dominated trees. Other studies [78, 79] reported higher basal area increments in thinned stands than in unthinned ones. Furthermore, regardless of dimension, trees increase growth after thinning, though larger individuals have higher growth rates than the smaller ones. This can be, at least partially, explained by the higher position in the canopy of the taller trees, and thus more vigor, less competition for light, higher leaf area, and more developed root system [78, 79].

3.4 Models to support decision-making processes

The study and modeling of biometric characteristics at tree and stand level for *Pseudotsuga menziesii* is extensive, with over 1000 results in the Scopus database related to the species name and associated models, but only a few entries pertain to Portugal. This review focused on relevant information for growing the species in Portugal and considered equations for tree stem volume, tree biomass, diameter, and height growth models and site index models.

3.4.1 Volume and biomass equations

A summary of the available volume equation models is provided in **Table 2**, along with the reference studies.

The equations to estimate tree volume with bark (v, m^3) presented in **Table 2** use diameter at breast height (d, cm) or tree basal area (g, m^2) and tree height (h, m) as regressors. The model by Diniz [9] was developed using data from trees with diameters ranging from 15 to 95 cm, collected from stands in the Estrela, Padrela, Marão, Gerês, and Soajo mountains, with a predominance of trees from the first two locations. Anjos [80] tested the development of volume equations for stands located in different sites and found that the average volumes did not differ between sites, except for one stand located in Serra da Padrela (Padrela-1), which had significantly higher values than the other stands. The diameter ranges were from 5 to 60 cm, depending on the site.

Reference	n	Region	Model	s_{yx} (m^3)
[9]	417	Sub-Atlantic Zone of Northern Portugal	$v = 0.45926gh$	—
[80]	105	Bornes	$v = 1.476 \times 10^{-4} d^{2.319}$	0.072
[80]	105	Bornes	$v = 9.910 \times 10^{-5} d^{1.936} h^{0.664}$	0.068
[80]	105	Marão	$v = 2.360 \times 10^{-4} d^{2.212}$	0.053
[80]	105	Marão	$v = 7.45 \times 10^{-5} d^{1.992} h^{0.718}$	0.045
[80]	105	Nogueira	$v = 3.289 \times 10^{-4} d^{2.001}$	0.061
[80]	105	Nogueira	$v = 2.427 \times 10^{-4} d^{1.790} h^{0.408}$	0.060
[80]	33	Padrela-1	$v = 9.183 \times 10^{-4} d^{2.025}$	0.043
[80]	33	Padrela-1	$v = 6.84 \times 10^{-5} d^{1.629} h^{1.198}$	0.034
[80]	70	Padrela-2 and 3	$v = 3.119 \times 10^{-3} d^{1.660}$	0.056
[80]	70	Padrela-2 and 3	$v = 7.93 \times 10^{-5} d^{1.536} h^{1.249}$	0.041
[80]	105	Soajo	$v = 4.140 \times 10^{-4} d^{2.21}$	0.063
[80]	105	Soajo	$v = 2.72 \times 10^{-5} d^{1.789} h^{1.305}$	0.048

Table 2.
Existing stem volume equations developed for the species using data collected from Pseudotsuga menziesii stands in Portugal.

For an overview of available volume and biomass models for the species in Europe, a search was conducted on the GlobAllomeTree database (http://www.globallometree.org/; [81, 82]). Two volume equations were identified for *Pseudotsuga menziesii* in France and five biomass equations in the Netherlands. In the compilation by Zianis et al. [83], 15 models for the species are listed, including volume equations from Belgium, the Netherlands, Norway, and Romania, as well as biomass equations with references to the Netherlands, Italy, and the United Kingdom. To the authors' best knowledge, there are no equations developed for estimating stem biomass for this species in Portugal. To address this gap, volume estimates can be combined with information on wood density, similar to the approach described for *Pinus radiata* in Portugal by Fonseca et al. [84].

3.4.2 Growth models, simulators, and web products

In the context of growth models for species, it is important to acknowledge the contributions of Diniz [9], as well as the works of Luis [20] and Fontes [7]. Their research has played a fundamental role in advancing the understanding of growth patterns of *Pseudotsuga menziesii* in Portugal. Diniz [9] developed and presented the first provisional Yield Table for the sub-Atlantic zone of northern Portugal. Later, Luis [20] comprehensively studied height growth, diameter, basal area, and volume, as well as the evolution of stem shape and longitudinal profile for the species. It also assesses site quality and competition in stands. Fontes [7, 85] modeled the dominant height growth and found that the estimates are similar to those obtained for the UK and the Netherlands, but slightly lower than those defined for Belgium, France, and Germany. The Fontes [7, 86] took into account location factors to model the quality index. This allowed the identification of the best areas for the species' development in the country.

Various advancements contribute to informed decision-making. Although there are currently no established simulators for the species in Portugal, there are already web-based tools that can assist in defining scenarios and forecasting growth. Examples of these tools are the models integrated into the Computer-Aided Projection for Strategies In Silviculture (CAPSIS, http://www.inra.fr/capsis platform [87]), specifically SimCoP, Organon, Gymnos, FCBA *Pseudotsuga menziesii*, and Luberon 2.

4. Wood traits

Knowledge of the anatomical, physical, and mechanical properties of wood is essential for its correct use and determines its final application. Although there is considerable variability in the characteristics of wood from the same species, influenced by the growth conditions of the tree, its age, origin, and even its position within the tree [88, 89], it is important to understand the pattern and extent of this variability because its use is directly related to some of its characteristics. This section summarizes the main characteristics of *Pseudotsuga menziesii* wood, with particular emphasis on the wood of this species growing in Portugal (Sections 4.2 and 4.3). The wood anatomy of this species reveals structural characteristics that make it suitable for a wide range of applications. It is known for its good strength properties, stability, and attractive appearance with a distinctive orange heartwood [90]. It is one of the most important construction woods in the world and is also used for veneers, fibers, particle boards, and in the pulp industry [91–93].

4.1 Wood anatomy

Macroscopically, *Pseudotsuga menziesii* wood has clearly visible growth rings, with a well-defined, dark-colored latewood, and often a sharp transition from earlywood to latewood [91, 94]. Its sensitivity to growth conditions results in distinct growth rings that reflect climate fluctuations, making it an excellent species for dendroclimatological studies [95].

The wood has a distinctive heartwood, brownish-red in color, and a light-colored sapwood with yellowish tones, typically consisting of 10–20 growth rings that ensure the conduction of sap in the tree [92]. The heartwood is moderately resistant to biodegradation [96] due to the accumulation of extractives in this part of the trunk, which can reach a content of about 6–12% [97]. *Pseudotsuga menziesii* wood is considered difficult to impregnate with preservatives, often requiring incisions to allow penetration into the wood [98]. This reduced permeability is also attributed to the accumulation of extractives, presumably due to decreased porosity and increased hydrophobicity [96]. The grain is predominantly straight but can sometimes be wavy or spiral [93]. The texture is moderately coarse, due to the marked contrast between earlywood and latewood zones [91].

Microscopically, the main structural element of *Pseudotsuga menziesii* wood, like any other conifer, is the tracheid, which are long cells generally 100 times longer (1–10 mm) than they are wide [99]. Longitudinal tracheids are prismatic cells with closed ends, constituting more than 90% of the volume of conifer woods [100]. The difference between earlywood and latewood tracheids lies in the radial diameter and the thickness of cell walls [100], with earlywood tracheids having larger diameters and thinner walls (**Figure 4**).

The rays are heterogeneous and uniseriate, and therefore only visible under a microscope. When accompanied by horizontal resin ducts, they are fusiform rays. They typically consist of 8–15 cells in height but can occasionally reach up to 25 cells [94]. These are parenchymal cells with oval sections and thick walls with many pits, as well as marginal horizontal tracheids with thin walls and numerous small bordered pits. In cross-fields, there are 2–4 pinoid-type pits [94].

Pseudotsuga menziesii wood has longitudinal and radial resin ducts, which are small in cross-sectional size and usually only visible under magnification. The longitudinal ducts are associated with the latewood (**Figure 4**) or the transition zone between earlywood and latewood [91, 94]. The epithelial cells surrounding the resin duct secrete the resin, which is deposited inside the canal (**Figure 5**).

4.2 Wood biometry

The length of fibers (in hardwoods) and tracheids (in conifers) is a biometric characteristic of woody tissue, related to the biological processes occurring in the tree during wood formation, which influences the properties of pulp for paper or fiber panels [89]. Indeed, fiber morphology—shape and size—as well as cell wall architecture, directly affect its flexibility, plasticity, and strength, and indirectly, the type of bonds they form among themselves. This largely determines the physical and mechanical properties of the products made from them. Longer fibers lead to better bonding by increasing the number of cell-to-cell contacts [101], an important aspect for paper or fiberboard production, and greater strength in solid wood when used, for example, in beams [102]. Therefore, knowledge of fiber/tracheid length is essential in determining the characteristics of products obtained from this raw material.

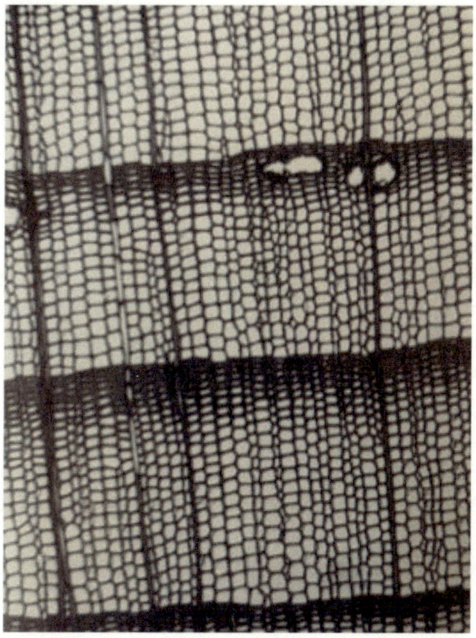

Figure 4.
Growth rings of Pseudotsuga menziesii *(40×) where the differences between earlywood and latewood can be observed. (Source: Schweingruber [94]).*

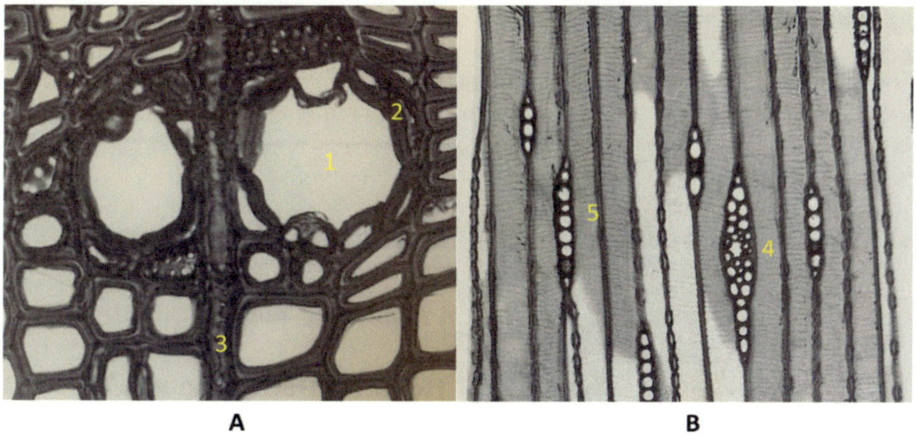

Figure 5.
Appearance of the resin ducts of Pseudotsuga menziesii*. (A) (40×)—Longitudinal resin ducts (2) with epithelial cells (1), showing ray cells (3). (B) (250×)—Fusiform resin duct (4), with uniseriate rays (5) visible. (Source: Schweingruber [94]).*

There are different patterns of variability in tracheid length, which can be within the tree, within and between growth rings from pith to bark, between individual trees of the same species, within a site, and between sites with different characteristics [89]. The structure of the wood varies depending on the tree's growth [103].

The presence of juvenile and mature wood is one of the main sources of variation in wood, with this difference being more pronounced in conifers compared to hardwoods [104]. In the case of *Pseudotsuga menziesii*, the transition between juvenile and mature wood occurs within the first 10–20 years of the tree's age [92]. Louzada and Fonseca [105] found that the transition from juvenile to mature wood happens around 15 years of age.

In conifers, it is common to observe short tracheids near the tree center, followed by a rapid increase in length during the juvenile zone, with a subsequent decrease [100, 104]. Ištok et al. [103], in a study on the tracheid length of *Pseudotsuga menziesii* growing in three locations in Croatia, where they measured 5760 tracheids from different trees, concluded that all tested sources of variation (annual growth rings, trees, location, and annual growth ring × zone interaction) significantly affected tracheid length. However, the results suggested no relationship between the tracheid length of earlywood and latewood with growth parameters (earlywood width, latewood width, and growth ring width) across the three sites. For Portugal, Louzada and Fonseca [105] reached similar conclusions. In a study with 25 trees, they found that diameter growth classes had a very low influence on fiber length, which could be considered negligible for practical purposes. In the same work, they stated that the variation in fiber length is dominated by the influence of growth rings, i.e., physiological age, and the variation of fibers within the rings (**Figure 6**). The variation due to individual trees and their interaction with growth rings, although highly significant, was considered of little interest for technological utilization and genetic improvement objectives.

Despite the significant variability that this characteristic can exhibit, several studies present average tracheid lengths. Ištok et al. [103] report that earlywood tracheids have an average length of 2.47 mm, while latewood tracheids are slightly longer, averaging 2.77 mm in length. Louzada and Fonseca [105] studied the differences between juvenile and mature wood tracheids, finding average lengths of 2.37 mm for the first 15 growth rings corresponding to juvenile wood and 4.34 mm for mature wood. The differences were not as pronounced in the work of Fabisiak and Moliński [106], which reported average tracheid lengths of 3.02 mm for juvenile wood and 3.61 mm for mature wood.

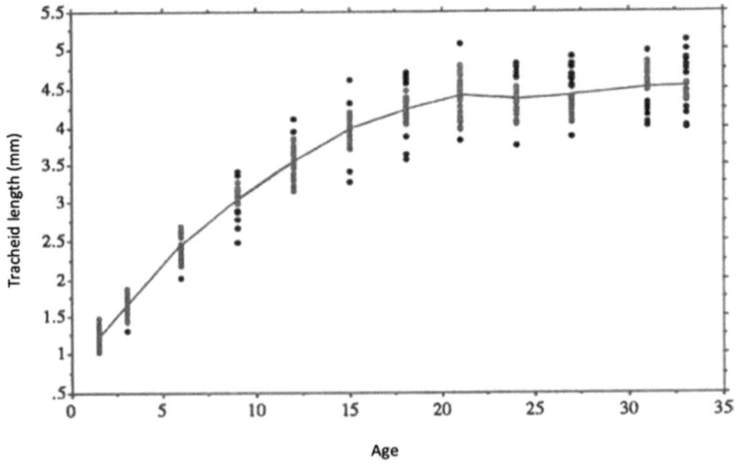

Figure 6.
Radial variation in fiber length, showing the average values for 25 Pseudotsuga menziesii *trees, along with the dispersion of individual tree values [105].*

4.3 Wood density

In the field of wood science, there are few areas of more global interest to wood quality than density. As well as being a direct measure of the amount of woody material in a given volume, it is closely linked to numerous important technological properties and characteristics that are essential for the production and utilization of forest products. As such, it is the most revealing index for understanding the fundamental characteristics of wood.

However, while in most materials density corresponds to a relatively constant property, defined by mass per unit volume, in the case of wood it is highly variable, since it can vary considerably, not only from one species to another, but also within the same species depending on the location, the tree, the height level, and age of the tree, the characteristics of the year in which the growth ring is formed and, even within this, the time of its formation. In general, conifers have a tendency to increase in density from the pith to the periphery. In the particular case of *Pseudotsuga menziesii*, it has been observed that density is high near the pith, decreases in the following rings, and then increases to reach its maximum value near the cambium. Much of the variation in density between growth rings is due to the fact that the radial variation in the density of the initial and final wood, considered separately, is not identical, as well as the variation in the proportion of initial and final wood present in the different rings.

According to Elliott [107], the density variation curves show that the initial wood rapidly decreases in density in the first few years from the pith, significantly influencing the density values of the juvenile wood rings. After this, they maintain a minimum density with a small fluctuation or increase slightly in the last growth rings. On the other hand, the density of the final wood shows a pattern of variation from the pith to the periphery that is very similar to that shown for the average density. Thus, the density of the growth rings of adult wood is extremely influenced by the values of the final wood. As for the percentage of final wood, it seems to be directly related to age, especially in the early years, and inversely to tree height [89].

The effect of location on wood density has attracted the attention of numerous researchers. Erickson and Harrison [108] in a study of *Pseudotsuga menziesii* stated that the accelerated growth of 20-year-old trees caused a significant decrease in density in the first few years, but it recovered after 6 years. There was no significant change in density in uniformly growing trees over the same period. Keller and Thoby [109], in a study of the same species, stated that, strong growth does not cause a decrease in density. Although Krajnc et al. [110] found differences in the average width of the growth ring of *Pseudotsuga menziesii* trees from three different sites, these differences were not seen in the average density. Louzada and Fonseca [105], in a study in Portugal on *Pseudotsuga menziesii*, concluded that, as long as the effect of age is not confused in the relationship between density and growth rate, the relationship between these variables is not significant, i.e., that for a given age, density is not dependent on growth rate. Site quality only affects density negatively and significantly when juvenile wood is analyzed. Juvenile wood is mentioned in this study as a highly significant source of density variation for this species.

The fact that the density of *Pseudotsuga menziesii* wood is independent of its growth opens up good prospects for the management of stands of this species, since it is not because the tree grows more that the quality of its wood is poorer, as has so often been the case with other coniferous wood. In average terms, the density of this

Property	Value
Fiber saturation moisture, HSF (%)	27
Shrinkage (%)	
Volumetric	10.7–12.4
Tangential	6.9–7.6
Radial	3.8–4.8
Shrinkage coefficient (%)	0.46
Anisotropy (T/R)	1.6
Modulus of rupture, MOR (MPa)	82–91
Modulus of elasticity, MOE (GPa)	10.3–13.4
Hardness (N)	3160

Table 3.
Average values and variation intervals for physical and mechanical properties of Pseudotsuga menziesii. (Source: [91, 98]).

species, determined at 12% humidity, is around 0.480 g cm^{-3} and can be considered a light wood [91]. Alden [98] refers to higher values of 0.540–0.610 g cm^{-3}, depending on whether the trees are from the coast or the interior of the United States of America. Louzada and Fonseca [105] found average basic density values near the pith (juvenile wood) of 0.460 g cm^{-3} and near the cambium (adult wood) of 0.500 g cm^{-3} in *Pseudotsuga menziesii* growing in the north of Portugal.

4.4 Other physical properties and mechanical properties

Thanks to its anatomical structure, *Pseudotsuga menziesii* wood has excellent mechanical strength, including high resistance to compression, bending, and shear. These properties make the wood suitable for a wide variety of applications, from construction to the manufacture of furniture and joinery products. **Table 3** summarizes the main physical properties, apart from density, and mechanical properties of this species. Wood, as a biological material, shows great variability in its characteristics, both intrinsic to the trees and species themselves and external to the growing conditions. As such, the values presented here are indicative and based on work mainly outside Portugal.

5. Conclusions

The characteristics of the *Pseudotsuga menziesii* species and its ability to adapt beyond its natural range have made it very valuable for afforestation programs in Europe. In Portugal, this non-native species is recognized by the national authorities as an important option for afforestation in the central and northern regions. In order to manage the species sustainably, a thorough understanding of its ecology and silviculture is essential. Effective management, based on comprehensive information about the species, is essential to ensure that the species fulfills its intended ecological, social, and economic roles.

This review aimed to bring together a comprehensive knowledge of the species, building on pioneering research carried out in Portugal, focusing on specific aspects of the species' adaptability, growth and wood properties, and updating the information to support improved silvicultural and management practices and use of the species. There are simulators and tools available on the web that integrate growth models developed for the species in other countries. A line of research to be developed will be the creation or adaptation to the Portuguese context of a simulator to test silvicultural models and predict growth under different management scenarios and facilitate decision-making.

Acknowledgements

This work was partially funded by the FCT—Fundação para a Ciência e a Tecnologia to Center for the Research and Technology of Agro-Environmental and Biological Sciences (CITAB), under project UIDB/04033/2020 (https://doi.org/10.54499/UIDB/04033/2020), to Forest Research Centre (CEF), within project UIDB/00239/2020, and to MED—Mediterranean Institute for Agriculture, Environment and Development under the Project UIDB/05183/2020 (https://doi.org/10.54499/UIDB/05183/2020).

Conflict of interest

The authors declare no conflict of interest.

Author details

Teresa Fidalgo Fonseca[1,2,3]*, Maria Emília Silva[1,2], José Luís Lousada[1,2], Jaime Sales Luis[1] and Ana Cristina Gonçalves[4]

1 Department of Forestry Sciences and Landscape Architecture (CIFAP), University of Trás-os-Montes and Alto Douro, Vila Real, Portugal

2 Centre for the Research and Technology of Agro-Environmental and Biological Sciences (CITAB), University of Trás-os-Montes and Alto Douro, Vila Real, Portugal

3 Forest Research Centre (CEF), School of Agriculture University of Lisbon, Lisboa, Portugal

4 MED – Mediterranean Institute for Agriculture, Environment and Development & CHANGE – Global Change and Sustainability Institute, Institute of Research and Advanced Education (IIFA), Department of Rural Engineering, University of Évora, Évora, Portugal

*Address all correspondence to: tfonseca@utad.pt

References

[1] Ronch FD, Caudullo G, Rigo D. *Pseudotsuga menziesii* in Europe: Distribution, habitat, usage and threats. In: European Atlas of Forest Tree Species. Luxembourg: Publication Office of the European Union; 2016. pp. 147-146

[2] Ukrainetz NK, Kang KY, Aitken SN, Stoehr M, Mansfield SD. Heritability and phenotypic and genetic correlations of coastal Douglas-fir (*Pseudotsuga menziesii*) wood quality traits. Canadian Journal of Forest Research. 2008;**38**(6):1536-1546

[3] Spiecker H, Lindner M, Schuler J, editors. Douglas-Fir: An Option for Europe. Joensuu: European Forest Institute, European Forest Institute; 2019. 121 p. (What science can tell us)

[4] Kownatzki D, Kriebitzsch WU, Bolte A, Liesebach H, Schmitt U, Elsasser P. Zum Douglasienanbau in Deutschland: Ökologische, Waldbauliche, Genetische und Holzbiologische Gesichtspunkte des Douglasienanbaus in Deutschland und den Angrenzenden Staaten aus Naturwissenschaftlicher und Gesellschaftspolitischer Sicht. Braunschweig: Johann Heinrich von Thünen-Inst; 2011. 67 p. (Landbauforschung Sonderheft)

[5] Ponette Q, Ranger J, Ottorini JM, Ulrich E. Aboveground biomass and nutrient content of five Douglas-fir stands in France. Forest Ecology and Management. 2001;**142**(1-3):109-127

[6] Spiecker H, Schuler J. Introduction. In: Spiecker H, Lindner M, Schuler J, editors. Douglas-Fir—An Option for Europe. Vol. 11. Joensuu: European Forest Institute; 2019. (What Science Can Tell Us)

[7] Fontes L. The Performance, Constraints and Potential of Douglas-Fir [*Pseudotsuga menziesii* (Mirb.) Franco] in Portugal. Oxford, UK: University of Oxford; 2002

[8] ICNF. 6° Inventário Florestal Nacional: 2015 Relatório Final. Lisboa: ICNF; 2019. 284 p. Available from: https://www.icnf.pt/api/file/doc/c8cc40b3b7ec8541

[9] Diniz D. Estudo do Crescimento da *Pseudotsuga menziesii* (Mirb) Franco no Norte de Portugal. Lisboa: Instituto Superior de Agronomia; 1969

[10] Goes A. Contribuição para o Estudo do Fomento da *Pseudotsuga menziesii* (Mirb) Franco no Norte do País. Lisboa: Instituto Superior de Agronomia; 1976

[11] Anonymous. Portaria n.° 57/2019. Aprova o Programa Regional de Ordenamento Florestal de Trás-os-Montes e Alto Douro (PROF TMAD), Approves the Regional Forest Management Programme for Trás-os-Montes and Alto Douro (PROF TMAD). Vol. 29/2019. Série I de 2019-02-11. Diário da República; 2019

[12] Anonymous. The Gymnosperm Database [Internet]. 2022. Available from: https://www.conifers.org/index.php

[13] Sillett SC, Van Pelt R, Freund JA, Campbell-Spickler J, Carroll AL, Kramer RD. Development and dominance of Douglas-fir in North American rainforests. Forest Ecology and Management. 2018;**429**:93-114

[14] Correia AV, Oliveira AC. Principais Espécies Florestais com Interesse para Portugal: Zonas de Influência Atlântica. Lisboa: Direcção-Geral das Florestas;

2003. 322 p. Available from: https://www.icnf.pt/florestas/arborizacoes/projetosrjaar

[15] Thomas FM, Rzepecki A, Werner W. Non-native Douglas fir (*Pseudotsuga menziesii*) in Central Europe: Ecology, performance and nature conservation. Forest Ecology and Management. 2022;**506**:119956

[16] Lavender DP, Hermann RK. Douglas-Fir: The Genus Pseudotsuga. Oregon: Forest Research Laboratory; 2014. 352 p

[17] Van Loo M. Douglas-fir distribution in Europe. In: Spiecker H, Lindner M, Schuler J, editors. Douglas-Fir—An Option for Europe. Vol. 11. Joensuu: European Forest Institute; 2019. pp. 21-25. (What Science Can Tell Us)

[18] Alizoti P, Bastien JC, Chakraborty D, Klisz MM, Kroon J, Neophytou C, et al. Non-native Forest tree species in Europe: The question of seed origin in afforestation. Forests. 2022;**13**(2):273

[19] Fontes L, Savill P, Luis JS, Harris S. Identification of the origin of Portuguese Douglas-fir [*Pseudotsuga menziesii* (Mirb.) Franco] provenances. Silvae Genetica. 2003;**52**(5-6):266-273

[20] Luis JFS. Crescimento e Competição em Povoamentos de *Pseudotsuga menziesii* (Mirb.) Franco em Portugal. Vila Real, Portugal: Universidade de Trás-os-Montes e Alto Douro; 1989

[21] Correia A, Oliveira ÂC. Principais Espécies Florestais com Interesse para Portugal, Zonas de Influência Atlântica. Lisboa: Direcção-Geral das Florestas; 2003. 187 p. (Estudos e Informação)

[22] Nicolescu VN, Mason WL, Bastien JC, Vor T, Petkova K, Podrázský V, et al. Douglas-fir (*Pseudotsuga menziesii* (Mirb.) Franco) in Europe: An overview of management practices. Journal of Forest Research. 2023;**34**(4):871-888

[23] Charlet de Sauvage J, Bugmann H, Bigler C, Lévesque M. Species diversity and competition have minor effects on the growth response of silver fir, European larch and Douglas fir to drought. Agricultural and Forest Meteorology. 2023;**341**:109664

[24] Thurm EA, Biber P, Pretzsch H. Stem growth is favored at expenses of root growth in mixed stands and humid conditions for Douglas-fir (*Pseudotsuga menziesii*) and European beech (*Fagus sylvatica*). Trees. 2017;**31**(1):349-365

[25] Vitali V, Forrester DI, Bauhus J. Know your neighbours: Drought response of Norway spruce, silver fir and Douglas fir in mixed forests depends on species identity and diversity of tree neighbourhoods. Ecosystems. 2018;**21**(6):1215-1229

[26] Lévesque M, Saurer M, Siegwolf R, Eilmann B, Brang P, Bugmann H, et al. Drought response of five conifer species under contrasting water availability suggests high vulnerability of Norway spruce and European larch. Global Change Biology. 2013;**19**(10):3184-3199

[27] Vitali V, Büntgen U, Bauhus J. Silver fir and Douglas fir are more tolerant to extreme droughts than Norway spruce in South-Western Germany. Global Change Biology. 2017;**23**(12):5108-5119

[28] Lévesque M, Rigling A, Bugmann H, Weber P, Brang P. Growth response of five co-occurring conifers to drought across a wide climatic gradient in Central Europe. Agricultural and Forest Meteorology. 2014;**197**:1-12

[29] Rais A, Kuilen JWG, Pretzsch H. Growth reaction patterns of tree height,

diameter, and volume of Douglas-fir (*Pseudotsuga menziesii* [Mirb.] Franco) under acute drought stress in Southern Germany. Eur J Forest Res. 2014;**133**(6):1043-1056

[30] Boudru M. Forêt et Sylviculture. Le Traitement des Forêts. Vol. 2. Gembloux: Presses Agronomiques de Gembloux; 1989. 344 p

[31] Burns RM, Honkala BH, editors. Silvics of North America. Vol. 1. Conifers/Washington: USDA, Forest Service; 1990. 675 p

[32] Niinemets Ü, Valladares F. Tolerance to shade, drought, and waterlogging of temperate northern hemisphere trees and shrubs. Ecological Monographs. 2006;**76**(4):521-547

[33] Schütz JP, Ammann PL, Zingg A. Optimising the yield of Douglas-fir with an appropriate thinning regime. European Journal of Forest Research. 2015;**134**(3):469-480

[34] Frei ER, Moser B, Wohlgemuth T. Competitive ability of natural Douglas fir regeneration in central European close-to-nature forests. Forest Ecology and Management. 2022;**503**:119767

[35] Mason W, Edwards C, Hale S. Survival and early seedling growth of conifers with different shade tolerance in a Sitka spruce spacing trial and relationship to understorey light climate. Silva Fennica. 2004;**38**(4):357-370. Available from: https://www.silvafennica.fi/article/404

[36] Stokes V, Kerr G, Connolly T. Underplanting is a practical silvicultural method for regenerating and diversifying conifer stands in Britain. Forestry: An International Journal of Forest Research. 2021;**94**(2):219-231

[37] Savill PS. The Silviculture of Trees Used in British Forestry. 2nd ed. Cambridge, MA: CABI; 2013. 280 p

[38] Alves AAM, Pereira JS, Correia AV. Silvicultura. A Gestão dos Ecossistemas Florestais. Lisboa: Fundação Calouste Gulbenkian; 2012. 597 p

[39] Bindewald A, Miocic S, Wedler A, Bauhus J. Forest inventory-based assessments of the invasion risk of *Pseudotsuga menziesii* (Mirb.) Franco and Quercus rubra L. in Germany. Eur J Forest Res. 2021;**140**(4):883-899

[40] Kohnle U, Klädtke J, Chopard B. Management of Douglas-fir. In: Spiecker H, Lindner M, Schuler J, editors. Douglas-Fir—An Option for Europe. Vol. 11. Joensuu: European Forest Institute; 2019. pp. 73-83. (What Science Can Tell Us)

[41] Harrington TB, Harrington CA, DeBell DS. Effects of planting spacing and site quality on 25-year growth and mortality relationships of Douglas-fir (*Pseudotsuga menziesii* var. *menziesii*). Forest Ecology and Management. 2009;**258**(1):18-25

[42] Eberhard BR, Eckhart T, Hasenauer H. Evaluating strategies for the management of Douglas-fir in Central Europe. Forests. 2021;**12**(8):1040

[43] Louro V, Cabrita P. Pseudotsuga: Contribuição para o Conhecimento da sua Cultura em Portugal. Vol. 208. Direcção-Geral das Florestas; 1989. 128 p. (Estudos e Informação)

[44] Hsu HW, Stuke M, Bakker JD, Kim SH. A time-to-event analysis for temperature dependence of seed germination in four conifers: Ecological niche and environmental gradients. Forest Ecology and Management. 2024;**562**:121972

[45] Riou-Nivert PH. Des Itinéraires Sylvicoles Diversifiés pour le Douglas. Paris: Centre national de la Propriété Forestière; 2020. p. 27. Available from: https://www.cnpf.fr/sites/socle/files/2022-01/itineraires_douglas_bd%5B1%5D.pdf

[46] Gonçalves AC. Thinning: An overview. In: Gonçalves AC, editor. Silviculture. London: IntechOpen; 2021. pp. 41-58. Available from: https://www.intechopen.com/books/silviculture/thinning-an-overview

[47] Schütz JP. Sylviculture 1. Principes d'Éducation des Forêts. Lausanne: Presses Polytechniques et Universitaires Romandes; 1990. 245 p

[48] Lanier L, Badré M, Delabraze P, Dubourdieu J, Flammarion JP. Précis de Sylviculture. Nancy: ENGREF; 1986. 468 p

[49] Luis JFS, Monteiro M, do L. Dynamics of a broadleaved (*Castanea sativa*) conifer (*Pseudotsuga menziesii*) mixed stands in Northern Portugal. Forest Ecology and Management. 1998;**107**(1-3):183-190

[50] Nunes L, Gower S, Monteiro M, Lopes D, Rego F. Growth dynamics and productivity of pure and mixed *Castanea sativa* Mill. and *Pseudotsuga menziesii* (Mirb.) Franco plantations in northern Portugal. iForest. 2014;**7**(2):92-102

[51] Schütz JP, Pommerening A. Can Douglas fir (*Pseudotsuga menziesii* (Mirb.) Franco) sustainably grow in complex forest structures? Forest Ecology and Management. 2013;**303**:175-183

[52] Malcolm DC, Mason WL, Clarke GC. The transformation of conifer forests in Britain Ð regeneration, gap size and silvicultural systems. Forest Ecology and Management. 2001;**151**:7-23

[53] Wheeler A, Franklin JF, Wessell SJ. Ecological silviculture in Douglas-fir–Western hemlock ecosystems. In: Palik BJ, D'Amato AW, editors. Ecological Silvicultural Systems: Exemplary Models for Sustainable Forest Management. Hoboken, USA: Wiley; 2024. pp. 40-52

[54] Schütz JP, Götz M, Schmid W, Mandallaz D. Vulnerability of spruce (*Picea abies*) and beech (*Fagus sylvatica*) forest stands to storms and consequences for silviculture. Eur J Forest Res. 2006;**125**(3):291-302

[55] Cameron AD, Hands MOR. Developing a sustainable irregular structure: An evaluation of three inventories at 6-year intervals in an irregular mixed-species stand in Scotland. Forestry. 2010;**83**(5):469-475

[56] Kerr G, Snellgrove M, Hale S, Stokes V. The Bradford–Hutt system for transforming young even-aged stands to continuous cover management. Forestry: An International Journal of Forest Research. 2017;**90**(4):581-593

[57] Mason WL. Are irregular stands more windfirm? Forestry. 2002;**75**(4):347-355

[58] Louro G, Marques H, Salinas F. Elementos de apoio à elaboração de projectos florestais. Lisboa: Direcção-Geral das Florestas; 1999. 126 p. (Estudos e Informação)

[59] Sales Luis JF, Fonseca TF. The allometric model in the stand density management of *Pinus pinaster* Ait. in Portugal. Ann For Sci. 2004;**61**(8):807-814

[60] Drew TJ. Stand density management: An alternative approach and its application to Douglas-fir plantations. Forest Science. 1979;**25**(3):518-532

[61] Heiderman RR, Kimsey MJ. A species-specific, site-sensitive maximum stand density index model for Pacific Northwest conifer forests. Canadian Journal of Forest Research. 2021;**51**(8):1166-1177

[62] Reineke LH. Perfecting a stand-density index for even-aged forests. Journal of Agricultural Research. 1933;**46**(7):627-638

[63] Augusto L, Dupouey JL, Ranger J. Effects of tree species on understory vegetation and environmental conditions in temperate forests. Annals of Forest Science. 2003;**60**(8):823-831

[64] Bärmann L, Kaufmann S, Weimann S, Hauck M. Future forests and biodiversity: Effects of Douglas fir introduction into temperate beech forests on plant diversity. Forest Ecology and Management. 2023;**545**:121286

[65] Podrázský V, Martiník A, Matějka K, Viewegh J. Effects of Douglas-fir (*Pseudotsuga menziesii* [Mirb.] Franco) on understorey layer species diversity in managed forests. Journal of Forest Science. 2014;**60**(7):263-271

[66] Wohlgemuth T, Gossner MM, Campagnaro T, Marchante H, Van Loo M, Vacchiano G, et al. Impact of non-native tree species in Europe on soil properties and biodiversity: A review. NB. 2022;**78**:45-69

[67] Burton JI, Ares A, Olson DH, Puettmann KJ. Management trade-off between aboveground carbon storage and understory plant species richness in temperate forests. Ecological Applications. 2013;**23**(6):1297-1310

[68] Duguid MC, Ashton MS. A meta-analysis of the effect of forest management for timber on understory plant species diversity in temperate forests. Forest Ecology and Management. 2013;**303**:81-90

[69] Cremer M, Kern NV, Prietzel J. Soil organic carbon and nitrogen stocks under pure and mixed stands of European beech, Douglas fir and Norway spruce. Forest Ecology and Management. 2016;**367**:30-40

[70] Glatthorn J, Appleby S, Balkenhol N, Kriegel P, Likulunga LE, Lu JZ, et al. Species diversity of forest floor biota in non-native Douglas-fir stands is similar to that of native stands. Ecosphere. 2023;**14**(e4609):16

[71] Roach WJ, Simard SW, Defrenne CE, Pickles BJ, Lavkulich LM, Ryan TL. Tree diversity, site index, and carbon storage decrease with aridity in Douglas-fir forests in Western Canada. Front For Glob Change. 2021;**4**:682076

[72] Thurm EA, Pretzsch H. Improved productivity and modified tree morphology of mixed versus pure stands of European beech (*Fagus sylvatica*) and Douglas-fir (*Pseudotsuga menziesii*) with increasing precipitation and age. Annals of Forest Science. 2016;**73**(4):1047-1061

[73] Plaga BNE, Bauhus J, Pretzsch H, Pereira MG, Forrester DI. Influence of crown and canopy structure on light absorption, light use efficiency, and growth in mixed and pure *Pseudotsuga menziesii* and Fagus sylvatica forests. European Journal of Forest Research. 2024:479-491. DOI: 10.1007/s10342-023-01638-w

[74] Pretzsch H, Schütze G. Effect of tree species mixing on the size structure, density, and yield of forest stands. Eur J Forest Res. 2016;**135**(1):1-22

[75] Goes. A Floresta Portuguesa: Sua Importância e Descrição das Espécies de Maior Interesse. Lisboa: Portucel; 1991

[76] Amoroso MM, Turnblom EC. Comparing productivity of pure and mixed Douglas-fir and western hemlock plantations in the Pacific Northwest. Canadian Journal of Forest Research. 2006;**36**(6):1484-1496

[77] Trouvé R, Bontemps JD, Collet C, Seynave I, Lebourgeois F. Growth partitioning in forest stands is affected by stand density and summer drought in sessile oak and Douglas-fir. Forest Ecology and Management. 2014;**334**:358-368

[78] Acquah SB, Marshall PL, Eskelson BNI, Moss I, Barbeito I. Growth responses to thinning from below in uneven-aged interior Douglas-fir dominated stands. Canadian Journal of Forest Research. 2024;**54**(5):486-499

[79] O'Hara KL, Narayan L, Cahill KG. Twelve-year response of coast redwood to precommercial thinning treatments. Forest Science. 2015;**61**(4):780-789

[80] Anjos M d R. Tabelas de volume de simples e dupla entrada e estudo da rapidez de crescimento em povoamentos de Pseudotsuga menziessi a norte do Douro [Dissertation]. Vila Real: Universidade de Trás-os-Montes e Alto Douro; 1987

[81] GlobAllomeTree - Assessing Volume, Biomass and Carbon Stocks of Trees and Forests [Internet]. Available from: https://www.fao.org/sustainable-forest-management/toolbox/tools/tool-detail/en/c/217974/

[82] Henry M, Bombelli A, Trotta C, Alessandrini A, Birigazzi L, Sola G, et al. GlobAllomeTree: International platform for tree allometric equations to support volume, biomass and carbon assessment. iForest. 2013;**6**(6):326-330

[83] Zianis D, Muukkonen P, Mäkipää R, Mencuccini M. Biomass and stem volume equations for tree species in Europe. Silva Fenn, Monogr. 2005;**2005**(4):1-63

[84] Fonseca T, Costa NMR, Pacheco Marques C, Luis Louzada J, Cristina Gonçalves A. Exploring the biometric traits and potential of radiata pine (*Pinus radiata* D. Don) as a non-native species for sustainable forest systems in Portugal. In: Fonseca T, Cristina Gonçalves A, editors. Conifers—From Seed to Sustainable Stands. London, UK: IntechOpen; 2024. Available from: https://www.intechopen.com/chapters/1165355

[85] Fontes L. Modelling dominant height growth of Douglas-fir (*Pseudotsuga menziesii* (Mirb.) Franco) in Portugal. Forestry. 2003;**76**(5):509-523

[86] Fontes L. Modelling the Douglas-fir (*Pseudotsuga menziesii* (Mirb.) Franco) site index from site factors in Portugal. Forestry. 2003;**76**(5):491-507

[87] Dufour-Kowalski S, Courbaud B, Dreyfus P, Meredieu C, De Coligny F. Capsis: An open software framework and community for forest growth modelling. Annals of Forest Science. 2012;**69**(2):221-233

[88] Wilhelmsson L, Arlinger J, Spångberg K, Lundqvist SO, Grahn T, Hedenberg Ö, et al. Models for predicting wood properties in stems of *Picea abies* and *Pinus sylvestris* in Sweden. Scandinavian Journal of Forest Research. 2002;**17**(4):330-350

[89] Zobel BJ, Van Buijtenen JP. In: Timell TE, editor. Wood Variation: Its Causes and Control. Heidelberg: Springer-Verlag; 1989. 363 p

[90] Dellus V, Scalbert A, Janin G. Polyphenols and colour of Douglas fir heartwood. Holzforschung. 1997;**51**(4):291-295

[91] de Carvalho A. Madeiras Portuguesas: Estrutura Anatómica, Propriedades, Utilizações. Vol. II. Lisboa: Direção Geral das Florestas; 1997. Available from: https://books.google.pt/books?id=XZ4QoAEACAAJ

[92] Henin JM, Pollet C, Schmitt U, Blohm JH, Koch G, Melcher E, et al. Technical properties of Douglas-fir wood. In: Douglas-Fir—An Option for Europe. Joensuu, Finland: European Forest Institute; 2019

[93] Walker A, editor. The Encyclopedia of Wood: A Tree-by-Tree Guide to the world's Most Versatile Resource. New York: Facts on File; 2005. 192 p

[94] Schweingruber FH. Anatomy of European Woods—An Atlas for the Identification of European Trees, Shrubs and Dwarf Shrubs. Bern Stuttgart: P. Haupt; 1990

[95] Balanzategui D, Nordhauß H, Heinrich I, Biondi F, Miley N, Hurley AG, et al. Wood anatomy of Douglas-fir in eastern Arizona and its relationship with Pacific Basin climate. Frontiers in Plant Science. 2021;**12**:702442

[96] Donaldson LA, Singh A, Raymond L, Hill S, Schmitt U. Extractive distribution in *Pseudotsuga menziesii*: Effects on cell wall porosity in sapwood and heartwood. IAWA Journal. 2019;**40**(4):721-740

[97] Taylor AM, Renée Brooks J, Lachenbruch B, Morrell JJ, Voelker S. Correlation of carbon isotope ratios in the cellulose and wood extractives of Douglas-fir. Dendrochronologia. 2008;**26**(2):125-131

[98] Alden HA. Softwoods of North América. Madison, WI: U.S Department of Agricultural, Forest Service, Forest Products Laboratory; 1997. Report No.: Gen. Tech. Rep. FPL-GTR-102

[99] Wiedenhoeft A, Eberhardt T. Structure and function of wood. In: Wood Handbook Wood as an Engineering Material. Madison, WI, USA: U.S. Department of Agriculture, Forest Service, Forest Products Laboratory. p. 20. (General Technical Report)

[100] Panshin AJ, De Zeeuw C. Textbook of Wood Technology: Structure, Identification, Properties, and Uses of the Commercial Woods of the United States and Canada. 4th ed. New York: McGraw-Hill; 1980. 722 p. (McGraw-Hill series in forest resources)

[101] Via BK, Stine M, Shupe TF, So CL, Groom L. Genetic improvement of fiber length and coarseness based on paper product performance and material variability—A review. IAWA J. 2004;**25**(4):401-414

[102] Van Leeuwen M, Hilker T, Coops NC, Frazer G, Wulder MA, Newnham GJ, et al. Assessment of standing wood and fiber quality using ground and airborne laser scanning: A review. Forest Ecology and Management. 2011;**261**(9):1467-1478

[103] Ištok I, Sedlar T, Orešković G, Jambreković B. The variations in tracheid length of *Pseudotsuga menziesii* (Mirb.) Franco wood in relation to cambium age, site, and growth. Forests. 2023;**14**(6):1165

[104] Zobel BJ, Sprague JR. In: Timell TE, editor. Juvenile Wood in Forest Trees. Berlin, Heidelberg: Springer Berlin Heidelberg; 1998. (Springer Series in Wood Science). Available from: http://link.springer.com/10.1007/978-3-642-72126-7

[105] Louzada J, Fonseca F. Influência do Crescimento em Diâmetro (dap) e da Qualidade do Local na Variação do

Comprimento das Fibras em *Pseudotsuga menziesii* Mirb. (Franco). Vila Real, Portugal: Universidade de Trás-os-Montes e Alto Douro; 1991. 28 p. (Série Pedagógica UTAD)

[106] Fabisiak E, Moliński W. The density and length of tracheids of Douglas fir (*Pseudotsuga menziesii* Franco) in relations to the biosocial position of given tree in the stand. Folia Forestalia Polonica. 2002;**33**:25-31

[107] Elliott GK. Wood Density in Conifers. Oxford, England: Commonwealth Forestry Bureau; 1970. Report No.: 8

[108] Erickson H, Harrison A. Douglas-fir wood quality studies part I: Effects of age and stimulated growth on wood density and anatomy. Wood Science and Technology. 1974;**8**:207-226

[109] Keller R, Thoby M. Liaisons entre l'état juvénil et l'état adulte pour quelques caractères technologiques et auxométriques chez el Douglas (*Pseudotsuga menziesi*). Annales des Sciences Forestières. 1977;**34**(3):175-203

[110] Krajnc L, Gričar J, Jevšenak J, Hafner P, Brus R. Tree rings, wood density and climate–growth relationships of four Douglas-fir provenances in sub-Mediterranean Slovenia. Trees. 2023;**37**(2):449-465

Chapter 8

Integration of Waste from Invasive Alien Species *Phytolacca americana* with Eucalyptus Commercial Pulp for Papermaking: A Review on the Evolution of Physic-Mechanical and Chemical Paper Properties

Antía Iglesias, Ángeles Cancela, Juan Picos and Ana Soler

Abstract

The paper industry contributes significantly to the transformation of native forests into monocultures of fast growing exotic species such as *Eucalyptus globulus*, reducing biodiversity. This expansion, driven by the demand for cellulose-based products, threatens ecosystems by displacing native flora and causing soil impoverishment. To address these issues, the study explores the possibility of reusing waste fibers obtained after cleaning areas affected by a major presence of invasive species such as *Phytolacca americana*, as a sustainable alternative to traditional wood fibers in paper production. The research focused on Ribadumia riparian forest in Galicia, Spain, and involved the entire process of obtaining and processing *Phytolacca americana* pulp. It assesses the physical and mechanical properties of the resulting paper, using advanced techniques such as SEM visualization and physical paper tests. Results confirmed the suitability of *Phytolacca americana* fibers for paper production, with certain properties showing improvement and/or maintenance over those from commercial paper. Blends of *Phytolacca americana* and commercial bleached *Eucalyptus globulus* pulp demonstrated comparable quality, suggesting the feasibility of replacing commercial fibers like *Eucalyptus globulus* with those derived from the waste of invasive species. This approach could not only contribute to reduce monocultures and promote biodiversity, but to apply zero-waste dynamics in usually pollutant industries.

Keywords: invasive alien species, paper production, biomass reutilization, non-woody pulp integration, biodiversity loss

1. Introduction

Paper and pulping industries are extremely related to environmental and ecosystem health. The paper industry is not only the fifth sector most pollutant, generating an amount of 220–380 m^3 of highly colored and potentially toxic wastewater [1], but it is responsible for most monoculture production, consuming 40% of the world's commercially cut timber. Apart from this, paper production remains one of the most relevant industries globally, acquiring a total amount of 2.519.089 € in past 2018 only in Spain, and as stated by the As Papel Association, in 2022, Spain's overall production reached 6.3 million tons, with 1.5 million tons designated for pulp and 6.3 million tons for paper and cardboard. This situation positioned Spain as sixth in paper and fifth in pulp production within the EU (European Union). According to preliminary CEPI (Confederation of European Paper Industries) reports in 2021, the paper industry has increased by 5.8% in relation to the previous year [2].

Those wicked situations contribute to the deterioration of natural ecosystems surrounding factory mills and pulpwood plantations. Paper mills are usually located near wood plantations in order to decrease costs; natural forests were usually used for papermaking industry in the past, but, nowadays, the use of natural forests is being substituted by plantations of fast-growing species, usually exotic, such as Eucalyptus and Pines. Planted forests were, in 2014, 7% of the total global forest area, or 264 million ha [3].

According to the Annual Forest Statistics of 2018, 33% of timber is destinated to pulp production, the highest proportion of wood commercial uses. In comparison, in the case of the sector, 38% of wood is destined for bioenergy, and 21% for pulp production [4]. At the north of Spain, specifically in Galicia, there is a volume of timber areas of $125m^3/ha$ in front of 189 m^3/ha of forest surface. Compared with this, the south of Spain presents an average of 20 m^3/ha – 50 m^3/ha [4]. A third part of total forest plantation in Galicia is destinated to commercial deforestation, regarding the felling of tree areas for commercial and industrial purposes and interests. The three principal arboreal species in Spain are from the Pinus family, but the fourth one, even exotic, is *Eucalyptus globulus*, occupying 79.645.351 m^3 *c.c.* in all Spanish surface [5], part of the forest area destinated to production, an 8% of the total forest area, specifically 619.918 ha.

Eucalyptus globulus plantations proliferation in Galicia, Spain, are mainly derived from ENCE (Energy and Cellulose Paper Mill), Pontevedra's paper mill, which in 2022 spent 236.000.000€ in timber from individuals [6], covering 95% of demand needs. Other studies demonstrated that a third part of the Galician Forest has been planted by humans, with commercial ends.

Eucalyptus globulus species are not classified as invasive exotic species in Spain, according to the Invasive Exotic Species Catalog. However, this exotic species has demonstrated expansive behavior altering local ecosystems. Among other conditions, its leaf's chemical composition, which intervenes in soil, increasing its pH and avoiding biodiverse flora proliferation [7], derives in preventing local species from growing. Even nowadays, several studies [8, 9] demonstrated high possibilities for *Eucalyptus globulus* industry by-products (such as leaves and bark) reutilization for antioxidant extraction and papermaking; main problem remains in high commercial demands that are translated into monocultures incrementation [6].

Furthermore, the problematic relationship between monocultures and biodiversity flora maintenance, in addition to forest urbanization and anthropogenic factors as agriculture and land management, usually results in ecosystem diversity loss, favoring exotic and invasive species proliferation [10]. Local vegetation in riparian areas is being naturally substituted by invasive species such as those selected for this study, *Phytolacca americana* [11]. The impact of invasive alien species in natural ecosystems disturbing its balance through nutrient and resource competition, disease transmission, and intertwining in soil and light conditions, provoking the reconfiguration of the ecosystem functioning [2, 12]. The major presence of invasive plant species in managed environments tends to have consequences in the ecological integrity of the ecosystem, altering the natural composition of species and habitat [13, 14].

According to the European Commission, 41 plant species pose a threat to European natural ecosystem's biodiversity. In Regulation (EU) 1143/2014, Invasive alien species are presented as one of the main threats to biodiversity and related to ecosystem services, which are being intensified by global trade, transport, tourism, and climate change, consisting of a cross-border challenge affecting the whole Union [11]. When referring to invasive alien species, it is relevant to confirm anthropogenic introduction and naturalization in different ecosystems than its original habitat; by human means, these species were usually introduced for cultivation or for ornamental means and derived into ecosystem degradation [11]. In affected ecosystems, rapid restoration and mitigation of Invasive species is crucial for flora maintenance; for this reason, several management activities are being performed all around affected areas, such as the one selected for this study, in Ribadumia, Galicia. Derived from this management initiative, promoted by the city council and governmental decisions, amounts of fiber waste are being generated, which were suitable to be reintroduced into production systems.

In order to help mitigate and reduce paper industry contradictions with ecosystem biodiversity and maintain of ecological integrity of natural forests, this study proposes the reutilization of invasive alien species *Phytolacca americana* plant waste, after cleaning initiatives, as a substitution for *Eucalyptus globulus* commercial pulp in pulping and papermaking industry. The substitution of first raw materials for secondary raw matter helps mitigate waste production and waste burns and reduce price volatility, availability, and import dependency [15].

The need for creating and designing sustainable products is an increasing concern, becoming one of the most relevant issues nowadays. According to the European Commission Circular Economy Action Plan [16], the anticipated global consumption of materials like biomass, fossil fuels, metals, and minerals is set to double within the next 40 years. Concurrently, annual waste generation is forecasted to surge by 70% by 2050 [17].

The selected area of study and recollection for this research was the riverbank area of Umia River in Ribadumia, Galicia, Spain, which is affected by the presence of several invasive alien species due to anthropogenic factors [10]. *Phytolacca americana* (**Figure 1**) has been identified as one of the most present alien species in riparian forests, and its controversial effects have been studied by experts [18].

Selected species *Phytolacca americana* is a perennial plant of the Polygonaceae family, native to Eastern North America, commonly known as American pokeweed, poke sallet, or inkberry. It is a large herbaceous species, reaching up to 3 meters. Its stem (**Figure 2**) is cylindrical, fleshy, and hollow inside, with purple or red spots, having a woody constitution at the base; it is easily identifiable due to its reddish/fuchsia coloration [10]. The root of Phytolacca is thick, with multiple heads and a strong system of taproots [19]. The leaves are alternate, simple, lanceolate, and bitonal, darker on the upper side. The maximum length of the leaves is 25 cm. Its fruit is fleshy and intensely purple, highly toxic, like all parts of the plant [20].

Phytolacca americana is one of the most common herbaceous invasive alien species in Mediterranean areas [21]. Nationally, it is included under Royal Decree 630/2013 in the Spanish Catalog of Invasive Alien Species; at the European level, it is regulated by Regulation (EU) 1143/2014 of the European Parliament and of the Council of 22

Figure 1.
(left) Phytolacca americana *in the riverbank area of Umia river, Spain.*

Figure 2.
(right) Detail of stem and fruit of Phytolacca americana.

October 2014 on the prevention and management of the introduction and spread of invasive alien species; and internationally, under the Convention on Biological Diversity (CBD) 1992, as well as the Convention on the Conservation of European Wildlife and Natural Habitats (Bern Convention, 1979) (Life, Med, WET Rivers, IAS, PHYAME/EPI/FL032).

Currently, the distribution of this species has significantly expanded, being naturalized in various regions of the world beyond its natural habitat in North America [19]. From Canada to Argentina in the Americas or since the 17th century in Europe, where it began to be cultivated around the Mediterranean. References to the cultivation of this species in areas of North Africa and Southern Europe date back to the mid-17th century [19]. Similarly, this plant has naturalized in other parts of the world, from Turkey, Asia, India, China, Taiwan, Japan, Indonesia, and Australia, to New Zealand. It is one of the most widely distributed species globally and is, in numerous locations, considered an IAS [19, 21].

The plant's life cycle in the Mediterranean climate occurs between April and October, disappearing during the colder months. The plant begins to sprout as the average temperatures rise after winter, and the first plants can be found in April or May [19].

According to [19], it can be found in uncultivated vineyards, orchards, cultivated fields, ruderal areas, or even urban areas. The fruits ripen during the warmest seasons, and it is from the last months of autumn and the first of winter that the plant disappears completely, although the root remains from 1 year to the next. It can be found growing annually in the same areas [21]. In the riverbank area of Ribadumia, it can be found isolated from other individuals of its species, growing in the middle of other species such as *Tradescantia fluminensis*. But it can also be found in residual spaces between agricultural land, forming big groups of individuals dominating the spaces.

In this study, fibers from *Phytolacca americana* were mixed with commercial bleached pulp from Eucalyptus, commonly used in the paper industry. Raw matter was donated to the Forestry Engineering School of the University of Vigo by the ENCE Research Laboratory in Marin, Pontevedra.

Therefore, the present research aims to analyze the potential of fibers from the invasive alien species *Phytolacca americana* as a replacement for commercially used fibers derived from monocultures of trees such as *Eucalyptus globulus*. By focusing on the properties of laboratory-tested handsheets, we examine various proportions of raw material, starting with 100% *Phytolacca americana* and progressing through a series of mixtures—100%, 90%, 70%, 50%, 30%, 10%, and 0% in relation to *Eucalyptus globulus* commercial pulp. This stepwise approach allows for a comprehensive comparison of the resultant paper sheet properties in terms of air resistance, drainability, grammage, and thickness. The findings from this research aim to provide valuable insights into the feasibility of using *Phytolacca americana* fibers as a sustainable alternative to traditional monoculture-sourced fibers for the papermaking industry and extended to printing applications. Thereby contributing to the reduction of the environmental impact associated with controversial commercial pulp origins, such as monocultures, this research initiative search to advance into biodiversity maintenance and vegetal waste management in areas affected by the presence of invasive alien species.

2. Materials and methods

For this study, it has been used and applied specific instruments and processes. As follows, it is possible to find the materials and methods involved in its development.

2.1 Harvesting and delignification process

Harvesting of raw matter from *Phytolacca americana* was carried out in June of 2022 in a specific area of 2 km in the river Umia riparian forest space between Cabanelas and Ribadumia, Spain. In order to proceed with the de-lignification process, raw matter was divided into leaves and stems, cut into pieces of a maximum of 20 cm, and dried in the oven at the Wood Chemistry Laboratory at the Forestry Engineering School at a temperature of 105°C for 24 h. After dehydration, stems and leaves were weighed and stored in a cooking pot of 12 L of capacity. A 20% of its weight was added in Caustic SOSA (NaOH) content, and distilled water was added until full covering of the material, an average quantity of 5 L. As observed in previous

Figure 3.
From left to right, delignified raw matter from Phytolacca americana *(1) leaves, (2) small branches and leaves mixture, and (3) stem material.*

research [22, 23], optimized times and quantities are a ratio of 5:1 liquid-to-solid mix and 90 minutes of cooking time. Temperature was selected by maintaining the content at the boiling point. Internal conformation of matter, which presents strong and long fibers in the stem area, suggested the addition of 30 minutes extra for delignification process, performing a total time of 120 min at boiling temperature.

After delignification, the black bleaches resulting from the process, were filtered and stored in special containers. It is possible to neutralize its pH with sawdust in order to use it as compost material for further occasions, as we can observe in **Figure 3**, raw delignified matter presents completely different structures when referring to leaves and stem material. While stems keep their original strength, leaves are almost decomposed in a homogeneous mix.

Once raw matter was cleaned with distilled water, it was dried at the hoven during 24 h at 105°C, for further pulping and refining processes.

Previous studies performed with *Phytolacca americana* demonstrated that more suitable results of papermaking were obtained with stem material, so in this study, just stem matter was used for blending with commercial pulp due to similar conditions. In the publication [22], *Phytolacca americana* leaves pulp material obtained a drainability degree of 63°Schopper-Riegler (SR) and stem matter oscillated in between 21°SR and 34.5°SR depending on refining times (60 seconds or 120 seconds). On the other hand, commercial pulp with 60 seconds of refining time obtained 27.5°SR or 43°SR for 120 seconds, which is closely to the drainability obtained with Phytolacca stem material. It is for these reasons that it has been decided for this study to specifically use raw matter from the Phytolacca stem, as it appears to demonstrate similar properties and behavior with *Eucalyptus globulus* commercial pulp.

2.2 Pulping and refining process

Stem material from *Phytolacca americana*, once dried, was divided into small portions of 30 g and rehydrated with 1 L of distilled water, according to UNE-EN-ISO 5263-1:2005 [24] indications for commercial pulp. Then, commercial pulp dried material was summitted in the same process, rehydrated with 1 L of distilled water. The refining process was performed separately.

First, unbleached matter from *Phytolacca americana* stem was disintegrated in a laboratory disintegration beater, obtaining a consistency of 10%, ready to be refined, following indications set up in UNE-EN ISO 4119:1996 [25]. Mixture was refined for 2 minutes in a PFI laboratory beater mill at 600 and 1200 rpm according to UNE-EN-ISO 5264-2:2011 [26]. Refined material was then weighed, and distilled water was added until 300 g of material was reached.

The same process was performed with commercial bleached pulp from Eucalyptus. First, disintegrated and then refined for 2 minutes. Resultant pulp was measured, and the matter was adjusted with distilled water until obtention of needed quantities.

In order to perform efficient laboratory sheets and physic-mechanical tests, raw matter was divided into proportions of a mixture of 10%, 30%, 50%, 70%, and 90% of raw matter from Phytolacca in relation to commercial pulp material. After calculation, the raw matter was mixed and stored for pulping test development before laboratory sheets formation. Related information with 100% *Phytolacca americana* handsheets and 100% commercial pulp handsheet properties obtained in previous studies [22] was used to have a global vision of physical characteristic evolution.

2.3 Physic-mechanical properties analysis of pulp and Hansheets

After performing the refining process separately, each pulp matter was weighed to obtain its corresponding part in order to perform each test. To achieve each proportion quantity for optimizing 30 g of raw matter from Phytolacca and 30 g of raw matter of commercial pulp the following equation was calculated:

$$Q = \frac{Percentage\ n \times raw\ matter\ mass}{(\Sigma\, percentage)} \tag{1}$$

where
Q is the raw matter quantity to be used.
Percentage n is the relative percentage to be calculated.
Raw matter mass is 30 g of Phytolacca or commercial pulp.
Σ percentage is the summation of all percentages to be calculated.

The result is presented as grams of mass to be used in each mixture test in order to maintain the final mass constant according to the study [25]. The proportion of distilled water added to the experiment has been adapted to maintain proportions of consistent pulp dissolution. The refining time was 120 seconds with an initial raw matter of 30 g, The following pulp quantity for the mixture of *Phytolacca americana* and commercial pulp are presented in **Table 1**.

The calculated pulp mixture was adapted to consistency needs following UNE-EN-ISO 5267-1:1999 [27] in order to perform the drainability Schopper-Riegler (°SR) test. Prepared samples of selected proportions were disintegrated and reformulated with distilled water to obtain a consistency of 0.5% to perform the test with the Schopper-Riegler apparatus.

2.3.1 Shopper-Riegler drainability property and calculation process

As specified in Ref. [27], the drainability index refers, in the context of papermaking, to the feasibility and facility of how water can be removed from a fiber suspension. To this end, the calculation of the drainage index involves measuring the

Mixture proportion	Pulp quantity for mixture (g)	
	Phytolacca americana	Commercial pulp
10–90%	1.2	10.8
30–70%	3.6	8.4
50–50%	6	6
70–30%	8.4	3.6
90–10%	10.8	1.2

Table 1.
Pulp quantity to be used for mixture proportions for pulping tests.

drainage capacity of a fiber cake formed on a specific apparatus called Schopper-Riegler, which measures in this same scale. Results are expressed in Schopper-Riegler degrees (°SR). A higher index means a more tighter surface, so a denser and compact fiber. On the contrary, lower °SR used to imply longer and thinner fibers with bigger holes in between.

After drainability tests, laboratory sheets were made with each pulp mixture. A conventional sheet former, ubicated at the Wood Chemistry Laboratory of the Forestry Engineering School, with a diameter of 16 cm, was used with blended pulps, following UNE-EN-ISO 5269-1:2005 [28] indications, and ambient conditions were measured and set according to UNE-EN-ISO 187:1990 [29] standards. In order to achieve an average and significant result, each test was performed in triplicates. Obtained handsheets can be observed in **Figure 4**, where it is possible to notify color and texture variations due to fiber mixture evolution.

With the intention to better understand paper resistance and properties, the were the following tests on the laboratory sheets: Thickness and medium thickness (μm) and grammage (g/m^2) according to UNE-EN ISO 536:2012 [30]; specific volume (cm^3/g) and specific density (g/cm^3) According to standards in UNE-EN ISO 534:2012 [31]; Air resistance was calculated with a Gurley test apparatus, expressed in second (s), and air permeability according to Gurley (μm/(Pa.s)) according to standard UNE-EN ISO 5636-5:2015 [32]. All tests were performed in triplicates, following related standards.

Figure 4.
Laboratory sheets were performed with different pulp proportions. Correspondence, from left to right: 100% of Phytolacca fibers, 90% of Phytolacca fibers, and 10% of commercial pulp; 70% - 30%; 50% - 50%; 30% - 70%; 10% -90%; and 100% of commercial pulp.

2.3.2 Grammage property and calculation process

The grammage is known as one of the most representative properties of paper when referring to the creative industry. This affects not only to the paper utilization but to its resistance and durability. The grammage is defined as the property that measures the mass per unit area of a sheet [30]. For its determination, each sample must be weighed on a calibrated scale, obtaining an average value that will allow us to calculate the final grammage. Due to the specific test conditions we are performing, the hand sheet diameter is constant at 160 mm, with a radius of 80 mm, obtaining an area of 0.02010 m^2. These data are relevant as the formula for calculating the grammage depends on paper diameter and surface, calculated in 49.72 constant for 160 mm diameters.

Then, grammage is calculated as follows:

$$G = W\,x\,S \tag{2}$$

where
G is the grammage expressed in (g/m^2)
W is the weight expressed in g
S is the surface, equivalent to 49.74 m^2

2.3.3 Thickness and medium thickness properties and calculation process

The thickness of a sheet is defined as the distance between two surfaces of the material [30]. Usually expressed in micrometers (μm), to calculate this, it is necessary to use a specific machine whose accuracy allows for specific slight thickness variations. Positioned between two surfaces, the paper sheet is tight in at least five different points. This matrix of measurement points has to be distributed over the surface of the hand sheet to evaluate the complete material. After the calculation of the average, we can observe the thickness of the paper, which will affect further specific volume and density properties.

2.3.4 Specific volume and density properties and calculation process

The apparent specific volume of the sheet is the volume per unit mass, expressed in cubic centimeters per gram, and is calculated from the thickness and grammage of a single sheet as indicated [31]. Meanwhile, the apparent density is the mass per unit volume, calculated from the thickness of a single sheet or in relation to the apparent specific volume of a single sheet [31].

To calculate the specific volume expressed in cm^3/g, it is as follows:

$$Vs = \frac{Ts}{G} \tag{3}$$

where
Vs is specific volume
Ts is specific thickness, calculated from the average of a single handsheet.
G is the grammage of the laboratory sheet.

Furthermore, the specific density, expressed in g/cm^3, as the inversion of specific volume, should be calculated as follows:

$$Ds = \frac{G}{Ts} \tag{4}$$

where
Ds is specific density
Ts is specific thickness, calculated from the average of a single handsheet.
G is the grammage of the laboratory sheet.

2.3.5 Gurley air resistance and permeance property and calculation process

Air resistance is the time required for a specific volume of air to pass through a unit area under a unit pressure. It should be expressed as "air resistance Gurley" and measured in seconds. For its calculation, the air is compressed by the weight of a vertical cylinder floating in a liquid using the Gurley apparatus [32]. The time it takes for a given volume of air to pass through the sample indicates the Gurley air resistance in s.

On the other hand, air permeance is the average airflow rate passing through a unit area under a specific pressure difference, expressed in (μm/(Pa.s)) [32]. For its calculation it is needed to apply the following equation:

$$Pe = \frac{135.3}{t} \tag{5}$$

where
Pe is the air permeance
t is the time

The equation is based on an average pressure difference of 1.22 Kpa, a test area of 642 $mm^2$and a volume of 100 ml of air passing through the sample, measured at ambient pressure.

2.4 Chemical composition and optical properties analysis

After calculating physic-mechanical properties, the chemical composition and evolution of handsheets were analyzed when developing the specified blends.

The chemical composition of performed tests was analyzed using SEM visualization at the CACTI (Help Center for Scientific and Technological Research) of the University of Vigo. Electron Microscope JEOL6100 was used in order to observe surface images of the internal structure of samples and chemical composition evaluation through variations in pulp concentrations. Images of optical microscope Nikon SZM1500 were studied to better appreciate texture and color properties of papers.

As it is possible to observe in **Figure 5**, the main composition of commercial pulp from *Eucalyptus globulus* made paper is based on carbon, oxygen, magnesium, silicon, calcium, sodium, alum, and phosphorus [22]. On the contrary, 100% *Phytolacca americana* fibers made laboratory paper demonstrated a composition based on carbon, oxygen, calcium, phosphorus and silicon (**Figure 6**) [22].

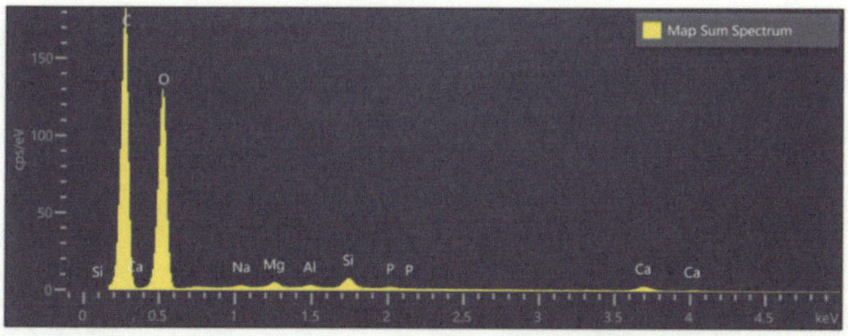

Map Sum Spectrum				
Element	Line Type	Weight %	Weight % Sigma	Atomic %
C	K series	47.93	0.16	56.15
O	K series	47.50	0.16	41.78
Mg	K series	0.57	0.03	0.33
Si	K series	1.29	0.03	0.65
Ca	K series	1.93	0.06	0.68
Na	K series	0.25	0.03	0.15
Al	K series	0.28	0.03	0.15
P	K series	0.26	0.03	0.12
Total		100.00		100.00

Figure 5.
Spectrum of chemical composition of 100% commercial bleached Eucalyptus globulus *pulp, laboratory paper.*

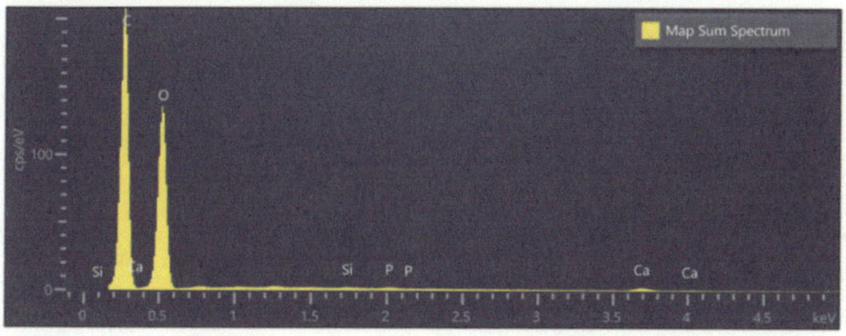

Map Sum Spectrum				
Element	Line Type	Weight %	Weight % Sigma	Atomic %
C	K series	49.69	0.15	57.25
O	K series	48.75	0.15	42.17
Ca	K series	1.22	0.06	0.42
P	K series	0.23	0.03	0.10
Si	K series	0.11	0.02	0.05
Total		100.00		100.00

Figure 6.
Spectrum of chemical composition of 100% Phytolacca americana *fiber's laboratory paper.*

In **Figure 7**, we observe how the internal structure of both papers present similarities, composed of long and thin fibers that have been measured with Image J software. Fibers presented in paper made of 100% *Phytolacca americana* pulp, demonstrated an average thickness of 8.67 μm, with thinner fibers of 1.969 μm and thicker one's of 18.89 μm. The fiber length is around 200.49 μm. In comparison, fibers that form commercial pulp handsheets are an average of 10.56 μm thick, presenting more constant and stable fibers balancing between 5.099 and 13.99 μm. Besides its thickness constancy, fibers are longer than 727.23 μm, being intertwined, forming a homogeneous surface structure.

In **Figure 8**, it is possible to appreciate, from a lateral point of view, how papers of 100% *Phytolacca americana* (right) and 100% commercial pulp (left) are composed. As commented before, internal composition of Phytolacca papers is less ordinated than those made from commercial pulp.

Chemical analysis presented in this section allows us to observe the main composition of pure papers without mixing materials. As it will be demonstrated in the

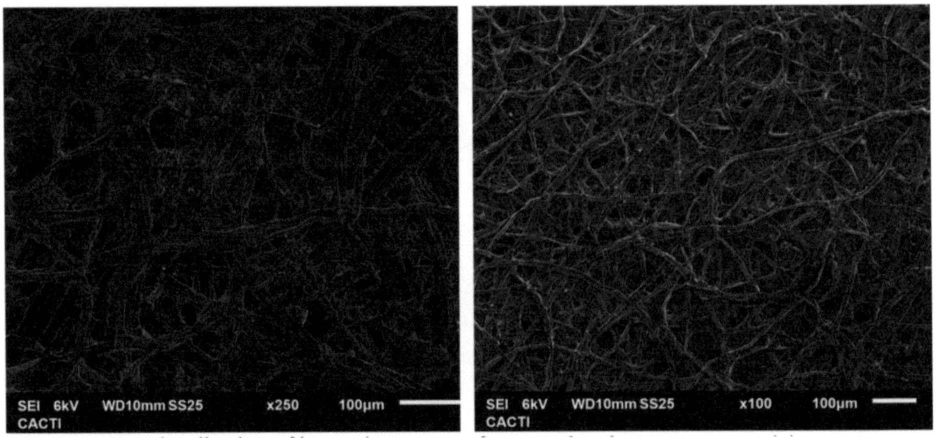

Figure 7.
SEM visualization of internal structure of 100% Phytolacca americana *laboratory paper (right) and SEM visualization of internal structure of 100% commercial bleached pulp handsheet. (left).*

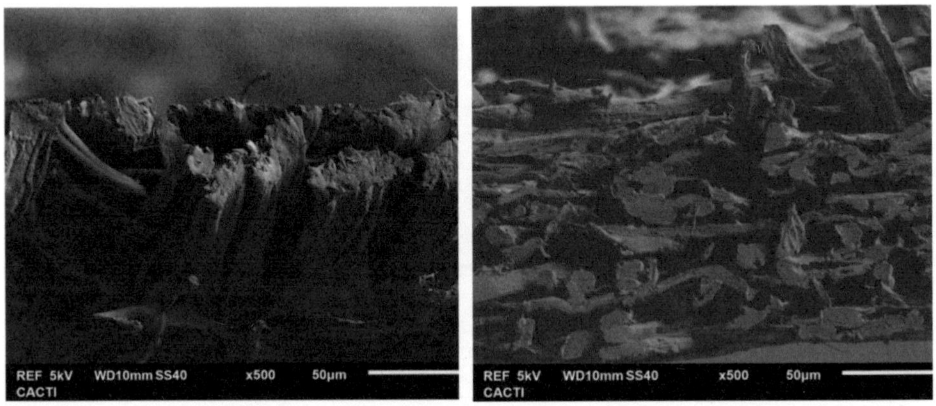

Figure 8.
Transversal section of internal structure of 100% Phytolacca americana *laboratory paper (right) and Transversal section of internal structure of 100% commercial bleached pulp handsheet (left).*

following sections, these properties and compositions will be affected and suffer an evolution when blending raw matter, obtaining different results depending on the quantity.

2.5 Evaluation of physical, mechanical and morphological properties

Physic-mechanical and morphological characteristics were evaluated in accordance with the mentioned ISO standards, and statistical analyses were integrated into all measurements. The properties and tests were conducted and calculated in triplicate, with the data being averaged.

Mechanical properties were tested in triplicate and carried out using PFI correspondences of 1 loop = X10 in modern equipment, where 60s was deemed equivalent to 1200 rpm [22, 23]. In order to avoid misunderstandings, PFI-related data are given in seconds (s) instead of rpm.

3. Results

3.1 Physic-mechanical properties

Experiments with different proportions of pulp blend have been named as test, corresponding test 0: 100% *Phytolacca americana* fibers pulp; 1: 90% Phytolacca fiber 10% commercial *Eucalyptus globulus* bleached pulp; 2: 70% Phytolacca fiber 30% commercial *Eucalyptus globulus* bleached pulp; 3: 50% Phytolacca fiber 50% commercial *Eucalyptus globulus* bleached pulp; 4: 30% Phytolacca fiber 70% commercial *Eucalyptus globulus* bleached pulp; 5: 10% Phytolacca fiber 90% commercial *Eucalyptus globulus* bleached pulp; and 6: 100% commercial *Eucalyptus globulus* bleached pulp. In **Table 2** it is possible to see the different pulp blends numered by performed tests. This information will be referenced this way (Test 0, Test 1, Test 2...) in the following figures and tables of the document.

As it has been explained before, pulp drainability has been tested following Shopper-Riegler apparatus indications, obtaining an average degree of 36.14°SR. Higher Schopper-Riegler degrees are achieved in the extremes of proportions, being 100% *Phytolacca americana* and 90–10% mixture, in addition to 100% commercial bleached pulp with higher drainability degrees (**Figure 9**). On the contrary, even with a small

Tests	Phytolacca matter	Eucalyptus commercial pulp
0	100	0
1	90	10
2	70	30
3	50	50
4	30	70
5	10	90
6	0	100

Table 2.
Relation of performed tests according to proportion blends between Phytolacca americana *matter and* Eucalyptus globulus *commercial pulp.*

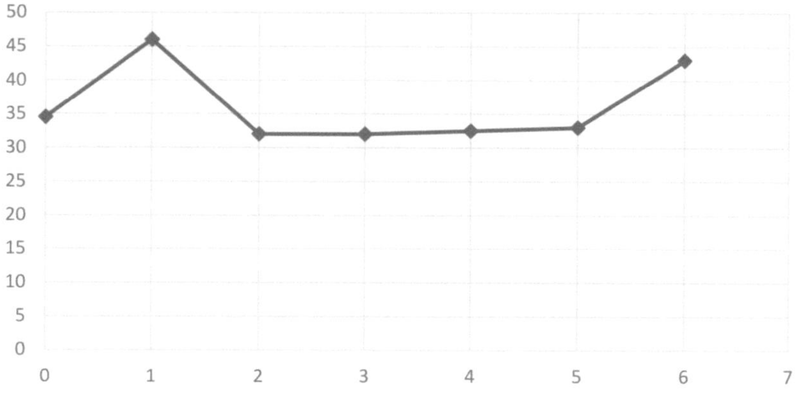

Figure 9.
Drainability transformation in performed mixture test. Correspondence of Test 0 = 100% of Phytolacca fibers, 1 = 90% of Phytolacca fibers and 10% of commercial pulp; 2 = 70% - 30%; 3 = 50% - 50%; 4 = 30% - 70%; 5 = 10% -90%, and 6 = 100% of commercial pulp.

descensus of drainability, it remains constant in intermediated blends, which indicates a homogeneous mixture and properties compensation of both raw materials.

The physical properties relation between weight and thickness is represented in **Figure 10**. It is possible to observe how thickness remains constant, with a variation of 0.0205 mm; but in relation to handsheet weight, we can observe how commercial pulp apports heavier fibers, augmenting it as we increment commercial pulp in the mixture in decrement of Phytolacca fibers. Lower grammages are represented by *Phytolacca americana* 100% content and 90% with an average of 23.53 and 33.49 g/m^2in front of higher grammages of tests 5 and 6, which obtain an average of 64.47 and 56.7 g/m^2, respectively. Medium tests with a balanced proportion of raw matter demonstrated an average of 52.39 g/m^2 of grammage.

In **Figure 11**, it is possible to observe grammage evolution in performed tests, deriving to a clear augment of grammage when adding commercial pulp to the mixture. Both raw matters become integrated, achieving a consistent surface. As can be observed in **Table 3**, in specific volume (cm^3/g) and specific density (g/cm^3) evolution, we can appreciate the same tendency, concluding higher specific density when adding commercial pulp, in decrement of specific volume (**Figure 12**).

The last physic-mechanical property studied in this research was air permeability and air resistance, measured in (μm/(Pa.s)) and seconds (s) by using the Gurley apparatus in 300 cm^3 measure. In **Figure 13**, it is possible to observe how air permeability decreases drastically when incrementing commercial pulp to the handsheet, contrary with air resistance, which increments softly. It is possible to suppose the relation between fiber structures of different raw matter, finding a more consistent and homogeneous surface upon commercial pulp, which increments the paper's capacity to resist air transmission. Higher air resistance times are obtained with tests 5 and 6 (9.2 and 13.15 s), corresponding with higher contents of commercial pulp. Air

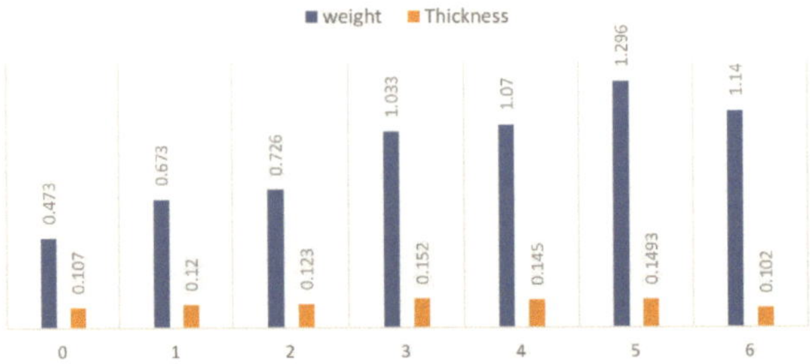

Figure 10.
Graphical representation of the evolution of weight (g) and thickness (mm) in performed tests.

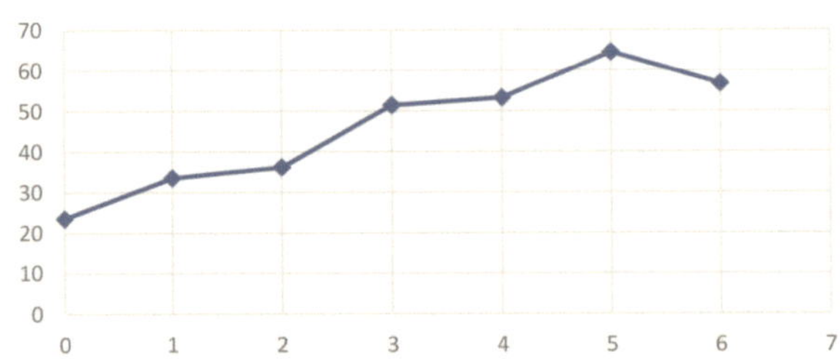

Figure 11.
Evolution of grammage (g/m^2) in performed tests.

Tests	Specific volume (cm³/g)	Specific density (g/cm³)
0	4.56	0.22
1	3.58	0.28
2	3.41	0.29
3	2.96	0.34
4	2.73	0.37
5	2.32	0.43
6	1.79	0.56

Table 3.
Specific volume and specific density tests results.

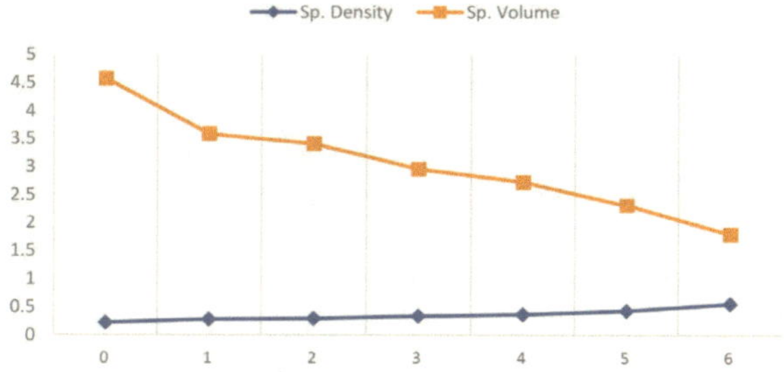

Figure 12.
Inverse relation of specific density and volume along tests.

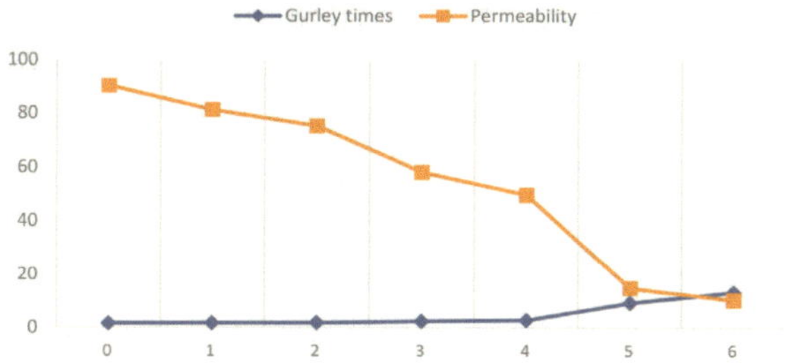

Figure 13.
Air permeability (µm/(Pa.s)) *and air resistance (s) evolution in performed tests.*

permeability for these cases is the lowest obtained, being 14.7 and 10.29 µm/(Pa.s), respectively. In the case of tests 0 and 1, where *Phytolacca americana* concentration is higher, Gurley times are reduced to 1.5 and 1.66 s, while air permeability is 90.2 and 81.18 µm/(Pa.s). Medium steps, corresponding with 50% of both raw matters, present an average of 2.33 seconds and 57.98 µm/(Pa.s).

3.2 Internal structure integration

The internal composition of samples suffered an evolution when changing proportions of materials, but this difference is not significant when referring to surface structure due to the similitudes of fiber length and thickness. In **Figure 14**, it is possible to appreciate how main surface remains consistent, being composed of long

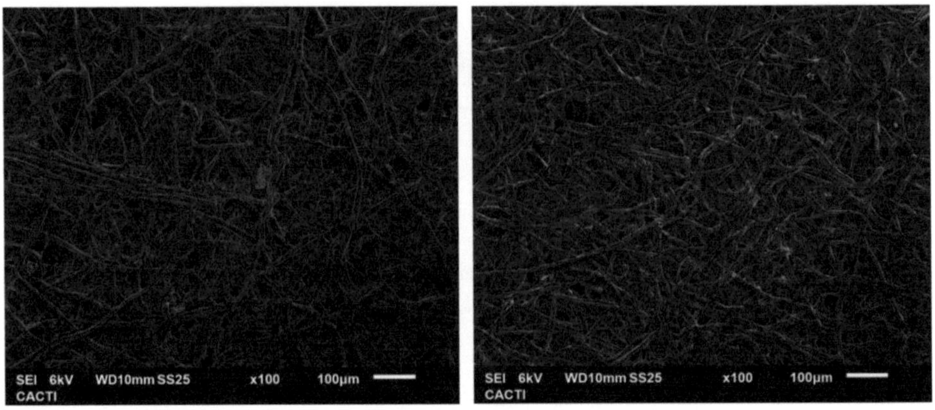

Figure 14.
Electron microscope images. Internal structure of handsheet performed in test 1 (90% Phytolacca and 10% commercial pulp) (left) and in test 5 (10% Phytolacca and 90% commercial pulp) (right).

fibers intertwined, creating a homogeneous surface in both cases. In **Figure 14** left, composed mainly of *Phytolacca americana* stem fibers, fibers thickness is not as consistent as in **Figure 14** right, but remains formatting a uniform surface.

In **Figure 15**, we observed how internal structure of proportions of test 2 and 4 remains similar while changing pulp blend. In **Figure 15** (left), we observe how predominant Phytolacca's fibers derive in a less uniform surface, while in **Figure 15** (right), where there is a higher presence of commercial pulp, surface remains homogeneous. In case of intermediated compositions of test 3, 50% of both materials, it can be appreciated how the laboratory sheet remains formed by the same kind of fibers (**Figure 16**). In **Figure 16**, right, we observe a lateral section view of test 3 handsheet, and it is possible to observe how different fibers intertwin and form a dense structure suitable to be used for any printing, writing, or cutting ends.

In **Figure 17**, it is possible to observe how color and texture of different blends evolve when changing the proportions of materials. From left to right, *Phytolacca americana* content decreases, and commercial pulp increases. Yellow color present in

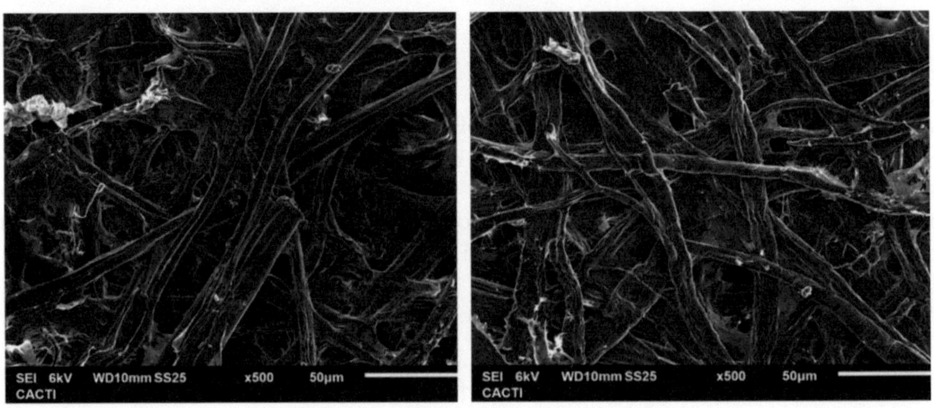

Figure 15.
Electron microscope image. Internal structure of handsheet formed by 70–30% of Phytolacca and commercial pulp (left) and surface view of internal structure of 30–70% Phytolacca material and commercial pulp (right).

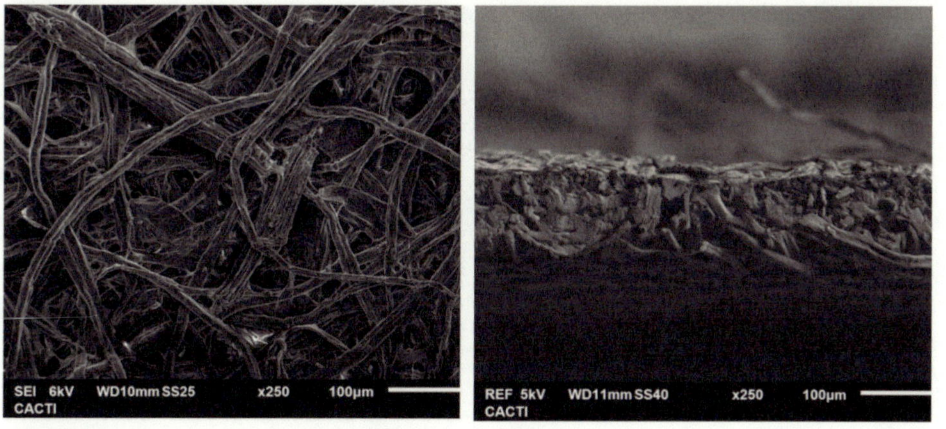

Figure 16.
Electron microscope image. Internal structure of third test, formed by 50% of each material (left) and Lateral section view of internal structure of 50% of each material (right).

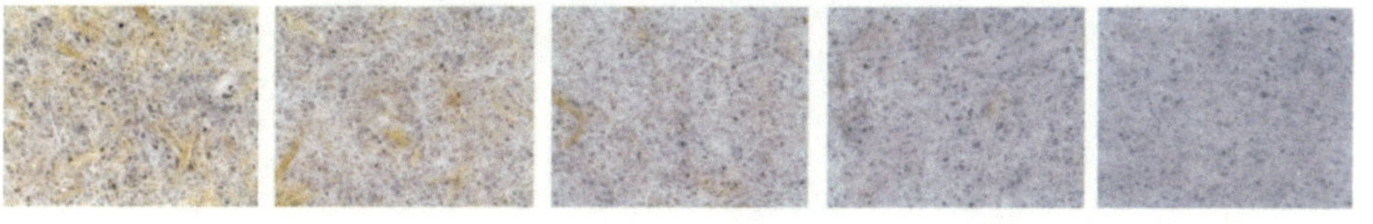

Figure 17.
Color variations when changing proportions in laboratory sheets. Imagen obtained from optical microscope Nikon SZM1500.

unbleached pulp from Phytolacca gives texture and color to the laboratory sheets, adding extra value to the proposed materials.

3.3 Chemical composition evolution

Internal composition of samples has been studied with SEM visualization. As it has been observed, the main composition of both materials demonstrated carbon and oxygen as principal components. Differing from commercial pulp paper, *Phytolacca americana* composition presents fewer chemical compounds as it is not bleached, and this variation can be observed in **Figures 18** and **19**, representing tests 2 and 4 (70% Phytolacca and 30% commercial pulp – 30% Phytolacca and 70% commercial pulp).

As we observe in **Figures 18** and **19**, composition remains almost constant in both tests, and in **Figures 20** and **21**, we can appreciate the slight evolution when mixing both materials along different tests. The main components remain carbon and oxygen, which exchange quantities after test 2, when the major content of oxygen is present in the decrement of carbon. A higher proportion of commercial pulp derives from a higher content of oxygen. On the contrary, a higher content of Phytolacca material leads to a higher carbon proportion in the mixture. In **Figure 19**, we can appreciate how calcium and silicon, both secondary most present components evolve when changing the proportions of Phytolacca and commercial pulp. Both component's percentage tends to be 0 when augmenting commercial pulp material. Silicon disappears in test 5 when Phytolacca mater is just 10% of the mixture.

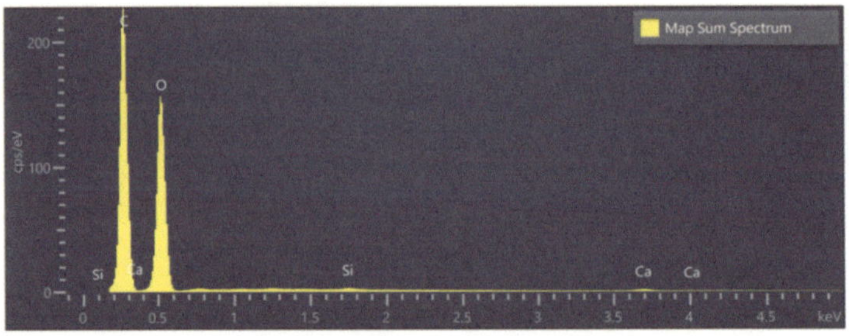

Map Sum Spectrum				
Element	Line Type	Weight %	Weight % Sigma	Atomic %
C	K series	49.16	0.15	56.53
O	K series	49.98	0.15	43.15
Ca	K series	0.68	0.05	0.23
Si	K series	0.18	0.02	0.09
Total		100.00		100.00

Figure 18.
Spectrum of composition in test 2.

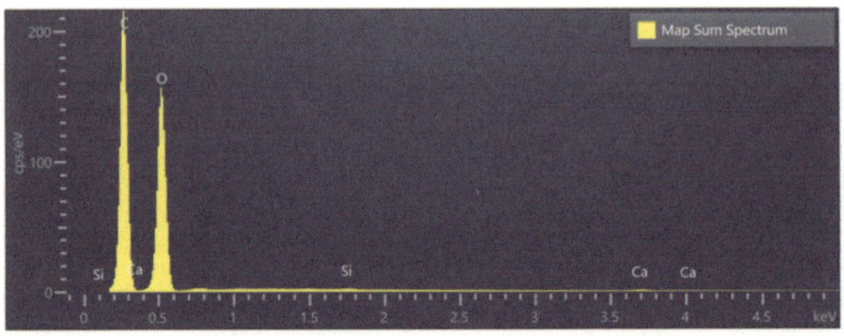

Map Sum Spectrum				
Element	Line Type	Weight %	Weight % Sigma	Atomic %
C	K series	48.57	0.12	55.86
O	K series	50.88	0.12	43.93
Ca	K series	0.44	0.04	0.15
Si	K series	0.11	0.02	0.06
Total		100.00		100.00

Figure 19.
Spectrum of composition in test 4.

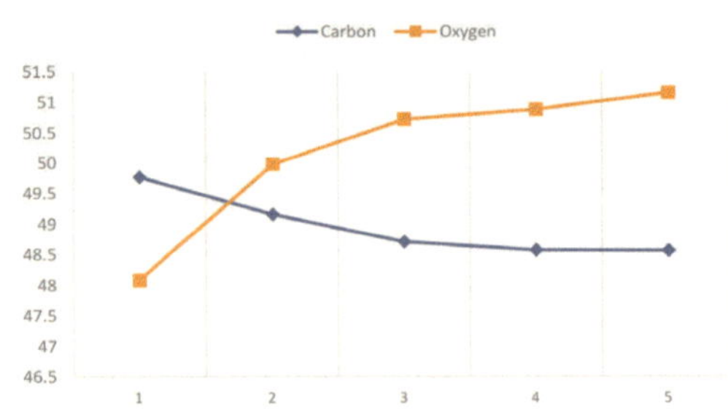

Figure 20.
Evolution of percentage of carbon and oxygen content when changing proportions of raw matter.

As it has been demonstrated, physic-mechanical, chemical, and optical properties of developed handsheets suffered evolutions when changing material proportions. But it is relevant to mention that the difference does not reflect on the quality or resistance of material, concluding that its blend is suitable to be applied in industry.

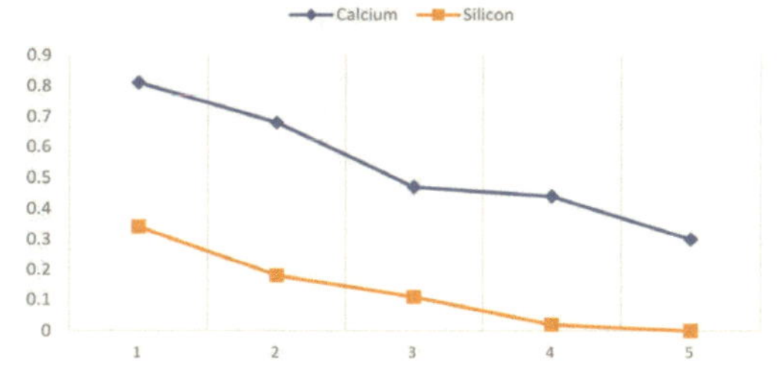

Figure 21.
Evolution of percentage of calcium and silicon content when changing proportions of raw matter.

When referring to physical properties, higher grammages are obtained with higher commercial pulp content in relation to *Phytolacca americana* material, but it remains higher with a mixture of 10–90% of Phytolacca and commercial pulp, than with 100% commercial pulp. Furthermore, paper-specific volume has been incremented when adding higher quantities of Phytolacca, and air resistance keeps constant when mixing material in the same proportions. Regarding the chemical composition of performed handsheets, there are no significant changes that will affect the biocomposability of the material; and in relation to optical properties, adding *Phytolacca americana* to commercial bleached *Eucalyptus globulus* pulp proposed value-added results, obtaining textures and colors suitable to be of interest in creative and packaging industries.

4. Discussion

The results obtained throughout this study demonstrate the viability of integrating the raw material from the invasive exotic species *Phytolacca americana* into paper production in varying proportions, thereby reducing the necessary amount of commercial *Eucalyptus globulus* pulp, which typically comes from monocultures. Specifically, we can observe that the drainability index of papers made with the mixture (in different intermediate proportions) remains between 30 and 35°SR (36.14°SR), while, according to [22], paper made with 100% commercial *Eucalyptus globulus* pulp with the same refining time of 120 seconds achieves values of 43°SR, indicating a slight variation in the pulp's drainage capacity. Therefore, we can say that approximately, the drainability values of the pulps are effective for commercial production. Other studies [23] show a grammage of 65.2 for paper made from bleached *Eucalyptus globulus* pulp; in this study, grammages between 20.39 and 77.26 were recorded for tests 1–5. This demonstrates that the properties of commercial paper regarding grammage remain stable in the intermediate tests, 3 and 4, where we find values of 48.08 and 54.55. On the other hand, the thickness of the samples, as previously indicated, undergoes minimal variation, remaining between 0.11 and 0.16, increasing with the addition of *Phytolacca americana* fiber. Previous studies [22, 23] demonstrate that the average thickness of commercial paper remains between 0.10 and 0.11. This property directly affects the apparent density and volume, which again experience slight variations that do not significantly affect the structural characteristics of the paper.

It is important to highlight that the slight increase in grammage, as well as the color variations contributed by the selected raw material, makes the resulting papers more valuable and particularly interesting for printing and working in creative sectors, which can benefit from the narrative potential of the material and its secondary origin.

The property with the greatest variation compared to other results obtained with commercial pulp [23] is in the Gurley air resistance and the material's air permeance. While paper made from commercial pulp presents a compact structure that prevents air passage for a longer duration, as we introduce pulp from *Phytolacca americana*, the times decrease from 3.58 to 2.51 seconds. The air permeance evolves from 81.18 to 14.70. Therefore, we can say that *Phytolacca americana* fibers affect the paper's structure, resulting in a less homogeneous constitution, which reduces its resistance to air passage.

Finally, as it was shown, the chemical composition of samples suffered an evolution when incrementing *Phytolacca americana* matter, presenting higher contents of oxygen in decrement of carbon and reducing the main composition of paper to essentials as it was avoided to use chemicals and the bleached pulp decreases in tests 3–6.

5. Conclusions

Conducted research on the utilization of *Phytolacca americana* fibers in combination with commercial pulp material has demonstrated significant potential for the paper industry. Value-added products demonstrate great possibilities for artistic applications and environmental communication purposes as they offer a variety of color and texturized papers along with proportions variations. The experiments undertaken have successfully optimized blending ratios, showcasing the feasibility of incorporating waste biomass from the invasive alien species *Phytolacca americana* into paper production. This approach not only promotes circular economy principles but also contributes to waste reduction initiatives, thereby mitigating the need for organic waste incineration.

The developed paper exhibits promising optical and structural properties, with varying proportions yielding diverse physical and mechanical characteristics suitable for a range of applications. Intermediated proportions of 50% of both materials demonstrate maintaining initial properties of commercial pulp paper with added values of color, texture, and fiber reutilization. The presented results confirm the feasible possibility of replacing commercial bleached pulp for proposed vegetal waste material without variation of properties, neither chemical, physic-mechanical nor optical. By harnessing waste biomass, particularly from invasive alien plant species, sustainable papermaking practices are promoted, aligning with biodiversity conservation goals and environmental stewardship principles. This innovative approach offers an eco-friendly alternative, addressing pressing environmental challenges within the papermaking sector, contributing to the possibility of reduction of wood demand for the paper industry, which will relay into monocultures reduction and local biodiversity maintenance.

As the industry faces increasing environmental pressures, embracing innovative, and sustainable methodologies is imperative for reducing ecological footprints. By promoting the adoption of such approaches, the papermaking sector can transition toward a more environmentally responsible and resource-efficient future, contributing to global efforts to combat climate change and preserve biodiversity. Simultaneously, a new paper surface with great printing and decorative ends is proposed for the creative industry, opening a huge range of possibilities in order to embrace environmental awareness and care.

Acknowledgements

This research was supported by the Xunta de Galicia through the predoctoral grant 290. We are grateful for the funding that made this work possible. Additionally, we would like to extend our thanks to IntechOpen for providing the opportunity to publish the results of our research.

This chapter was initially published as a preprint: Iglesias, A., Cancela, A., Picos, J., Soler, A. Sustainable innovation in Pulp and Paper Industry: Invasive alien species *Phytolacca americana* waste integration to commercial pulp. Physic-mechanical properties analysis [Internet]. *Research Square Journal*. 2024. Available from: https://doi.org/10.21203/rs.3.rs-4148576/v1

Conflict of interest

The authors declare no conflict of interest.

Author details

Antía Iglesias[1*], Ángeles Cancela[2], Juan Picos[3] and Ana Soler[1]

1 Draw Department, University of Vigo, Pontevedra, Spain

2 Chemical Engineering Department, University of Vigo, Pontevedra, Spain

3 Natural Resources and Environment Engineering Department, University of Vigo, Pontevedra, Spain

*Address all correspondence to: antia.iglesias@uvigo.es

References

[1] Badar S, Farooqi I. Pulp and paper industry manufacturing process, wastewater generation and treatment. In: Malik A, Grohmann E, editors. Environ-mental Protection Strategies for Sustainable Development. Strategies for Sustainability. Dordrecht: Springer; 2012

[2] Vrabič-Brodnjak U, Klemen M. Invasive alien plant species for use in paper and packaging materials. Fibers. 2022;**10**(11):94

[3] Area MC. Wood Fibres for Papermaking. Shawbury, Shrewsbury, Shropshire, United Kingdom: Smithers Pira; 2014

[4] MITECO. Producción, consumo y comercial exterior de la madera y sus productos. Vol. 1. Madrid, Spain: Anuario de estadística forestal; 2018. pp. 1-12

[5] MITECO. Estructura Forestal. Vol. 1. Madrid, Spain: Ministerio para la Transición Ecológica y el Reto Demográfico; 2018. pp. 4-10

[6] Merino A, Gil A. La fiebre del eucalipto: Especulación y fuego en Galicia. EOM & European Data Journalism. 2023. Available from: https://www.europeandatajournalism.eu/es/cp_data_news/eucalyptus-fever-speculation-and-wildfire-in-galicia/

[7] Falkiner R, Smith C. Changes in soil chemistry in effluent-irrigated *Pinus radiata* and *Eucalyptus grandis* plantations. Australian Journal of Soil Research. 1997;**35**(1):131

[8] Álvarez X, Cancela A, Merchán Y, Sánchez A. Anthocyanins, phenolic compounds, and antioxidants from extractions of six eucalyptus species. Applied Sciences. 2021;**11**(21):9818

[9] Iglesias A, Cancela Á, Álvarez X, Sánchez Á. Anthocyanins and total phenolic compounds from pigment extractions of non-native species from the Umia river basin: *Eucalyptus globulus*, *Tradescantia fluminensis*, and *Arundo donax*. Applied Sciences. 2023;**13**:5909. DOI: 10.3390/app13105909

[10] Valero E, Picos J, Álvarez X. Characterization of riparian forest quality of the Umia river for a proposed restoration. Ecological Engineering. 2014;**67**:216

[11] European Commission. Regulation (EU) N° 1143/2014 of the European Parliament and of the council on prevention and management of the introduction and spread of invasive alien species. Official Journal of the European Union. 2014;**1**:186/10

[12] Ordóñez C, Duinker PN. Ecological integrity in urban forests. Urban Ecosystems. 2012;**15**:863-877

[13] Conway MT, Almas A, Coore D. Ecosystem services, ecological integrity, and native species planting: How to balance these ideas in urban forest management? Urban Forestry & Urban Greening. 2019;**41**:1

[14] Goodwin S, Shriver G. Using a bird community index to evaluate national parks in the urbanized national capital region. Urban Ecosystems. 2014;**17**:979

[15] European Parliament. Circular economy: Definition, importance and benefits. 2023. Available from: https://www.europarl.europa.eu/topics/en/article/20151201STO05603/circular-economy-definition-importance-and-

benefits#:~:text=To%20protect%20the%20environment,total%20annual%20greenhouse%20gas%20emissions

[16] European Commission. Communication from the Commission to the European Parliament, the Council, the European Economic and Social Committee and the Committee of the Regions a New Circular Economy Action Plan for a Cleaner and more Competitive Europe. COM/2020/98. Brussels: European Commission; 2020

[17] Kaza S, Yao L, Bhada-Tata P, Van Woerden F. What a waste 2.0: A global snapshot of solid waste management to 2050. In: Urban Development. Washington, DC: World Bank; 2018. Available from: http://hdl.handle.net/10986/30317

[18] Álvarez-Bermúdez X. Modelo conceptual de la eutrofización y proliferación de cianobacterias. Un caso de estudio en el embalse de A Baxe [thesis]. Universidad de Vigo; 2015

[19] Balogh L, Juhasz J. American and Chinese Pokeweed (*Phytolacca americana* L., *Phytolacca esculenta van Houtte*). In: Zoltán Bota-Dukát Z, Balogh L, editors. The Most Important Invasive Plants in Hungary. Vácrátót: Institute of Ecology and Botany, Hungarian Academy of Sciences; 2008

[20] Acedo C, Llamas F. Catálogo de plantas alóctonas en la provincia de León (NW España). Journal Estudia Botánica. 2006;**25**:63-96. ISSN: 0211-9714

[21] Pepe M, Crescente MF, Varone L. Effect of water stress on physiological and morphological leaf traits: A comparison among the three widely-spread invasive alien species *Ailanthus altissima*, *Phytolacca americana*, and *Robinia pseudoacacia*. Plants. 2022;**11**(7):899

[22] Iglesias A, Cancela Á, Soler Baena A, Sánchez Á. Characterization of cellulose derived from invasive alien species plant waste for application in the papermaking industry: Physic-mechanical, optical, and chemical property analysis. Applied Sciences. 2023;**13**:11568-11588

[23] Kapun T, Zule J, Fabjan E, Brigita H, Brigita H, Miha G, et al. Engineered invasive plant cellulose fibers as resources for papermaking. European Journal of Wood and Wood Products. 2022;**80**:501-514. DOI: 10.1007/s00107-021-01779-y

[24] UNE-EN-ISO 5263-1. Pulps - Laboratory Wet Disintegration - Part 1: Disintegration of Chemical Pulps. United Kingdom: International Organization for Standardization; 2005

[25] UNE-EN-ISO 4119. Pulps Determination of Stock Concentration. United Kingdom: International Organization for Standardization; 1996

[26] UNE-EN-ISO 5264-2. Pulps. Laboratory Beating – Part 2: PFI Mill Method; 2011

[27] UNE-EN-ISO 5267-1. Pulps. Determination of Drainability – Part 1: Schopper-Riegler Method; 2001

[28] UNE-EN-ISO 5269-1. Pulps. Preparation of Laboratory Sheets for Physical Testing – Part 1: Conventional Sheet-Former Method; 2005

[29] UNE-EN-ISO 187. Paper, Board and Pulps – Standard Atmosphere for Conditioning and Testing and Procedure for Monitoring the Atmosphere and Conditioning of Samples; 1990

[30] UNE-EN-ISO 536. Paper and board. Determination of grammage; 2013

[31] UNE-EN-ISO 534. Paper and Board. Determination of Thickness, Density and Specific Volume; 2012

[32] UNE-EN-ISO 5636-5. Paper and Board. Determination of Air Permeance (Medium Range). Part 5: Gurley Method; 2015